EDEXCEL A LEVEL

KU-010-023

PHYSICS

Mike Benn
Graham George

DYNAMIC LEARNING

HODDER EDUCATION
AN HACHETTE UK COMPANY

In order to ensure that this resource offers high-quality support for the associated Edexcel qualification, it has been through a review process by the awarding body to confirm that it fully covers the teaching and learning content of the specification or part of a specification at which it is aimed, and demonstrates an appropriate balance between the development of subject skills, knowledge and understanding, in addition to preparation for assessment.

While the publishers have made every attempt to ensure that advice on the qualification and its assessment is accurate, the official specification and associated assessment guidance materials are the only authoritative source of information and should always be referred to for definitive guidance.

Edexcel examiners have not contributed to any sections in this resource relevant to examination papers for which they have responsibility.

No material from *Edexcel A Level Physics Year 1 Student's Book* will be used verbatim in any assessment set by Edexcel.

Endorsement of *Edexcel A Level Physics Year 1 Student's Book* does not mean that the book is required to achieve this Edexcel qualification, nor does it mean that it is the only suitable material available to support the qualification, and any resource lists produced by the awarding body shall include this and other appropriate resources.

Although every effort has been made to ensure that website addresses are correct at time of going to press, Hodder Education cannot be held responsible for the content of any website mentioned in this book. It is sometimes possible to find a relocated web page by typing in the address of the home page for a website in the URL window of your browser.

Hachette UK's policy is to use papers that are natural, renewable and recyclable products and made from wood grown in sustainable forests. The logging and manufacturing processes are expected to conform to the environmental regulations of the country of origin.

Orders: please contact Bookpoint Ltd, 130 Milton Park, Abingdon, Oxon OX14 4SB. Telephone: +44 (0)1235 827720. Fax: +44 (0)1235 400454. Lines are open 9.00a.m.–5.00p.m., Monday to Saturday, with a 24-hour message answering service. Visit our website at www.hoddereducation.co.uk

©Mike Benn, Graham George 2015

First published in 2015 by

Hodder Education,

An Hachette UK Company

Carmelite House, 50 Victoria Embankment, London EC4Y 0DZ

Impression number	4
Year	2016

All rights reserved. Apart from any use permitted under UK copyright law, no part of this publication may be reproduced or transmitted in any form or by any means, electronic or mechanical, including photocopying and recording, or held within any information storage and retrieval system, without permission in writing from the publisher or under licence from the Copyright Licensing Agency Limited. Further details of such licences (for reprographic reproduction) may be obtained from the Copyright Licensing Agency Limited, Saffron House, 6–10 Kirby Street, London EC1N 8TS.

Cover photo © Russell Kightley/Science Photo Library

Illustrations by Aptara

Typeset in 11/13 pt Bembo by Aptara, Inc.

Printed in Italy

A catalogue record for this title is available from the British Library.

ISBN 978 1471 807527

Contents

Introduction

Welcome to *Edexcel A level Physics Year 1 Student's book*. The Edexcel specification has been developed from the best of the Edexcel concept-led and the Salters Horners context-led approaches. Although the book has been written specifically to cover the concept approach to the specification, it also makes a most valuable resource for the context approach as it is illustrated throughout by contextual examples and practice questions. The authors both have vast experience of teaching, examining and writing about physics. Both have examined for Edexcel at a senior level for over 30 years.

As the title suggests, this book is designed primarily as a resource for the first year of the A-level course. It does, however, encompass all the material you need if you are only pursuing your studies of physics to AS level.

A key aspect of the text is the emphasis on practical work. Although you do not have a practical examination as such, questions on practical work pervade both AS papers. If you are doing A level, you will also have a third paper comprising practical-based and synoptic questions. There is also an internally assessed *Practical Endorsement* at A level, for which you need to start a portfolio of work in Year 1. All the *Core Practicals* in the specification are described here in detail and in such a way that students can carry out the experiments in a laboratory environment. Each experiment has a set of data for the reader to work through, followed by questions similar to those you will be asked in the examination. Questions within written examination papers will aim to assess the knowledge and understanding that students gain while carrying out practical activities, both within the context of the 16 core practical activities, as well as in novel practical scenarios. In addition, for A level, the completion of the 16 core practical activities can provide evidence of competence for the Science Practical Endorsement. The core practicals are also intended to provide students with opportunities to undertake investigative work, therefore the core practical experiments described in this book must be considered as examples of the sort of activity that could be undertaken.

Many other experiments – under the heading of *Activity* – are also described, together with data and questions. In addition, Chapter 2 is a guide to practical work which explains how you should approach practical work and contains some exercises for you to try. Before carrying out *any* practical activity, teachers *must* identify any hazards and assess any risks. This can be done by consulting a model (generic) risk assessment provided by CLEAPSS to subscribing authorities.

Emphasis is also placed on practice questions. The text is abundantly illustrated by *Examples* which are accompanied by answers to enable you to check your progress. There are then *Test Yourself questions* for you to try (answers for these can be accessed using the QR codes found in the *Free online resources* section at the end of this book) and at the end of each chapter there are *Exam Practice Questions*. These are graded in terms of difficulty (⬤ = AS/A level grade E–C, ⬤ = AS grade C–A/A level grade C and ⬤ = AS grade A/A level grade C–A). In the *Exam Practice Questions* the mark allocation for each part is shown, as in the examination. The answers give an *indication* of how the marks might be awarded but not in the same detail that there would be in an actual mark scheme. Examples of mark schemes are given in Chapter 19.

Most students using this book will have completed a GCSE course in physics or combined science. The *Prior Knowledge* boxes at the start of each chapter list some of the material that you may have covered at GCSE .These boxes should serve as a useful reminder of your earlier work before you start each topic. It should be emphasised, however, that all of the basic concepts are explained fully in each chapter and that previous knowledge is not an essential pre-requisite.

Throughout the book there are *Key Terms* highlighted in the margin that you need to learn. There are also numerous *Tips*. These may be reminders, for example to use SI units, warnings to avoid common errors or hints about short cuts in performing calculations.

At the end of the book, Chapter 19 (*Preparing for the exams*) is a valuable guide on revision and exam technique. As you need to put these principles into practice from day one, you are strongly advised to read through this *before* you start your course (although you probably won't be able to attempt the questions). You should then re-visit Chapter 19 from time to time. The same goes for Chapter 18 (*Maths in physics*) where you will find an outline of the mathematical requirements for AS and A level physics, together with lots of simple (and not so simple!) examples for you to try.

The authors have enjoyed writing this book – we hope you enjoy reading it and find it, along with the supporting material, a valuable resource to help you with your studies. Good luck!

Get the most from this book

Welcome to the **Edexcel A level Physics Year 1 Student's Book**! This book covers Year 1 of the Edexcel A level Physics specification and all content for the Edexcel AS Physics specification.

The following features have been included to help you get the most from this book.

Prior knowledge

This is a short list of topics that you should be familiar with before starting a chapter. The questions will help to test your understanding.

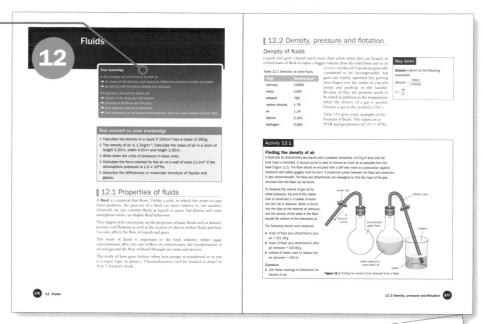

Test yourself questions

These short questions, found throughout each chapter, are useful for checking your understanding as you progress through a topic.

Tips

These highlight important facts, common misconceptions and signpost you towards other relevant topics.

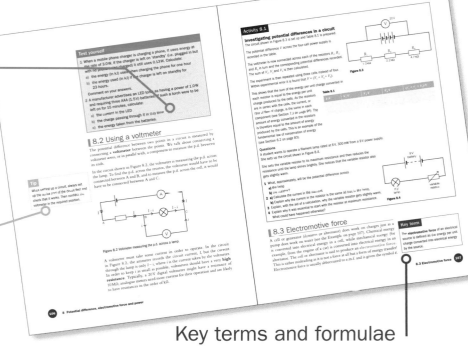

Key terms and formulae

These are highlighted in the text and definitions are given in the margin to help you pick out and learn these important concepts.

Examples

Examples of questions and calculations feature full workings and sample answers.

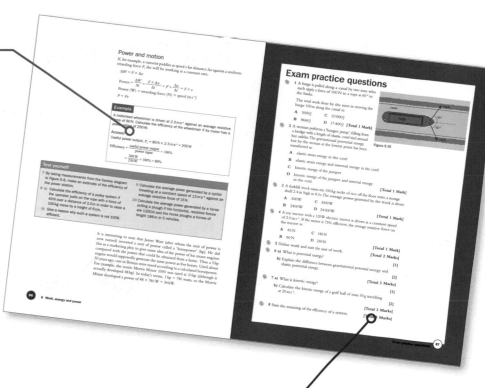

Exam practice questions

You will find Exam practice questions at the end of every chapter. These follow the style of the different types of questions you might see in your examination, including multiple-choice questions, and are colour coded to highlight the level of difficulty. Test your understanding even further, with Maths questions and Stretch and challenge questions.

Activities and Core practicals

These practical-based activities will help consolidate your learning and test your practical skills. Edexcel's Core practicals are clearly highlighted.

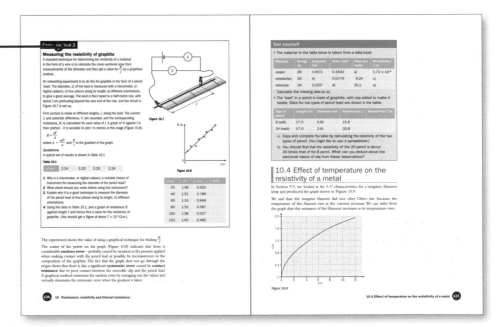

Dedicated chapters for developing your **Maths** and **Practical skills** and **Preparing for your exam** are also included in this book.

Acknowledgements

The publishers would like to thank the following for permission to reproduce copyright material:

Photo credits

p. 44 © Rui Vieira/PA Archive/Press Association Images; **p. 48** © Jamie McDonald/Getty Images; **p. 50** © Edward Kinsman/Science Photo Library; **p. 110** *l* © Mark Bassett / Alamy, *c* © Sebalos – Fotolia, *r* © Mark Boulton / Alamy; **p. 138** *t* © David J. Green / Alamy, *c* © sciencephotos / Alamy; **p. 151** *l* © 3dsculptor – Fotolia, *r* © manfredxy – Fotolia; **p. 153** Graham George 2015; **p. 163** *t* © Andrei Nekrassov – Fotolia, *b* © Wayne Higgins / Alamy; **p. 166** © Lyroky / Alamy; **p. 207** © Scott Camazine; **p. 239** © Peter Aprahamian/Sharples Stress Engineers Ltd/Science Photo Library; **p. 242** © Sovereign, ISM/Science Photo Library; **p. 261** *t* © Andrew Lambert Photography/Science Photo Library, *b* © Andrew Lambert Photography/Science Photo Library; **p. 264** *l* © Edward Kinsman/Science Photo Library, *c* © Edward Kinsman/ Science Photo Library, *r* © Giphotostock/Science Photo Library; **p. 276** © Giphotostock/Science Photo Library; **p. 283** © Paton Hawksley Education Ltd; **p. 287** *t* © imageegami – Fotolia, *b* © David Parker/Science Photo Library; **p. 287** © RGB Ventures / SuperStock / Alamy; **p. 294** © Giphotostock/Science Photo Library; **p. 295** © Andrew Lambert Photography/Science Photo Library

t = top, *b* = bottom, *l* = left, *c* = centre, *r* = right

1
Quantities and units

Prior knowledge

In this chapter you will need to:
→ use common measures and simple compound measures such as speed
→ substitute values into formulae and equations using appropriate units and rearrange equations in order to change the subject.

The key facts that will be useful are:
→ mass, length and time are examples of measurable physical quantities
→ kilogram (kg), metre (m) and second (s) are units of mass, length and time
→ speed is the distance covered per unit time and is measured in metres per second ($m\,s^{-1}$)
→ vector quantities have both size and direction.

Test yourself on prior knowledge

1 A man walks 1.6 km in 20 minutes. Calculate his average speed.

2 A sprinter runs 200 m at an average speed of $8.0\,m\,s^{-1}$. Calculate the time taken for her to complete the distance.

3 Acceleration can be calculated by dividing a change in speed by the time taken. State the unit of acceleration.

4 Speed is a scalar quantity and velocity is a vector. Explain the difference between the two.

1.1 Physical quantities, base and derived units

An elderly physicist was asked how much he had in the bank.

'How much of what?' he responded.

'Money, of course!'

'Fifteen million, three hundred thousand, one hundred and four,' he replied.

The physicist was not a rich man. He had quoted his balance in Turkish Lira which, at that time, had an exchange rate of 2.6 million to the pound (1.4 million to the American dollar).

The story has relevance to measurements in physics. It is meaningless to state that the size of a wire is 10; we must state the quantity that is measured (in this case, the length of the wire) and the unit (such as cm). In this chapter, and throughout this book, you will identify and use a number of **base quantities** (and their units) that are fundamental to all physical measurements. You will develop and use derived units for quantities for a wide range of physical properties.

Tip

Many students lose marks in examinations by failing to include the unit of a derived quantity! Always show the unit for all calculated quantities.

All measurements taken in physics are described as physical quantities. There are seven quantities fundamental to physics. These are mass, length, time, temperature, current, amount of substance and luminous intensity. All other quantities are derived from these base quantities – for example, speed is distance (length) divided by time.

SI units

A system of measurement is needed so that a comparison of the sizes can be made with the values of other people. Over the years, many different systems of units have been used. In the UK and US, pounds and ounces, degrees Fahrenheit and miles are still common measurements of mass, temperature and length, respectively. Scientists have devised an international system that uses agreed **base units** for the seven base quantities. These are termed **SI units** (abbreviated from the French Système International d'Unités).

The base units needed for AS (and A level) examinations are defined in Table 1.1.

Tip

Remember the convention that for quantities the symbols are *italicised*, and those for units are written in plain font. This convention is used throughout this book and the AS and A level examinations

Table 1.1 Base quantities and units.

Base quantity	Base unit	Symbol	Definition
length	metre	m	the distance travelled by electromagnetic radiation through a vacuum in a time of $\dfrac{1}{299\,792\,458}$ second
mass	kilogram	kg	the mass of a standard platinum–iridium cylinder held in Sèvres, France
time	second	s	$9\,192\,631\,770$ periods of the radiation emitted from an excited caesium-133 atom
current	ampere	A	the current that, when flowing in two infinitely long parallel wires placed one metre apart in a vacuum, produces a force per unit length of $2 \times 10^{-7}\,\mathrm{N\,m^{-1}}$
temperature interval	kelvin	K	$\dfrac{1}{273.16}$ of the thermodynamic temperature difference between absolute zero and the triple point of water
amount of substance	mole	mol	the amount of substance that contains the same number of elementary particles as there are atoms in 12 grams of carbon-12

Derived units

Many physical quantities are defined in terms of two or more base quantities. Some examples you may be familiar with are area, which is the product of two lengths, speed, which is the distance (length) divided by time, and density, which equals mass divided by volume (length cubed).

Table 1.2 includes some of the derived quantities and units that will be used in this book. Expanded definitions will be given in later chapters.

Table 1.2 Derived units.

Quantity	Definition	Derived unit	Base units
speed	$\dfrac{\text{distance}}{\text{time}}$		$m\,s^{-1}$
acceleration	$\dfrac{\Delta\text{velocity}}{\Delta\text{time}}$		$m\,s^{-2}$
force	$F = m \times a$	newton (N)	$kg\,m\,s^{-2}$
pressure	$p = \dfrac{F}{A}$	pascal (Pa)	$kg\,m^{-1}\,s^{-2}$
work (energy)	$W = F \times x$	joule (J)	$kg\,m^{2}\,s^{-2}$
power	$P = \dfrac{W}{t}$	watt (W)	$kg\,m^{2}\,s^{-3}$
charge	$Q = I \times t$	coulomb (C)	$A\,s$
potential difference	$V = \dfrac{W}{Q}$	volt (V)	$kg\,m^{2}\,A^{-1}\,s^{-3}$
resistance	$R = \dfrac{V}{I}$	ohm (Ω)	$kg\,m^{2}\,A^{-2}\,s^{-3}$

Tip

The Greek letter delta, Δ, is used to show a change in a quantity. So Δvelocity represents a change in velocity.

Example

Use the definitions from Table 1.2 to show that the pascal can be represented in base units as $kg\,m^{-1}\,s^{-2}$.

Answer

$$\text{Force (N)} = \text{mass (kg)} \times \text{acceleration (m\,s}^{-2})$$

$$= kg\,m\,s^{-2}$$

$$\text{Pressure (Pa)} = \frac{\text{force (kg\,m\,s}^{-2})}{\text{area (m}^{2})}$$

$$\Rightarrow \text{the unit (Pa)} = \frac{kg\,m\,s^{-2}}{m^{2}}$$

$$= kg\,m^{-1}\,s^{-2}$$

Prefixes

Many measurements are very much larger or smaller than the SI base unit. The thickness of a human hair may be a few millionths of a metre and the voltage across an X-ray tube hundreds of thousands of volts. It is often useful to write these as multiples or sub-multiples of the base unit. You will probably be familiar with the kilometre (1 km = 1000 m) and the millimetre (1 mm = 0.001 m). Table 1.3 gives the prefixes commonly used in the AS (and A level) courses.

Table 1.3 Standard prefixes.

Prefix	Symbol	Multiple	Example
pico	p	10^{-12}	$1\,pF = 10^{-12}\,F$
nano	n	10^{-9}	$1\,nA = 10^{-9}\,A$
micro	μ	10^{-6}	$1\,\mu V = 10^{-6}\,V$
milli	m	10^{-3}	$1\,mm = 10^{-3}\,m$
kilo	k	10^{3}	$1\,kW = 10^{3}\,W$
mega	M	10^{6}	$1\,M\Omega = 10^{6}\,\Omega$
giga	G	10^{9}	$1\,GHz = 10^{9}\,Hz$
tera	T	10^{12}	$1\,Tm = 10^{12}\,m$

Examples

1 A metal sphere has a radius of 3.0 mm and a mass of 0.96 g. Calculate the volume of the sphere and determine the density of the metal.

2 The resistance of a conductor is given by the equation

$$R = \frac{V}{I}$$

Calculate the potential difference, V, needed for a current, I, of 25 μA to flow through a 2.2 kΩ resistor. Give your answer in mV.

Answers

1 Volume $= \frac{4}{3}\pi r^3 = \frac{4}{3}\pi(3.0 \times 10^{-3}\,\text{m})^3 = 1.1 \times 10^{-7}\,\text{m}^3$

 Density $= \dfrac{\text{mass}}{\text{volume}} = \dfrac{9.6 \times 10^{-4}\,\text{kg}}{1.1 \times 10^{-7}\,\text{m}^3} = 8.5 \times 10^3\,\text{kg}\,\text{m}^{-3}$

2 Rearrange the equation and write the quantities in standard form.

 $V = IR = 25 \times 10^{-6}\,\text{A} \times 2.2 \times 10^3\,\Omega = 0.055\,\text{V} = 55\,\text{mV}$

Test yourself

1 State the units of force, work and power.

2 Momentum is defined as mass × velocity.

 a) Give the unit for momentum in base units.

 b) Show that the unit is equivalent to N s.

3 Convert

 a) 240 mg to kg

 b) 470 pF to μF

 c) 11 GHz to Hz.

4 The frequency of a vibrating wire is given by the equation

$$f = \frac{1}{2l}\sqrt{\frac{T}{\mu}}$$

 where T is the tension in the wire, l is its length and μ is the mass per unit length.

 a) Rearrange the equation with T as the subject.

 b) Show that the unit of T is the newton, N.

1.2 Scalar and vector quantities

If you travel from town A to town B, it is unlikely that you will follow a direct route between the towns (Figure 1.1).

Assume that town B is 10 km due east of A and that the journey by road is 16 km. On arrival at B, you will have travelled a **distance** of 16 km and you will have been **displaced** from your starting point by 10 km **due East**. Distance gives only a magnitude (size) of the journey's length, while displacement needs both magnitude and direction.

Quantities represented solely by their magnitude are called scalar quantities and those with both magnitude and direction are vector quantities.

Key term

Scalar quantities have magnitude only; **vector quantities** have both magnitude and direction.

If the journey takes two hours, the average **speed** $\left(\frac{\text{distance}}{\text{time}}\right)$ is $8\,\text{km}\,\text{h}^{-1}$, whereas the average **velocity** $\left(\frac{\text{displacement}}{\text{time}}\right)$ is $5\,\text{km}\,\text{h}^{-1}$ due East.

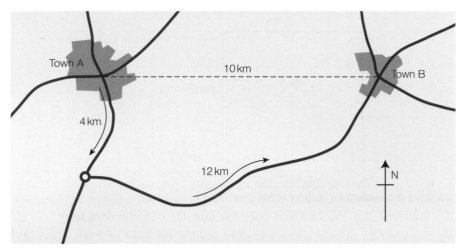

Figure 1.1 Distance and displacement.

When adding scalar quantities, normal arithmetic is applied. In the example above, the total distance travelled after returning to town A will be 16 km + 16 km, which equals 32 km. However, you would be back at your starting point, so your displacement (10 km due East plus 10 km due West) is zero.

Example

An aeroplane flies a distance of 1150 km from London to Oslo in 1 hour and 30 minutes. The return journey follows a different flight path and covers a distance of 1300 km in 2 hours. Calculate:

a) the average speed of the aircraft

b) its average velocity over the two trips.

Answer
a) Average speed $= \dfrac{\text{distance}}{\text{time}} = \dfrac{2450\,\text{km}}{3.5\,\text{h}}$

$\qquad\qquad = 700\,\text{km}\,\text{h}^{-1}$

b) Average velocity $= \dfrac{\text{displacement}}{\text{time}} = \dfrac{0\,\text{km}}{3.5\,\text{h}} = 0\,\text{km}\,\text{h}^{-1}$

Addition of vectors requires direction to be considered. It can be achieved with the help of scale drawings. If, after travelling to town B, you move 10 km due North to town C, your resultant displacement will be 14 km at an angle of 45 ° to AB (Figure 1.2).

Example

An oarsman rows a boat across a river with a velocity of $4.0\,\text{m}\,\text{s}^{-1}$ at right angles to the bank. The river flows parallel to the bank at $3.0\,\text{m}\,\text{s}^{-1}$. Draw a scale diagram to determine the resultant velocity of the boat.

Tip

Throughout this book, quantity algebra is used in all calculations. This means that when the values of quantities are substituted into an equation, the appropriate unit is also included. In later examples, where several quantities are given as a multiple or sub-multiple of the SI unit, it can be useful in ensuring that the final answer has the correct magnitude. Although this is a recommended practice, it is not a requirement in the examinations.

Tip

Vectors are usually differentiated from scalars by putting an arrow above the symbol, but they are usually represented as vector diagrams in most physics exams.

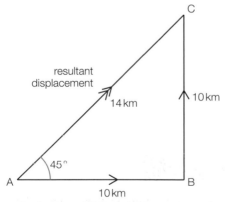

Figure 1.2 Vector addition.

Answer

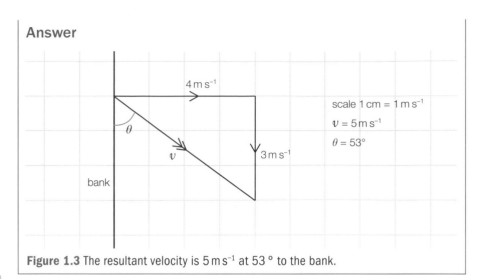

Figure 1.3 The resultant velocity is $5\,\mathrm{m\,s^{-1}}$ at 53 ° to the bank.

Table 1.4 Scalars and vectors.

Scalar	Vector
distance	displacement
speed	velocity
time	acceleration
mass	force
amount of substance	weight
temperature	momentum
charge	magnetic flux density
energy	electric field strength
potential difference	
resistance	

For the addition of two vectors at right angles, the laws of Pythagoras and trigonometry can be used to determine the resultant vector.

In the above example, the magnitude of the resultant velocity, v, is given by the expression:

$$v^2 = (4.0\,\mathrm{m\,s^{-1}})^2 + (3.0\,\mathrm{m\,s^{-1}})^2$$
$$v = 5.0\,\mathrm{m\,s^{-1}}$$

and the direction is found using the expression:

$$\tan\theta = \frac{4.0\,\mathrm{m\,s^{-1}}}{3.0\,\mathrm{m\,s^{-1}}}$$
$$\theta = 53\,°$$

These methods will be used in Chapter 5 to add forces and in the Year 2 Student's book to combine magnetic fields.

Table 1.4 gives a list of some scalar and vector quantities.

1.3 Vector addition and the resolution of vectors

Imagine a child is pulling a toy cart. The child exerts a force at an angle, θ, to the ground, as shown in Figure 1.4.

Figure 1.4

Part of the force can be thought to be pulling the cart horizontally and part can be thought to be lifting it vertically. We have seen that two vectors at right angles to each other can be added together to give a single resultant vector. It therefore follows that a single vector has the same effect as two **components** at right angles to each other.

The force exerted by the child has a horizontal component of F_H and a vertical component of F_V. Using trigonometry:

$$\cos\theta = \frac{F_H}{F}$$

and

$$\sin\theta = \frac{F_V}{F}$$

⇒ the horizontal component of the force, $F_H = F\cos\theta$ and the vertical component, $F_V = F\sin\theta$.

Tip

The component adjacent to θ is always $F\cos\theta$. You may find it easier to use the cosine function for both components – for example, if $\theta = 30°$, the horizontal component will be $F\cos 30°$ and the vertical component $F\cos 60°$.

Example

The Earth's magnetic field has a flux density of about $5.0 \times 10^{-5}\,T$ in the UK and enters the Earth at about 70° to the surface. Calculate the horizontal and vertical components of the field.

Answer

Horizontal component $= (5.0 \times 10^{-5}\,T) \times \cos 70°$

$$= 1.7 \times 10^{-5}\,T$$

Vertical component $= (5.0 \times 10^{-5}\,T) \times \sin 70°$

$$= 4.7 \times 10^{-5}\,T$$

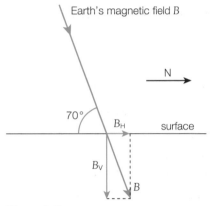

Figure 1.5

The study of projectiles in Chapter 3 is simplified by taking horizontal and vertical components of the initial velocity of the projectile. The vertical and horizontal motions can then be treated separately.

Test yourself

5 Make a list of four scalar quantities and four vector quantities.

6 A girl cycles 5.0 km due East and then 12 km due North. Calculate:

   a) the total distance she has travelled

   b) her displacement from the starting position.

   She then cycles a further 10 km 30° South of West (W 30° S).

   c) Draw a scale diagram to determine the displacement from her original position.

7 A stone is thrown with a velocity of 18 m s⁻¹ at an angle of 25° to the horizontal. Calculate:

   a) the initial horizontal component of the velocity

   b) the initial vertical component of the velocity.

Exam practice questions

1 a) Which of the following is a base quantity?

    A   Length   **C**   Second

    B   Metre   **D**   Speed   **[Total 1 Mark]**

b) Which of the following is a derived unit?

    A   Force   **C**   Newton

    B   Metre   **D**   Second   **[Total 1 Mark]**

c) Which of the following is a scalar quantity?

    A   Acceleration   **C**   Metre

    B   Force   **D**   Speed   **[Total 1 Mark]**

d) Which of the following is a vector quantity?

    A   Distance   **C**   Time

    B   Mass   **D**   Velocity   **[Total 1 Mark]**

2 Calculate the force applied to a surface of area $220\,\text{mm}^2$ to create a pressure of $10\,\text{MPa}$. Give your answer in kN.   **[Total 2 Marks]**

3 An aeroplane takes off with a velocity of $120\,\text{m s}^{-1}$ at an angle of $30°$ to the runway. Calculate the horizontal and vertical components of the velocity.   **[Total 2 Marks]**

4 The capacitance, C, of a capacitor is defined by the equation:

$$C = \frac{Q}{V}$$

where Q is the charge stored and V is the potential difference across the capacitor.

Use Table 1.2 to show that the unit of capacitance – the farad, F – may be represented as $\text{kg}^{-1}\,\text{m}^{-2}\,\text{A}^2\,\text{s}^4$ in base units.   **[Total 3 Marks]**

5 A hiker walks $5\,\text{km}$ due East and then $10\,\text{km}$ due South.

  a) Draw a scale diagram to calculate the displacement of the hiker from the starting position.   **[2]**

  b) Calculate the average speed and the average velocity of the hiker if the complete journey took 5 hours.   **[2]**

                                         [Total 4 Marks]

6 An oarsman rows a boat across a river aiming for a point on the other bank directly opposite his starting point and $40\,\text{m}$ away. He rows at a velocity of $2.0\,\text{m s}^{-1}$ towards this point, but finds that he reaches the other bank $10\,\text{m}$ downstream from the intended target. Sketch a vector diagram for his journey, and calculate the velocity of the boat between the start and finishing points.   **[Total 3 Marks]**

7 The rate of flow of a fluid (volume per second) through a pipe is given
by the equation

$$\frac{V}{t} = \frac{\pi p r^4}{8\eta l}$$

p is the pressure difference across the pipe, l is the length of the pipe, r
its radius and η is a property of the fluid known as its viscosity. Rearrange
the equation with η as the subject and show that the unit of viscosity may
be given as Pa s. **[Total 2 Marks]**

8 Two forces, F_1 and F_2, act on an object. F_1 is 6.6 N acting vertically
down and the resultant force (the sum of F_1 and F_2) on the object is
a horizontal force of 4.7 N. What is the value of F_2? **[Total 2 Marks]**

2 Practical skills

2.1 Introduction

Physics is a very practical subject, and experimental work should form a significant part of your AS and A level Physics courses. It is commonly acknowledged that it is easier to learn and remember things if you have actually done them rather than having read about them or been told about them. That is why this book is illustrated throughout by experiments, usually with a set of data for you to work through and questions to answer.

Nothing is like the 'real thing', however, so you should be carrying out many of these experiments for yourself in the laboratory. In particular you will need to pay careful attention to the experiments designated as *Core Practicals* in the Edexcel specification. These will form the basis of questions on practical work in the two AS papers and in Paper 3 of the A level examination. You may be asked to describe these experiments in the examination and/ or answer questions based on them. All eight AS *Core Practicals* are fully described in this Year 1 Student's Book, with a set of typical results and the sort of questions you may be asked in the examination.

Many other experiments are also described in detail. Some are included to illustrate and give you a better understanding of the theory, while some are particularly designed to help you develop your practical skills

The Edexcel specification identifies particular key practical skills that should be developed through teaching and learning and that will form the basis of practical assessment:

- independent thinking
- use and application of scientific methods and practices
- numeracy and the application of mathematical concepts in a practical context
- instruments and equipment.

Full details of the requirements for practical work, including a comprehensive appendix on *Uncertainties and practical work*, can be found in the specification posted on the Edexcel website: www.edexcel.com.

Before we look at these aspects of practical work, however, we need to be clear about certain terminology used in physics, particularly precision, accuracy and errors.

In the following sections, the bulleted lists in orange indicate in more detail the skills that will be assessed in the examinations.

2.2 Errors: accuracy and precision

The Edexcel specification emphasises that *it is important that the words used have a precise and scientific meaning as distinct from their everyday usage.* Edexcel has used the terminology adopted by the Association for Science Education, full details of which can be found in the specification. In the following text, these key terms are expressed in *italics*.

In everyday English, *accuracy* and *precision* have similar meanings, but in physics this is not the case. Their meanings are not the same and the difference must be understood.

An *error* is the difference between a measured result and the *true value*, i.e. the value that would have been obtained in an ideal measurement. With the exception of a fundamental constant, the true value is considered unknowable. An error can be due to *random* or *systematic* effects and an error of unknown size is a source of *uncertainty*.

When you repeat a measurement you often get different results. There is an *uncertainty* in the measurement you have taken. It is important to be able to determine the uncertainty in measurements so that its effect can be taken into consideration when drawing conclusions about experimental results.

A measured value is considered to be *accurate* if it is judged to be close to the true value – it cannot be quantified and is influenced by random and systematic errors.

The term *precision* denotes the consistency between values obtained by repeated measurements – a measurement is precise if the values 'cluster' together. Precision is influenced only by random errors and can be quantified by measures such as standard deviation. This is best illustrated by considering an archer shooting arrows at a target. In this situation, precision means getting all the shots close together and accuracy means getting them where they should be (the bull's eye!).

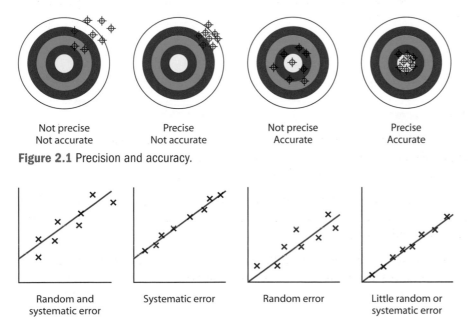

| Not precise
Not accurate | Precise
Not accurate | Not precise
Accurate | Precise
Accurate |

Figure 2.1 Precision and accuracy.

| Random and
systematic error | Systematic error | Random error | Little random or
systematic error |

Figure 2.2 Random and systematic errors.

A good archer may achieve high *precision*, but if the sights of the bow are not adjusted properly, the *accuracy* will be poor. This gives rise to a *systematic error*. A poor archer with correctly adjusted sights will scatter the arrows around the bull or, in terms of physics, will make *random errors*.

In physics experiments:

- *systematic errors* can be minimised by taking sensible precautions, such as checking for zero errors and avoiding parallax errors, and by drawing a suitable graph
- *random errors* can be minimised by taking the average of a number of repeat measurements and by drawing a graph that, in effect, averages a range of values.

Consider a mechanical micrometer screw gauge. Such an instrument has high *resolution* as it can measure to a very small interval – it can be read to 0.01 mm – which is the source of uncertainty in a single reading.

Furthermore, with careful use, it can make measurements with high *precision* – the measurements are *repeatable*, i.e. very similar when several readings are taken. However, its *accuracy* will depend on how uniformly the pitch of the screw has been manufactured and whether or not there is a zero error. You cannot do much about the former, although misuse, such as over tightening, can damage the thread. You should **always** check for zero error before using any instrument and make allowance for it if any is present. This is particularly true for digital instruments, which we often assume will be accurate and thus do not bother to check for any zero error. You should also understand that the accuracy of digital instruments is determined by the quality of the electronics used by the manufacturer.

> **Tip**
>
> An uncertainty for a single reading, or repeated readings that are the same, will be the *resolution* of the instrument. For repeated readings that are different, the uncertainty can be taken as being *half the range* of the readings.

Example

A student investigates the motion of a tennis ball rolling down a slope between two metre rules using the arrangement shown in Figure 2.3.

> **Tip**
>
> You will need to be able to rearrange an equation into the linear form
>
> $y = mx + c$
>
> For the proposed relationship, $s = \frac{1}{2}at^2$, it is important to understand that s and t are the variables and that a is a constant.
>
> A graph of s on the y-axis against t^2 on the x-axis should produce a linear graph of gradient $\frac{1}{2}a$. In this case, $c = 0$, so the graph goes through the origin.
>
> It is conventional to plot the *controlled* variable (s in this example) on the x-axis and so this is why the student decides to plot a graph of t^2 on the y-axis against s on the x-axis rather than the other way around.

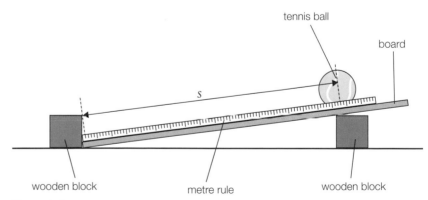

Figure 2.3

The student uses a digital stopwatch to find the time, t, that it takes for the ball to roll a distance, s, down the slope. She devises a technique in which she releases the ball from different markings on the metre rule and times how long it takes for the ball to roll down the slope and hit the block at the bottom. She records the results as shown in the table.

s/m	0.200	0.400	0.600	0.800	1.000
t/s	0.70	1.11	1.35	1.57	1.77
t²/s²					

The student thinks that s and t are related by the equation of motion:

$$s = \frac{1}{2}at^2$$

where a is the acceleration of the ball down the slope, which she assumes to be uniform. To test her hypothesis, she decides to plot a graph of t^2 against s.

1 Complete the table by adding values for t^2 and then plot a graph of t^2 against s.

2 Discuss the extent to which the graph confirms the student's hypothesis.

3 Determine the gradient of the graph and hence calculate a value for the acceleration of the ball down the slope.

4 Explain how the graph reduces the effects of random and systematic errors in the experiment.

5 Suggest the possible causes of these errors.

Answer

1 You should obtain a graph like that shown in Figure 2.4. Note that a large but convenient scale has been chosen for each axis and that the axes have been labelled with a forward slash between the physical quantity being plotted and its unit. This is a convention that you should adopt.

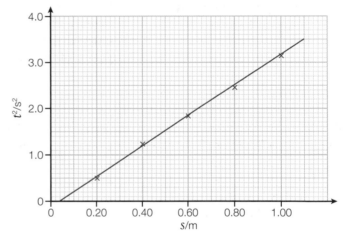

Figure 2.4

2 The equation of the line should be:

$$t^2 = \frac{2s}{a}$$

If this is correct, the graph should be a straight line of gradient $\frac{2}{a}$ passing through the origin. The graph **is** a straight line, but it does **not** go through the origin. This could be because there is an additional constant term in the equation or because there may be a small systematic error in the readings. The latter would suggest a value for s of about 0.04 m (4 cm) when $t = 0$ (see Question 5).

3 Using a **large triangle**, and remembering that the graph does not pass through the origin, you should find the gradient to be about 3.25 (s² m⁻¹).

The gradient is equal to $\frac{2}{a}$, so the acceleration is $\frac{2}{\text{gradient}} = 0.62\,\text{m s}^{-2}$.

Note that the answer is given to two *significant figures*. Although the timings are recorded to the resolution of the stopwatch, giving three significant figures for all except the first reading, the third figure (0.01 s) is over optimistic in view of likely human error in the timing in the order of 0.1 s. A final value stated to two significant figures is therefore more realistic. You will see that there is further justification for this when errors are discussed in more detail in Section 2.4.

Tip

The number of *significant figures* used depends on the *resolution* of the measuring instruments. A final value should usually be stated to the same number of significant figures as the instrument with the fewest significant figures in its reading.

4 The graph reduces the effect of *random* errors by using the straight line of best fit through the points to average out the five values. The gradient will still be equal to $\frac{2}{a}$ even if there is a *systematic* error, so drawing a graph both indicates whether there **is** a systematic error and enables allowances to be made for it.

5 The scatter of the points suggests a *random* error, particularly for the smaller values. This is probably due to timing errors, which are more apparent when the times are short (e.g. only about 1 s for the first two values).

The intercept of about 4 cm when $t = 0$ is caused by a *systematic* error by the student. She wrongly assumes that if she releases the bottom of the ball from, for example, the 20-cm mark on the rule, it will travel 20 cm before hitting the block. In reality, the ball travels about 4 cm less than this (see Figure 2.6 on page 17).

2.3 Planning, making measurements and recording data

You will be expected to:

- identify and control significant quantitative variables and plan approaches to take account of variables that cannot readily be controlled
- select appropriate equipment and measurement strategies in order to ensure suitably accurate results.

In the tennis ball experiment, the student identified the distance s and the time t as the two variables and measured them appropriately with a metre rule and digital stopwatch, respectively. She identified the acceleration, a, as being constant. To control this, she should ensure that the angle of the slope is constant – for example, by making sure that the runway does not 'bow' in the middle.

You will also be expected to:

- obtain accurate, precise and sufficient data
- record these data methodically using appropriate units and conventions
- present data in an appropriate scientific way
- plot and interpret graphs.

The student sensibly presented her data in a table and then plotted a suitable graph. The data were recorded with appropriate units, expressed in the conventional way, i.e. the symbol of the quantity being measured, followed by a forward slash and then the unit. For example t/s means the time, measured in seconds. The graph enabled an appropriate value for the acceleration to be determined by averaging several values, thereby reducing random error. We could also interpret from the graph that there appeared to be a systematic error in the measurements.

Mathematically, the forward slash means 'divided by'. Thus t/s actually means the time (in seconds) divided by seconds and so results in a number. The quantities in the table are therefore just numbers. The same convention is used when plotting graphs, so she has correctly labelled the axes as t^2/s^2 and s/m.

> **Tip**
>
> Always express measurements, whether in tables or on the axes of graphs, in the form *quantity*/unit, for example t^2/s^2.

> **Tip**
>
> You should understand the difference in meaning between italic s for the distance (a physical quantity) and upright s for seconds (a unit).

Unfortunately the student neither measured the distances correctly (which gave rise to a systematic error) nor took repeat readings (which would have improved the precision of her data). She also gave no indication that she had checked the stopwatch for zero error. In an experiment such as this, several repeat readings are essential as the times are very short and subject to human error in starting and stopping the stopwatch. The student should also have gone back and taken timings for extra values of s as five points are not really enough for a graph.

In general, measurements should be repeated, if practically possible, to provide a check against a misreading and to allow reductions in random errors by averaging two or more values. For example, when finding the diameter of a length of wire, the micrometer (or digital calipers) should first be checked for zero error. The diameter should then be measured at each end of the wire and at the centre (to check for taper), taking readings at right angles to each other at each point (to check for uniformity of cross-section). This is shown in Figure 2.5.

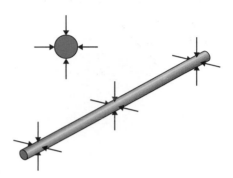

Figure 2.5 Measuring the diameter of a wire.

In some experiments, such as finding the current–voltage characteristics of a filament lamp, it would be completely **wrong** to take repeat readings, as the lamp will heat up over time and its characteristics will change. In such a situation you should **plan** for this by making sure that you take a sufficient number of readings the first time around.

You will further be expected to:

- consider margins of error, accuracy and precision of data
- evaluate results and draw conclusions with reference to measurement uncertainties and errors.

In the tennis ball experiment, the uncertainty in the length s is likely to be the accuracy with which the ball can be placed on the scale reading of the metre rule. This will be governed by the resolution of the scale (calibrated in mm) and the human error locating the point at which the ball touches the rule. The uncertainty is probably ±2mm. The uncertainty in the timing is most definitely **not** the resolution of the stopwatch (0.01 s). The uncertainty is governed by the human reaction time in starting and stopping the stopwatch. Although these effects tend to cancel out to some extent, the uncertainty would still be considered to be in the order of 0.1 s.

The accuracy of an experiment can be assessed best by expressing the uncertainties as **percentages**. In the tennis ball experiment, the percentage uncertainty in the measurement of, say, the mid-distance $s = 0.600\,\text{m}$ would be:

$$\frac{2\,\text{mm}}{600\,\text{mm}} \times 100\% = 0.3\%$$

The percentage uncertainty in the corresponding time, $t = 1.35\,\text{s}$, would be:

$$\frac{0.1\,\text{s}}{1.35\,\text{s}} \times 100\% = 7.4\%$$

As t^2 is being used, the percentage uncertainty is actually $2 \times 7.4\% = 14.8\%$. As $a = \frac{2s}{t^2}$, the percentage uncertainty in a is found by **adding** the percentage uncertainties in s and t^2: $0.3\% + 14.8\% = 15.1\%$. We would probably round this and say 15%, which is a relatively large percentage uncertainty.

> **Tip**
>
> The percentage uncertainty for a product (xy) or a quotient $\frac{x}{y}$ is found by **adding** the individual percentage uncertainties of x and y. As, for example, x^3 is $x \times x \times x$, the percentage uncertainty in x^3 will be three times the percentage uncertainty in x.

However, even taking as few as five readings and then plotting a graph reduces the uncertainty considerably. A rigorous treatment would be to draw the points on the graph with error bars and draw the steepest and shallowest best-fit straight lines. For simplicity, as there is not very much scatter of the plots on the graph, let us assume the percentage uncertainty in the value for the acceleration is in the order of 5%.

The uncertainty in the value of a can be found by taking 5% of its value, i.e. $\frac{5}{100} \times 0.62\,\text{m s}^{-2} = 0.03(1)\,\text{m s}^{-2}$

We can therefore write the value for the acceleration as $0.62 \pm 0.03\,\text{m s}^{-2}$.

Other skills that you may be assessed on include:

- correctly following instructions to carry out experimental techniques or procedures
- carrying out techniques or procedures methodically, in sequence and in combination, identifying practical issues and making adjustments when necessary
- using appropriate analogue apparatus and interpolating between scale markings
- safely using a range of practical equipment and materials, identifying safety issues and making adjustments when necessary.

These points are illustrated in the next two examples.

Example

When the student shows her teacher the results of the tennis ball experiment, he suggests that she develops the investigation further by comparing her value for the acceleration, a, with the theoretical value. He tells her that this is:

$$a = \frac{3}{5}g \sin \theta$$

where θ is the angle between the slope and the horizontal bench top.

1 Show that the angle should be about 6°.

2 Estimate the percentage uncertainty in attempting to measure this angle with a protractor and comment on your answer.

3 Draw a diagram to show how you could use a trigonometric method to determine $\sin \theta$ and comment on the advantage of this technique.

Answers

1 If $a = \frac{3}{5}g \sin \theta$

$$\Rightarrow \sin \theta = \frac{5a}{3g} = \frac{5 \times 0.62\,\text{m s}^{-2}}{3 \times 9.8\,\text{m s}^{-2}} = 0.105$$

$$\Rightarrow \theta = 6.05° \approx 6°$$

2 Using a protractor, the angle could probably not be measured to better than an uncertainty of ±1°, which gives a percentage uncertainty of:

$$\frac{1°}{6°} \times 100\% = 17\%$$

This is a large uncertainty, which suggests that using a protractor is not a suitable technique.

3 Figure 2.6 shows how a trigonometric technique could be used. It also shows why the student's measurements for s have a systematic error of about 4 cm.

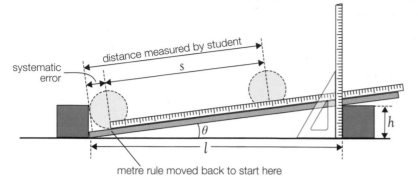

Figure 2.6

Using the technique adopted in Figure 2.6 to determine the value of s would enable readings to be taken methodically, quickly and accurately and would also allow several repeats of each value to be taken.

In the arrangement shown in Figure 2.6, the distance l was 1.000 m and the height h was 103 mm. These measurements give a value for $\tan \theta$ of:

$$\tan \theta = \frac{103 \text{ mm}}{1000 \text{ mm}} = 0.103$$

and hence a value for $\sin \theta$ of 0.102. This has been determined to three significant figures compared with a mere one significant figure using a protractor.

The advantage of this trigonometric method is that it involves two longish lengths, which can each be measured to an uncertainty of ±2 mm or better. This reduces the percentage uncertainty for $\sin \theta$ to about 1% or 2% at most.

There are no real safety issues associated with this experiment, although a possible hazard would arise if the tennis ball were allowed to bounce around the laboratory. The ball should therefore be caught at the end of each run and not allowed to fall off the bench.

Tip

You should always illustrate your work with **diagrams**. These should be drawn carefully, preferably in pencil, using a rule for straight lines. Distances should be marked precisely and any special techniques should be shown. In this case, for example, the use of a set square to ensure that the half-metre rule is vertical.

Example

A student planned to investigate a property of a prism. He wanted to find how the deviation, δ, produced by the prism depends on the angle of incidence, i, of a laser beam striking the prism.

The student placed the prism on a sheet of white paper and drew around it. He removed the prism and drew lines at angles of incidence of 30°, 40°, 50°, 60°, 70° and 80° using a protractor. He then replaced the prism and shone the laser beam along each of the lines in turn, taking particular care not to look directly into the laser beam. For each angle of incidence, he marked the direction of the emerging beam and determined the respective deviation. He recorded the data as shown in the table.

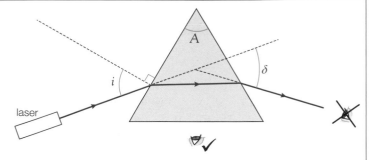

Figure 2.7

$i/°$	30.0	40.0	50.0	60.0	70.0	80.0
$\delta/°$	47.0	39.0	38.0	39.5	42.0	49.5

Note that the student took his measurements methodically by initially drawing angles of incidence at 10° intervals and recorded them in a suitable table with units. He attempted to measure the angles to a precision of 0.5° by interpolating between divisions on the protractor. This is reflected in the table by the student giving all of the angles to an appropriate number of significant figures.

1 Plot a graph of the deviation, δ, against the angle of incidence, i.

2 Use your graph to determine the angle of minimum deviation, δ_{min}.

3 Suggest extra readings that could be taken to improve the graph.

4 The student looked in a book on optics and found that the refractive index, n, of the material of the prism is given by the equation:

$$n = \frac{\sin \dfrac{A + \delta_{min}}{2}}{\sin \dfrac{A}{2}}$$

where A is the angle of the prism. The student measured A and found it to be 60.0°. Use this information to find a value for the refractive index.

Tip

Wherever realistically possible, try to interpolate between scale divisions, for example, 0.5 mm on a millimeter scale, 0.5° for a protractor and 0.5 °C or better on a mercury thermometer calibrated in degrees.

Answers

1 Your graph should look like that in Figure 2.8.

2 The minimum deviation is 37.9°.

3 The angle of minimum deviation could be determined with more certainty if extra readings were taken for $i = 45°$ and $i = 55°$.

4 $n = \dfrac{\sin \dfrac{A + \delta_{min}}{2}}{\sin \dfrac{A}{2}} = \dfrac{\sin \dfrac{60.0° + 37.7°}{2}}{\sin \dfrac{60.0°}{2}} = \dfrac{\sin 49°}{\sin 30°} = 1.51$

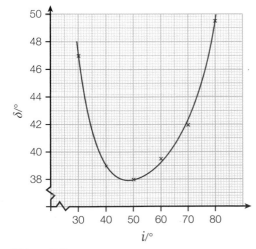

Figure 2.8

You should always take an appropriate number of measurements over as wide a range as possible. In the prism example, this is achieved by taking readings between 30° and 80° (readings with i less than about 30° are not achievable as total internal reflection occurs) and by subsequently taking extra readings at $i = 45°$ and $i = 55°$.

In consideration of safety, you should note that the student took particular care not to look directly into the laser beam. Even a laser pointer is a very hazardous device and could cause serious eye damage. It is not advisable to use lasers in a darkened room as your eye irises will be dilated. The utmost care should always be taken when working with lasers.

You always need to be aware of the potential dangers of the apparatus with which you are working and act accordingly. In particular, you will be doing a number of electrical experiments. Electricity should always be treated with the greatest respect – even at the low voltages with which you will be working.

2.4 Analysis and evaluation

You are required to develop and use a range of mathematical skills, details of which you can find on the Edexcel website. Your attention is also drawn to Chapter 18 of this book.

You will be expected to:

- process and analyse data using appropriate mathematical skills as exemplified in the mathematical appendix of the specification
- evaluate results and draw conclusions with reference to measurement uncertainties and errors.

It goes without saying that you should be able to produce a graph with sensible scales and with appropriately labelled axes, including correct units, and then plot points correctly and draw the straight line, or smooth curve, of best fit.

The graph in Figure 2.8 is an example of when the most suitable scale can be achieved by **not** starting the axes at the origin. You should always choose a scale that occupies at least half the graph paper in both the x and y directions (or else the scale could be doubled!) and one that avoids awkward scales such as scales having multiples of 3.

Points should be plotted accurately, with interpolation between the scale divisions, and marked with a neat cross or a small dot with a ring round it. In more advanced work the plots should be plotted as *error* bars. The length of the bars should reflect the uncertainty of a value in the region of the middle of the recorded measurements. When drawing the line of best fit, you should remember that not all functions in physics are linear. If your points clearly lie on a curve, you must draw a smooth curve through them. Furthermore, not every straight line will necessarily pass through the origin, even though you may be expecting it to (as in Figure 2.4 on page 13). If this is the case, you need to discuss possible reasons as to why the line does not pass through the origin.

> **Tip**
>
> First plot your points in pencil. If they look right, with no apparent anomalies, ink them in. Then draw your line in pencil. If you are not happy with your line, you can easily rub it out and have another go without also rubbing out the plotted points.

In the tennis ball experiment on page 12, we expected to get a straight line through the origin, but we found that there was a small intercept. We then discussed the possible reasons for this and came to the conclusion that a systematic error had arisen as the distance s had not been measured correctly.

In the second tennis ball experiment on page 16, we considered, quantitatively, the error likely to occur if a protractor was used to measure the angle of the slope. We concluded that an uncertainty of around 17% was unacceptable, so we improved the experiment by using a trigonometric method.

For the tennis ball experiment, we can conclude from the measurement of the angle of the slope that:

$$a = \frac{3}{5}g \sin \theta = \frac{3}{5} \times 9.81\,\text{m s}^{-2} \times 0.102 = 0.600\,\text{m s}^{-2}$$

The experimental value found for the acceleration was $0.62\,\text{m s}^{-2}$. The two values differ by:

$$\frac{(0.62 - 0.60)\,\text{m s}^{-2}}{0.61\,\text{m s}^{-2}} \times 100\% \approx 3\%$$

This is a very acceptable experimental error and so we can conclude that, within experimental error, the acceleration of a tennis ball rolling down a slope is given by the formula $a = \frac{3}{5}g \sin \theta$.

Note in the above experiment that the two values for the acceleration are both **experimental** values as the value of $\sin \theta$ had to be determined by taking measurements. Therefore the **average** value of the acceleration ($0.61\,\text{m s}^{-2}$) was used as the denominator.

Finally, you are expected to be able to:

- solve problems set in practical contexts and apply scientific knowledge to practical contexts
- identify uncertainties in measurements and use simple techniques to determine uncertainty when data are combined by addition, subtraction, multiplication, division and raising to powers.

These points are illustrated in the next example.

Tip

Remember:

- % difference between two experimental values is given by:

% difference = $\dfrac{\text{difference between values}}{\text{average of two values}} \times 100\%$

- % difference between an experimental value and a stated or known value is given by:

% difference = $\dfrac{\text{difference between values}}{\text{stated value}} \times 100\%$

Example

A technician needs to make up some leads. In an online catalogue he finds some PVC-insulated copper wire having the following manufacturer's specification:

Core: 1.5 mm²	Length: 10 m
Internal diameter: 2 mm	Weight: 231 g
External diameter: 3 mm	

Tip

Remember to distinguish carefully between the terms *mass* and *weight* when using them in physics. Manufacturers often use the everyday word 'weight' when it should really be 'mass'.

1 Comment on the error the manufacturer has made in stating one of the quantities in the specification.

2 The technician decides to check these data. Show that a core of 1.5 mm² would have a diameter of about 1.4 mm.

3 Concerned about this apparent discrepancy, he does some research and discovers that the core is actually made up of seven strands of copper wire each of diameter 0.52 mm arranged as shown in Figure 2.9.

a) Show that this gives a cross-sectional area for the copper close to the stated value of 1.5 mm².

b) Suggest why the manufacturer's statement that the diameter of the copper core is 2 mm is not unreasonable.

4 When the wire arrives he checks the diameter of the individual strands of copper wire using electronic digital vernier calipers having the following specification:

> Resolution: 0.01 mm   Repeatability: 0.01 mm
>
> Accuracy: 0.02 mm

Explain what is meant by each of these terms.

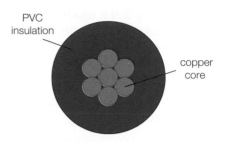

Figure 2.9 Cross-section of wire (magnified approximately 10 times).

5 He obtains the following measurements for the diameter of each of the seven single strands:

d/mm: 0.53, 0.54, 0.52, 0.53, 0.54, 0.53, 0.52

Using these data, together with the specification of the calipers,

a) determine the average area of cross-section these data give for a single strand

b) estimate the percentage uncertainty in this value

c) estimate the uncertainty this would give for the area of cross-section for the seven strands

d) comment on whether the stated specification of 1.5 mm² for this wire is valid.

6 The density of copper is 8930 kg m⁻³. Use this, with your answer to Question 5c, to calculate the mass of copper in the 10 m reel of wire.

7 To check this, the technician cuts off a 10 cm length of wire and strips off the PVC insulation. He finds that the mass of this length of copper is 1.3 g.

a) What value would this give him for the mass of copper in the cable?

b) Suggest two ways by which he could have got a more accurate value for the mass.

> **Tip**
>
> A measurement is *valid* if it measures what it is supposed to be measuring – this depends both on the method and the instruments.

Answers

1 The manufacturer has stated that the wire has a *weight* of 231 g when it should have been stated as a *mass* of 231 g. If the wire has a mass of 231 g, its weight would be

$$W = 0.231\,\text{kg} \times 9.81\,\text{N\,kg}^{-1} = 2.27\,\text{N}$$

2 From $A = \dfrac{\pi d^2}{4}$

$$\Rightarrow d^2 = \frac{4A}{\pi}$$

$$\Rightarrow d = \sqrt{\frac{4A}{\pi}} = \sqrt{\frac{4 \times 1.5\,\text{mm}^2}{\pi}} = 1.38\,\text{mm} \approx 1.4\,\text{mm}$$

3 **a)** For a single strand: $A = \dfrac{\pi d^2}{4} = \dfrac{\pi \times (0.52\,\text{mm})^2}{4} = 0.212\,\text{mm}^2$

For all seven strands: $A = 7 \times 0.212\,\text{mm}^2 = 1.49\,\text{mm}^2 \approx 1.5\,\text{mm}^2$

b) From Figure 2.9 it can be seen that the 'diameter' of the copper core can be considered to be three strand diameters, i.e. 3 × 0.52 mm = 1.56 mm. If the manufacturer is only stating values to the nearest millimetre (i.e. one significant figure), then a value of 2 mm is what would be quoted for the 'diameter' of the copper core.

4 A *resolution* of 0.01 mm means that this is the smallest measuring interval, i.e. the electronic readings are shown to 0.01 mm.

An *accuracy* of 0.02 mm means that the manufacturer cannot guarantee that the reading is any better than 0.02 mm from the true value.

A *repeatability* of 0.01 mm means that if you were to repeat exactly the same reading, the manufacture cannot guarantee that the readings will be the same – they could be 0.01 mm different.

5 a) Average diameter,

$$d = \frac{(0.53 + 0.54 + 0.52 + 0.53 + 0.54 + 0.53 + 0.52)\,\text{mm}}{7}$$

$$= 0.53 \pm 0.01\,\text{mm}$$

However, the manufacturer states an accuracy of 0.02 mm, so the uncertainty could be as much as (0.01 + 0.02) mm = 0.03 mm.

b) This gives a percentage uncertainty for

d of $\frac{0.03\,\text{mm}}{0.53\,\text{mm}} \times 100\% = 5.7\%$

The estimated percentage % uncertainty in d^2 will therefore be 2 × 5.7% = 11%.

c) For a single strand: $A = \dfrac{\pi d^2}{4} = \dfrac{\pi \times (0.53\,\text{mm})^2}{4}$

$$= 0.221\,\text{mm}^2$$

For all seven strands: $A = 7 \times 0.221\,\text{mm}^2 = 1.55\,\text{mm}^2$

As d is the average diameter of seven strands and the *percentage* uncertainty in the area of a single strand is 11%, the *percentage* uncertainty in the area of the seven strands will also be 11%. Uncertainty in area of seven strands

$$= \frac{11}{100} \times 1.55\,\text{mm}^2 = 0.17\,\text{mm}^2$$

The area should therefore be expressed as 1.55 ± 0.17 mm²

d) The specification states that the area is 1.5 mm². This value lies well within the experimental range of 1.55 ± 0.17 mm² and so the manufacturer's statement would appear to be valid.

6 Volume of copper,

V = area of seven strands × length of wire

$= 1.55 \times 10^{-6}\,\text{m}^2 \times 10\,\text{m} = 1.55 \times 10^{-5}\,\text{m}^3$

(Don't forget to convert mm² to m²!)

Mass of copper,

m = density × volume = 8930 kg m⁻³ × 1.55 × 10⁻⁵ m³

$= 0.138\,\text{kg}\ (= 138\,\text{g})$

As the % uncertainty in the area is 11%, the % uncertainty in m will also be 11% (assuming the density value is accurate).

Uncertainty in $m = \frac{11}{100} \times 138\,g = 15\,g$.

The mass can therefore be taken as $138 \pm 15\,g$.

7 a) If $10\,cm$ of bared copper wire has a mass of $1.3\,g$, then the mass of copper in a reel of length $10\,m$ will be $100 \times 1.3\,g = 130\,g$.

 b) He could have got a more accurate value by cutting off a longer length than $10\,cm$ or by using a balance having a greater resolution than $0.1\,g$.

If you are going on to take A level Physics, you will be internally assessed for a *Practical Endorsement*. There are no grades for this, just 'pass' or 'fail'. For this endorsement you will need to show practical competency in the skills highlighted in this chapter, together with evidence of the ability to:

- apply investigative approaches and methods to practical work
- keep appropriate records of experimental activities
- use appropriate software and/or tools to process data, carry out research and report findings
- use online and offline research skills, including websites and textbooks
- correctly cite sources of information.

As this Year 1 Student's book is written for the first year of an A level course, as well as for the AS examination, examples of these skills are included throughout the book in the *Core Practicals* and in the *Activities*. In addition, more advanced mathematical skills are needed for A level. These are also illustrated, where appropriate. There will be further examples in the Year 2 Student's book.

In Questions 1–3, you will be asked to make some simple measurements for yourself, using things that you can probably find at home. If you cannot do this, you can get the data from the first part of each answer and then work through the rest of the question.

1 You will need:

- an unopened 250 g packet of butter (or margarine)
- a mm scale, e.g. a 30 cm rule.

a) Take such measurements as are necessary to determine a value for the density of the 250 g pack of butter.

b) Estimate the percentage uncertainty in each of your measurements and hence discuss whether your value for the density would allow you to decide whether butter would float in water.

2 You will need:

- an unopened packet (500 sheets) of A4-sized 80 gsm printing paper (gsm is the manufacturer's way of writing 'grams per square metre' – that is, $g\,m^{-2}$)
- kitchen scales
- a mm scale, e.g. a 30 cm rule.

a) i) Use the kitchen scales to find the mass of the packet of paper.

   ii) Hence determine the mass of a single sheet of this A4 paper.

b) i) Take such measurements as are necessary to determine the area of a single sheet of this A4 paper.

   ii) Calculate an experimental value for the 'gsm' of the paper.

   iii) Calculate the percentage difference between your experimental value and the value given by the manufacturer. Comment on your answer.

c) i) Estimate the thickness of a single sheet of paper and hence determine a value for the density of the paper.

   ii) Describe how you could check the thickness of a single sheet of paper using a micrometer screw gauge or digital calipers.

3 You will need:

- ten 1p coins (all dated 1993 onwards)
- a mm scale, e.g. a 30 cm rule.

(If you do not live in the UK, you should use a small coin of your own currency. You will probably be able to find its mass on the internet.)

a) Take such measurements as are necessary to determine the average diameter, d, and thickness, t, of a 1p coin.

b) Calculate the volume, V, of a coin and hence the density of the material from which the coin is made given that the mass of a 1p coin is 3.56 g.

c) After 1992, 1p coins were made of copper-plated mild steel of density $7.8\,g\,cm^{-3}$.

   i) Determine the percentage difference between your value for the density and the value stated for mild steel.

   ii) Comment on your answer with reference to the uncertainties in your measurements.

d) Until 1992, 1p coins were made of brass (density $8.5\,g\,cm^{-3}$) and were 1.52 mm thick. Discuss whether your experiment would enable you to detect this difference.

4 In an experiment to find the acceleration of a mass falling freely under gravity, a student used an electronic timer to find the time, t, for a small steel sphere to fall vertically through a distance, h. The student recorded the following results:

h/cm	40	60	80	100	120
t/s	0.30	0.38	0.42	0.47	0.52

The student assumed that the relationship between h and t is $h = \frac{1}{2}gt^2$.

a) Show that a suitable graph to plot would be a graph of t^2 against h and that the gradient of this graph is $\frac{2}{g}$.

b) Tabulate values of h and t^2 and then plot a graph of t^2 against h, starting both scales at the origin.

c) Comment on the graph obtained.

d) Use your graph to determine a value for g.

e) Determine the percentage difference between the value you obtain for g and the accepted value of $9.81\,m\,s^{-2}$.

f) i) Comment on the way in which the student expressed the values of h.

   ii) Suggest how the measurements could have been improved.

Rectilinear motion

Prior knowledge

In this chapter you will need to be able to:
→ substitute numerical values into formulae and equations using appropriate units.
→ plot and draw graphs to show how distance and speed change with time
→ determine the gradient of linear graphs.

The key facts that will be useful are:
→ Velocity, acceleration and displacement are vector quantities.

→ average velocity $= \dfrac{\text{displacement}}{\text{time}}$

→ acceleration $= \dfrac{\text{change in velocity}}{\text{time}}$

→ A body in free fall on Earth accelerates downwards at $9.81\,\text{m s}^{-2}$.

Test yourself on prior knowledge

1 What is meant by a vector quantity?
2 Calculate the displacement of an object that travels with an average velocity of $4.5\,\text{m s}^{-1}$ for $8.0\,\text{s}$.
3 The velocity of a car changes from $36\,\text{km h}^{-1}$ to $18\,\text{km h}^{-1}$ in $8.0\,\text{s}$. Calculate the acceleration of the car in m s^{-2}.
4 Calculate the velocity of a brick $2.2\,\text{s}$ after it is accidentally dropped from the top of a building.
5 Sketch a displacement–time graph for a body moving at constant velocity.

3.1 Speed, velocity and acceleration

Rectilinear motion means motion along a straight path. We saw earlier that average speed is distance divided by time, and that velocity is a vector quantity defined as displacement divided by time. As all the motion to be studied in this chapter is rectilinear, it follows that velocity will be used throughout.

Average and instantaneous velocity

A sprinter accelerates for 20 metres and then maintains a uniform velocity for the remaining 80 metres of the 100-metre sprint. If the total time taken is $10.0\,\text{s}$, the average velocity, v_{av}, is found using:

$$v_{av} = \frac{\text{displacement}}{\text{time}} = \frac{100\,\text{m}}{10.0\,\text{s}} = 10.0\,\text{m s}^{-1}$$

During the first 20 m the runner's velocity is continuously increasing to a maximum of 12.0 m s⁻¹. If the displacement for a small period of time, say 0.01 s, during the acceleration, is 0.06 m, the velocity at this instant will be:

$$v = \frac{0.06\,\text{m}}{0.01\,\text{s}} = 6.0\,\text{m s}^{-1}$$

The instantaneous velocity is strictly defined as the velocity at an instant, which may be much smaller than 0.01 s but, for practical purposes, measurements of velocity taken over a time that is much shorter than the time of the overall measurements will be regarded as instantaneous.

Average velocity and instantaneous velocity are sometimes represented by the equations:

$$v_{\text{av}} = \frac{\Delta x}{\Delta t}$$

$$v_{\text{inst}} = \frac{\delta x}{\delta t}$$

where Δt represents an interval of time and δt represents a very small interval.

The instantaneous velocity is often given in the calculus notation $\dfrac{\text{d}x}{\text{d}t}$ that represents the value as δt tends to zero.

Activity 3.1

Measuring average velocity and instantaneous velocity

A trolley is released at the top of an inclined plane and allowed to run to the bottom of the plane. The distance, Δx, travelled by the trolley down the slope is measured, and the time, Δt, is measured using a stopclock. The average velocity is found by dividing the distance moved down the plane by the time taken.

An interrupter card cuts through the light beam directed at a light sensor (a 'light gate'), and the time is electronically recorded. The 'instantaneous' velocity at that position is calculated by dividing the length of the card, δx, by the time, δt, taken to cross the beam.

A sensor placed near the top of the runway can be used to find the initial velocity, u, and a lower sensor can give the final velocity, v.

A data-logging interface (such as Philip Harris' DL+) will measure the times at each light gate and also the time interval between the gates.

$$\text{average velocity} = \frac{\Delta x}{\Delta t}$$

$$\text{initial velocity, } u = \frac{\delta x}{\delta t_1}$$

$$\text{final velocity, } v = \frac{\delta x}{\delta t_2}$$

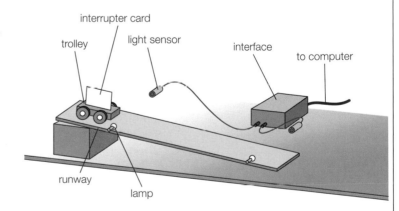

Figure 3.1 Measuring velocity and acceleration (lamp and sensor supports omitted for clarity).

If the acceleration down the slope is uniform, the average velocity will be $\dfrac{(u + v)}{2}$.

This can be compared with the measured average velocity:

$$\frac{\Delta x}{\Delta t} = \frac{(u + v)}{2}$$

The results of the experiment are shown in Figure 3.2. The time scale on the computer display has been broken to allow both timings to be observed.

Questions

The length of the interrupter card is 5.0 cm, and the distance between the sensors is 0.800 m.

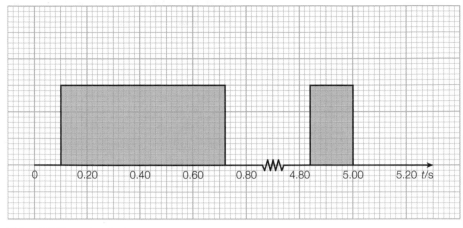

Figure 3.2 Computer display.

1 Use Figure 3.2 to determine:
   a) the time intervals during which the interrupter card cuts the beams at the top and the bottom of the ramp
   b) the time taken for the trolley to travel between the sensors.
2 a) Calculate the instantaneous velocity at the top and at the bottom of the slope.
   b) Use the values in part (a) to determine the average velocity of the trolley.
3 Calculate the average velocity of the trolley using the total time taken between the sensors.
4 Suggest a reason why the two values for the average velocity differ.

Acceleration

As the trolley moves down the slope, its velocity steadily increases. Any change in velocity indicates that the trolley is accelerating. The magnitude of the acceleration is a measure of the rate at which the velocity changes:

$$\text{acceleration, } a = \frac{\text{change in velocity}}{\text{change in time}} = \frac{\Delta v}{\Delta t}$$

The change in velocity is measured in metres per second (m s^{-1}), so acceleration has the unit m s^{-1} per second, which is written as m s^{-2}.

For uniform acceleration, as in the experiment to measure average and instantaneous velocity (in the Activity above), the acceleration is calculated using the equation:

$$a = \frac{(v - u)}{t}$$

Velocity and acceleration are vectors. If an object slows down, $(v - u)$ will be negative. This means that the acceleration is in the opposite direction to the velocities and the object has a negative value of acceleration.

Example

In an experiment, a trolley runs down an inclined plane. An interrupter card of length 20.0 cm cuts through light gates close to the top and the bottom of the slope. The following results were recorded from such an investigation.

Time to cut the top gate, $t_1 = 0.30$ s

Time to cut the bottom gate, $t_2 = 0.14$ s

Time to travel between the gates, $t = 0.50$ s

Calculate:

a) the velocity of the trolley at each gate

b) the acceleration of the trolley.

Answers

a) Velocity at top gate, $\quad u = \dfrac{(0.200\,\text{m})}{(0.30\,\text{s})} = 0.67\,\text{m s}^{-1}$

Velocity at bottom gate, $\quad v = \dfrac{(0.200\,\text{m})}{(0.14\,\text{s})} = 1.43\,\text{m s}^{-1}$

b) Acceleration, $\quad a = \dfrac{(v-u)}{t} = \dfrac{(1.43\,\text{m s}^{-1} - 0.67\,\text{m s}^{-1})}{0.50\,\text{s}} = 1.5\,\text{m s}^{-2}$

Equations of motion

The motion of an object moving at constant velocity, or accelerating uniformly, can be described by a set of equations known as the equations of motion. The following symbols represent the physical quantities involved in the equations:

s = displacement (m)

u = initial velocity (m s^{-1}) at $t = 0$ s

v = final velocity (m s^{-1})

a = acceleration (m s^{-2})

t = time (s)

The first equation is simply the definition of acceleration rearranged so that the final velocity, v, is the subject of the equation:

$$a = \frac{(v-u)}{t}$$

giving

$$v = u + at \qquad\qquad \textbf{Equation 1}$$

Average velocity is defined as total displacement divided by time. For uniform motion, the average velocity is $\dfrac{(u+v)}{2}$, so:

$$\frac{s}{t} = \frac{(u+v)}{2}$$

giving

$$s = \frac{(u + v)t}{2}$$ **Equation 2**

In order that any two quantities can be calculated if the other three are given, a further two equations can be obtained by combining Equations 1 and 2. The resulting expressions are:

$$s = ut + \frac{1}{2}at^2$$ **Equation 3**

$$v^2 = u^2 + 2as$$ **Equation 4**

You do not need to be able to perform these combinations, as Equations 3 and 4 will be included on the data sheet at the end of the AS (and A Level) examination papers.

Example

1 A train starts from rest at a station and accelerates at $0.2\,\text{m s}^{-2}$ for one minute until it clears the platform. Calculate the velocity of the train after this time, and the length of the platform.

2 The train now accelerates at $0.4\,\text{m s}^{-2}$ for the next 540 m. Calculate its final velocity and the time taken to travel this distance.

Answers

1 $u = 0\,\text{m s}^{-1}$, $a = 0.2\,\text{m s}^{-2}$, $t = 60\,\text{s}$, $v = ?$, $s = ?$

Using Equation 1:

$v = u + at$

$= 0 + 0.2\,\text{m s}^{-2} \times 60\,\text{s}$

$= 12\,\text{m s}^{-1}$

Using Equation 3:

$s = ut + \frac{1}{2}at^2$

$= 0 + \frac{1}{2} \times 0.2\,\text{m s}^{-2} \times (60\,\text{s})^2$

$= 360\,\text{m}$

2 $u = 12\,\text{m s}^{-1}$, $a = 0.4\,\text{m s}^{-2}$, $s = 540\,\text{m}$, $v = ?$, $t = ?$

Using Equation 4:

$v^2 = u^2 + 2as$

$= (12\,\text{m s}^{-1})^2 + 2 \times 0.4\,\text{m s}^{-2} \times 540\,\text{m}$

$= 576\,\text{m}^2\,\text{s}^{-2}$

giving

$v = 24\,\text{m s}^{-1}$

Using Equation 1:

$v = u + at$

$24\,\text{m s}^{-1} = 12\,\text{m s}^{-1} + 0.4\,\text{m s}^{-2} \times t$

giving

$t = 30\,\text{s}$

Test yourself

1 Explain why the quantities u, v, a and s must be vectors for rectilinear motion.

2 Calculate the acceleration of an object that increases in velocity from $2.0\,\text{m s}^{-1}$ to $5.0\,\text{m s}^{-1}$ in $0.5\,\text{s}$.

3 How long will it take a dog running at $4.0\,\text{m s}^{-1}$ to increase its velocity to $6.0\,\text{m s}^{-1}$ if it accelerates at $0.8\,\text{m s}^{-1}$?

4 How far will a car travel when it accelerates at $0.5\,\text{m s}^{-2}$ from rest for 20 s?

5 Calculate the final velocity of a cyclist who accelerates at $0.8\,\text{m s}^{-2}$ from an initial velocity of $5.0\,\text{m s}^{-1}$ over a distance of 12 m.

Tip

Always write down the values of s, u, v, a and t that you are given and then select the appropriate equation to obtain the unknown quantity.

Core practical 1

Measuring the acceleration of a free-falling object

For an object dropped from rest, the acceleration due to gravity can be calculated using the equations of motion:

$$s = ut + \frac{1}{2}at^2$$

or

$$v^2 = u^2 + 2as$$

As $u = 0\,\mathrm{m\,s^{-1}}$ and $a = g$, these equations become:

$$s \text{ (or } h) = \frac{1}{2}gt^2$$

and

$$v^2 = 2gs \text{ (or } 2gh)$$

Method 1

A ball-bearing is held by an electromagnet at height, h, above a trapdoor switch (Figure 3.3a). When the switch is thrown (downwards in the diagram), the circuit to the electromagnet is broken and the ball begins to fall. At that instant the stopclock starts timing. When the ball strikes the trapdoor, the lower circuit is broken and the clock is stopped.

The height is measured using a metre rule. The timing is repeated several times, and an average value of t is recorded.

The experiment is repeated for a range of different heights. A graph of t^2 against h is plotted. Rearranging $h = \frac{1}{2}gt^2$ gives $t^2 = \frac{2h}{g}$, so the gradient of the line will be $\frac{2}{g}$.

Method 2

A cylinder is dropped down a clear plastic tube so that it cuts through a light beam (Figure 3.3b). The length of the cylinder, l, and the height of the top of the tube, h, above the light gate are measured, and the time, t, for which the cylinder cuts the beam is recorded.

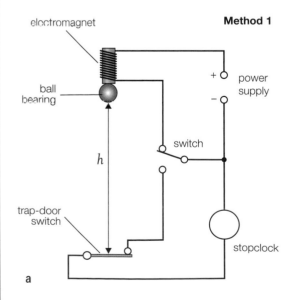

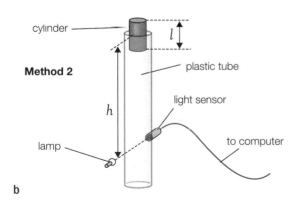

Method 1

Method 2

a

b

Figure 3.3 Two methods of measuring acceleration due to gravity.

The experiment is repeated several times, and an average value of t is obtained. The velocity, v, of the cylinder passing through the gate is calculated using $v = \dfrac{l}{t}$.

The experiment is repeated for a range of heights, and a graph of v^2 against h is plotted. As $v^2 = 2gh$, the gradient of the line will be $2g$.

A student performing experiments 1 and 2 obtained the results in Table 3.1.

Table 3.1 Results for Methods 1 and 2.

Method 1			Method 2 ($l = 0.050\,$m)		
h/m	t/s	t^2/s^2	h/m	t/s	v^2/(m s^{-1})^2
0.400	0.29		0.200	0.025	
0.800	0.40		0.400	0.018	
1.200	0.49		0.600	0.015	
1.600	0.57		0.800	0.013	
2.000	0.65		1.000	0.011	

(The values of t are the average of three readings.)

Method 3

If no specialist equipment is available, the principle of Method 1 can be used by measuring the time taken with a stopwatch for the object to be dropped over a range of heights measured with a metre rule. The times for each height should be taken several times and an average value calculated.

The data analysis is the same as for Method 2.

Questions

1 Copy and complete Table 3.1 to show the values of t^2/s^2 for Method 1 and v^2/(m s^{-1})^2 for Method 2. You may use a spreadsheet if you wish.
2 Plot a graph of t^2/s^2 against h/m for Method 1. Measure the gradient and calculate a value for the acceleration due to gravity.
3 Plot a graph of v^2/(m s^{-1})^2 against h/m for Method 2. Measure the gradient and calculate a value for the acceleration due to gravity.
4 Another student, performing Method 1, noticed that the graph did not pass through the origin, but was a straight line with a small intercept on the h axis. Suggest a reason for this.
5 Explain, with reference to the time measurements, why the uncertainty in the value of g is greater using Method 2.
6 Give one advantage of using Method 2.
7 A student using Method 3 has a manually operated stopwatch. She takes five readings for an object to fall from a height of 2.00 m into a sand tray.
   The recorded values are: 0.65 s, 0.61 s, 0.70 s, 0.63 s, 0.60 s
   a) Explain why five separate readings are taken at this height.
   b) Calculate the average time and determine the percentage uncertainty in this measurement.
   c) It is suggested that the experiment would be better performed from an upper storey of a tall building. Explain the advantages and disadvantages of such a change.

Free fall

In the absence of air resistance, all objects, whatever their mass, will fall freely with the same acceleration. Galileo Galilei tested this hypothesis by dropping different masses from the Leaning Tower of Pisa. Similarly, the astronauts on an Apollo Moon-landing mission showed that a hammer and a feather fall at the same rate on the Moon (where there is no air).

The acceleration of free fall on Earth, commonly termed the acceleration due to gravity, g, has a value of $9.8\,\mathrm{m\,s^{-2}}$.

When using the equations of motion for free-falling bodies, we need to be aware that displacement, velocity and acceleration are vector quantities and that the acceleration due to gravity always acts downward towards the Earth. If an object is thrown upwards, it will still be accelerating downwards at $9.8\,\mathrm{m\,s^{-2}}$. If the upward velocity is assigned a positive value, it follows that the acceleration must be negative. If the body was thrown downwards, its direction would be the same as that of g, so both can be given positive values.

Example

A ball is thrown vertically upwards with an initial velocity of $10\,\mathrm{m\,s^{-1}}$. Calculate the maximum height it will reach above its starting position, and the time it will take to reach this height.

Answer

At the instant the ball is at its maximum height, its velocity will be zero.

$u = +10\,\mathrm{m\,s^{-1}}$, $v = 0\,\mathrm{m\,s^{-1}}$, $a = -9.8\,\mathrm{m\,s^{-1}}$, $s = ?$, $t = ?$

Using $v^2 = u^2 + 2as$:

$0 = (+10\,\mathrm{m\,s^{-1}})^2 + 2 \times (-9.8\,\mathrm{m\,s^{-2}}) \times s$

$100\,\mathrm{m^2\,s^{-2}} = 19.6\,\mathrm{m\,s^{-2}} \times s$

$s = 5.1\,\mathrm{m}$

Using $v = u + at$:

$0 = +10\,\mathrm{m\,s^{-1}} + (-9.8\,\mathrm{m\,s^{-2}}) \times t$

$t = 1.0\,\mathrm{s}$

Test yourself

6 A football is dropped from the top of a building 30 m above the road below.

  a) Calculate the time the ball takes to reach the road.

  b) Calculate its velocity on impact with the road.

  c) State an assumption that you have made for these calculations.

7 A stone is thrown vertically upwards with an initial velocity of $8.0\,\mathrm{m\,s^{-1}}$ from a height of 1.8 m. Calculate:

  a) the stone's velocity after 0.50 s

  b) the maximum height it reaches

  c) the time taken for the stone to fall from the highest point to the ground.

3.2 Projectiles

Objects projected horizontally will still fall freely with a vertical downward acceleration of $9.8\,\mathrm{m\,s^{-2}}$.

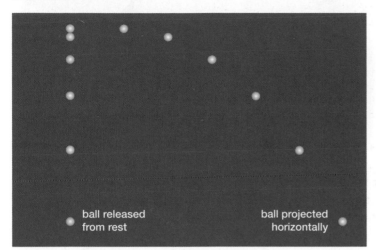

ball released from rest

ball projected horizontally

Figure 3.4 Falling spheres illuminated by a strobe lamp.

Figure 3.4 shows two balls – one released from rest and the other projected horizontally – photographed while illuminated by a strobe lamp. It is clear that the vertical acceleration of the projected ball is unaffected by its horizontal motion.

The horizontal distance travelled by projectiles can be found by considering the vertical and horizontal motions separately. In the vertical plane, the object will accelerate down at $9.8\,\mathrm{m\,s^{-2}}$, while the horizontal velocity remains constant (in the assumed absence of air resistance). The equations of motion can be applied to the vertical motion to ascertain the time spent by the object in free fall, and then the horizontal displacement is the product of the constant horizontal velocity and this time.

Example

A tennis ball is volleyed horizontally at a height of 1.5 m with a speed of $20\,\mathrm{m\,s^{-1}}$. Calculate the time taken by the ball to hit the ground and the horizontal distance travelled by the ball.

Answer

The instant the ball is struck, in addition to moving horizontally it will begin to fall downwards due to the gravitational force acting on it. The vertical motion of the ball is identical to that of a ball dropped from rest and falling to the ground.

In the vertical plane: $u = 0\,\mathrm{m\,s^{-1}}$, $a = 9.8\,\mathrm{m\,s^{-2}}$, $s = 1.5\,\mathrm{m}$, $t = ?$

Using $s = ut + \dfrac{1}{2}at^2$:

$$1.5\,\mathrm{m} = 0 + \frac{1}{2} \times 9.8\,\mathrm{m\,s^{-2}} \times t^2$$

giving

$t = 0.55\,\mathrm{s}$

In the horizontal plane: $u = 20\,\mathrm{m\,s^{-1}}$ (constant), $t = 0.55\,\mathrm{s}$, $s = ?$

$s = u \times t$

$\quad = 20\,\mathrm{m\,s^{-1}} \times 0.55\,\mathrm{s}$

$\quad = 11\,\mathrm{m}$

Activity 3.2

Monkey and hunter

A monkey hangs from the branch of a tree. A hunter aims his gun accurately at the monkey and fires. The sharp-eyed primate spots the bullet as it leaves the gun, releases its grip on the branch and falls to the ground. Does the monkey survive?

A laboratory model of the situation is shown in Figure 3.5. The 'gun' is clamped horizontally to the bench so that it is aimed directly at the 'monkey', which is held by the electromagnet 2 or 3 metres away. The 'bullet' is fired, breaking the circuit as it leaves the end of the barrel. The 'monkey' is released and falls freely to the ground.

For safety reasons all observers should be behind the 'gun' and protective glasses must be worn.

Question

Explain why the 'bullet' always hits the can, regardless of the distance from the 'gun'.

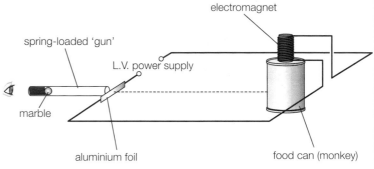

Figure 3.5 The 'monkey and hunter' scenario.

If an object is projected at an angle, the vertical and horizontal motion can still be treated separately by considering the components of the velocity in each plane.

Consider the motion of an object projected at an angle, θ, with an initial velocity, u.

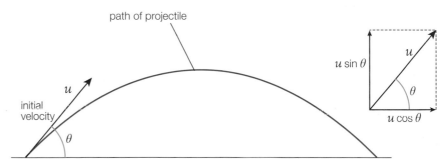

Figure 3.6 Projectile motion.

In the vertical plane, the initial velocity is $u \sin \theta$ (upwards) and the acceleration is $9.8 \, \mathrm{m \, s^{-2}}$ (downwards). As both are vectors, if a positive value is assigned to the initial velocity, the acceleration and the downward velocities and displacements will have negative values.

Neglecting air resistance, the horizontal component of the initial velocity, $u \cos \theta$, will remain constant throughout the motion.

Example

A football is kicked with a velocity of $12\,\text{m s}^{-1}$ at an angle of $30°$ to the ground.

Neglecting any effects of the air on the motion of the ball, calculate:

a) the vertical height reached by the ball

b) the time taken for the ball to rise to this height

c) the horizontal displacement of the ball when it falls to the ground.

Answers

In the vertical plane:

a) $u = 12\sin 30\,\text{m s}^{-1} = 6.0\,\text{m s}^{-1}$, $v = 0\,\text{m s}^{-1}$, $a = -9.8\,\text{m s}^{-1}$, $s = h = ?$

Using $v^2 = u^2 + 2as$:

$(0\,\text{m}^{-1})^2 = (6.0\,\text{m}^{-1})^2 + 2 \times (-9.8\,\text{m s}^{-2}) \times h$

$h = \dfrac{(6.0\,\text{m s}^{-1})^2}{19.6\,\text{m s}^{-2}} = 1.8\,\text{m}$

b) $u = 12\sin 30\,\text{m s}^{-1} = 6.0\,\text{m s}^{-1}$, $v = 0\,\text{m s}^{-1}$, $a = -9.8\,\text{m s}^{-2}$, $t = ?$

Using $v = u + at$:

$0\,\text{m s}^{-1} = 6.0\,\text{m s}^{-1} + (-9.8\,\text{m s}^{-2}) \times t$

$t = \dfrac{6.0\,\text{m s}^{-1}}{9.8\,\text{m s}^{-2}} = 0.61\,\text{s}$

In the horizontal plane:

c) $u = 12\cos 30 = 11\,\text{m s}^{-1}$, $t = 2 \times 0.61 = 1.2\,\text{s}$, $s = ?$

$s = u \times t$

$= 11\,\text{m s}^{-1} \times 1.2\,\text{s} = 13\,\text{m}$

Tip

The independence of the vertical and horizontal motion of bodies in free fall is an important concept. It allows you to use the equations of motion separately in each plane.

Test yourself

8 A coin is projected horizontally from a table top at the same time as an identical coin is dropped from the same point. Explain why they both strike the floor at the same time.

9 A cannonball is fired horizontally from the top of a vertical cliff, $60\,\text{m}$ above sea level, with a velocity of $200\,\text{m s}^{-1}$.

   a) How far from the cliff face does the cannonball enter the sea?

   b) What assumption have you made?

10 An arrow is fired at $40°$ to the horizontal with velocity $40\,\text{m s}^{-1}$. Calculate:

   a) the horizontal component of this velocity

   b) the vertical component of this velocity

   c) the time taken to hit a target in the same horizontal plane as the release point of the arrow

   d) the horizontal distance of the target from the archer.

3.3 Displacement–time and velocity–time graphs

Displacement–time graphs

Figure 3.7 shows the displacement–time graph for an object moving at constant velocity.

Constant, or uniform, velocity is calculated using the equation:

$$\text{velocity} = \frac{\text{displacement}}{\text{time}} = \frac{\Delta s}{\Delta t}$$

The velocity is represented by the **gradient** of a displacement–time graph.

The motion represented in Figure 3.8 is that of a trolley accelerating down an inclined plane. The gradient of the line gets steeper, which indicates an increase in velocity.

The instantaneous velocity is the gradient, $\frac{\delta s}{\delta t}$, of the graph at a point on the line.

To measure the small values of δs and δt would be very difficult and this would lead to large uncertainties in the measured velocity. The instantaneous velocity is more accurately measured by drawing a tangent to the graph line at the appropriate point. The gradient of this line is calculated using the larger values Δs and Δt.

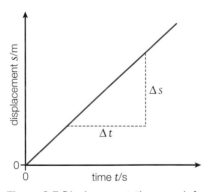

Figure 3.7 Displacement–time graph for constant velocity.

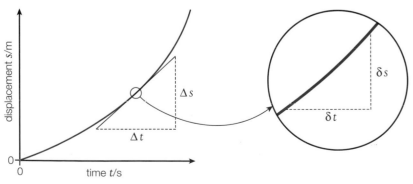

Figure 3.8 Determining the instantaneous velocity for accelerating object.

Activity 3.3

Measuring the displacement of a moving object

The transmitter of a motion sensor sends out pulses of ultrasound and infrared radiation that are picked up by the receiver. The distance of the transmitter from the sensor is continuously recorded. The sensor is interfaced with a computer with data-sampling software to measure the position of the trolley at fixed time intervals. The trolley may be pulled along the runway at constant velocity or allowed to accelerate down an inclined plane.

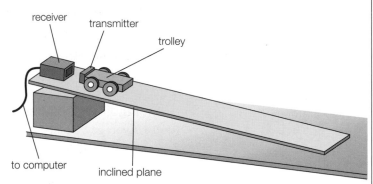

Figure 3.9 Measuring displacement using a motion sensor.

Displacement–time graphs can be drawn from the results or the graphs may be displayed on the computer.

The following readings were taken using the equipment shown in Figure 3.9.

Table 3.2 Experiment 1

t/s	0.20	0.40	0.60	0.80	1.00	1.20
s/m	0.120	0.250	0.404	0.538	0.681	0.822

Table 3.3 Experiment 2

t/s	0.10	0.20	0.30	0.40	0.50	0.60
s/m	0.078	0.165	0.250	0.332	0.448	0.600

Questions

1 On the same axes, plot displacement–time graphs for both experiments.
2 Describe the motion of the trolley in each experiment.
3 Use the graphs to determine the velocity of the trolley after 0.50 s in each case.

Velocity–time graphs

The results of the displacement–time experiment above can be used to illustrate how the velocity of the trolley changes as it moves along the runway. The gradient of the displacement–time graph is taken for a range of times, and a velocity–time graph is plotted. This is quite a tricky exercise, and one that is usually better left to the computer program.

Figure 3.10 Velocity-time graphs.

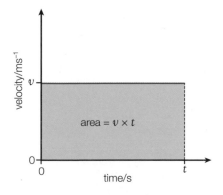

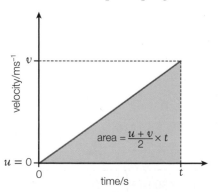

The graphs in Figure 3.10 represent (left) an object moving at constant velocity and (right) an object with uniform acceleration.

Acceleration is defined by the expression:

$$\text{acceleration} = \frac{\text{change in velocity}}{\text{change in time}} = \frac{\Delta v}{\Delta t}$$

The acceleration of an object is therefore equal to the **gradient** of a velocity–time graph. If the object is slowing down, the gradient will be negative. A negative value of acceleration indicates that the vector has the opposite direction to the velocity.

Finding displacement from velocity–time graphs

Average velocity is displacement divided by time. For uniform motion in a straight line, this leads to the equation:

$$s = \frac{(u + v)}{2} \times t$$

At constant velocity, the displacement is simply the product of velocity and time and will be the area under the horizontal line on the left-hand graph in Figure 3.10.

For the accelerating object (right-hand graph), the area under the line is $\frac{1}{2} \times$ base × height of the triangle, which is the same as the expression for displacement given above (since the value of u in this case is zero.)

For any velocity–time graph, the displacement is equal to the area between the line and the time axis.

Example

The velocity–time graph in Figure 3.11 represents the motion of a train as it travels from station A to station D, via B and C.

1 Describe the changes in the motion of the train.

2 Calculate:

  a) the acceleration from A to B

  b) the acceleration from C to D

  c) the total displacement from A to D.

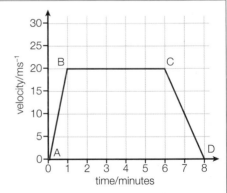

Figure 3.11 Velocity-time graph

Answers

1 The train accelerates uniformly from A to B, travels at $20\,\text{m s}^{-1}$ until it reaches C and then decelerates, uniformly, to D.

2 a) $\text{acceleration} = \text{gradient} = \dfrac{20\,\text{m s}^{-1}}{60\,\text{s}} = 0.33\,\text{m s}^{-2}$

  b) $\text{acceleration} = \text{gradient} = \dfrac{-20\,\text{m s}^{-1}}{120\,\text{s}} = -0.17\,\text{m s}^{-2}$

  c) displacement = area under graph

$$= \left(\frac{1}{2} \times 20\,\text{m s}^{-1} \times 60\,\text{s}\right) + (20\,\text{m s}^{-1} \times 300\,\text{s}) + \left(\frac{1}{2} \times 20\,\text{m s}^{-1} \times 120\,\text{s}\right)$$

$$= 7800\,\text{m}$$

Test yourself

11 Sketch a displacement–time graph for:

   a) a stationary object

   b) an object moving at constant velocity

   c) an object moving with uniform acceleration.

12 How would you determine the velocity of an object at a given time using a displacement–time graph?

13 Sketch a velocity–time graph for:

   a) an object moving at constant velocity

   b) an object moving with uniform acceleration.

14 How would you determine the acceleration at a given time using a velocity–time graph?

15 How would you determine the displacement of an object over a given time interval using a velocity–time graph?

Bouncing motion

The motion of a bouncing ball provides a good example of how motion is represented graphically.

Activity 3.4

Investigating the motion of a bouncing ball

An advanced motion sensor with the appropriate data logger is ideal for this investigation. The transmitter is attached to a basketball, or similar large ball, using Blu-tack. The receiver is clamped in position vertically above the transmitter, as shown in Figure 3.12. The motion sensor uses a mixture of infrared and ultrasonic signals to measure accurately the distance between the transmitter and the receiver.

The sensor is activated, and the ball is dropped and allowed to bounce two or three times. Practice is needed to ensure that the ball does not rotate and that the transmitter stays beneath the receiver throughout.

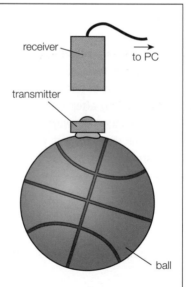

Figure 3.12 The set-up.

Tip

To describe the motion represented by a graph, always look at the gradient. Check if it is constant, positive, negative or zero. For a displacement–time graph, the gradient will give you the velocity at any time; for a velocity–time graph, the gradient gives you the acceleration.

The data can be displayed as a displacement–time, a velocity–time or an acceleration–time graph. The graphs in Figure 3.13 show the variations in displacement, velocity and acceleration on a common timescale. The initial displacement (the height above the floor) is positive, as are all upward values of displacement, velocity and acceleration; all downward values will be negative.

Questions

Explain the following features of the velocity–time graph.

a) The gradient between bounces is always the same.

b) The gradient during the bounce has a large positive value.

c) The area above and below the time axis is the same for a given bounce.

d) The velocity of the ball immediately after it has bounced is always smaller than that with which it strikes the ground.

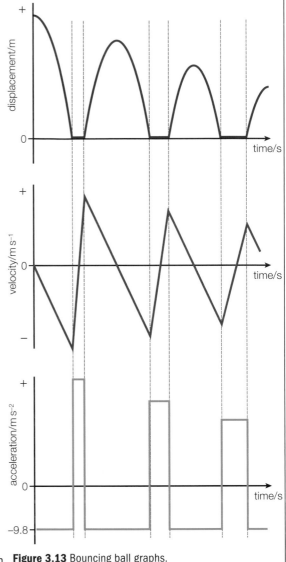

Figure 3.13 Bouncing ball graphs.

Exam practice questions

1 A train accelerated from rest to a velocity of $40\,\text{m s}^{-1}$ in a time of 1 minute and 20 seconds.

   a) What was the average acceleration of the train?

   A $0.33\,\text{m s}^{-2}$ **C** $2.0\,\text{m s}^{-2}$

   B $0.50\,\text{m s}^{-2}$ **D** $3.0\,\text{m s}^{-2}$ **[Total 1 Mark]**

   b) What was the distance travelled by the train?

   A $48\,\text{m}$ **C** $3200\,\text{m}$

   B $1600\,\text{m}$ **D** $4800\,\text{m}$ **[Total 1 Mark]**

2 The gradient of a displacement–time graph represents:

   A acceleration   **C** speed

   B distance   **D** velocity   **[Total 1 Mark]**

3 The area under a velocity-time graph represents

   A acceleration   **C** speed

   B displacement   **D** velocity   **[Total 1 Mark]**

4 Explain the difference between average velocity and instantaneous velocity.   **[Total 4 Marks]**

5 Write down the equations of motion.   **[Total 4 Marks]**

6 a) Define acceleration.   **[1]**

   b) The road–test information for a car states that it can travel from 0 to 60 mph in 8.0 s.

   $(1\,\text{mph} \approx 0.4\,\text{m s}^{-1})$

   i) Estimate the average acceleration of the car during this time.

   ii) Why is the acceleration unlikely to be uniform?   **[2]**

                                               [Total 3 Marks]

7 A cyclist travelling at $4.0\,\text{m s}^{-1}$ accelerates at a uniform rate of $0.4\,\text{m s}^{-2}$ for 20 s. Calculate:

   a) the final velocity of the cyclist

   b) the distance travelled by the cyclist in this time.   **[Total 2 Marks]**

8 A stone was dropped down a well. The splash was heard 2.2 s later. Calculate:

   a) the depth of the well

   b) the velocity of the stone when it hit the water.   **[Total 4 Marks]**

9 A ball was thrown vertically upwards with a velocity of $12\,\text{m s}^{-1}$ on release. Calculate:

    a) the maximum height from point of release reached by the ball

    b) the time taken for the ball to reach this height

    c) the velocity of the ball 2.0 s after it was released.   **[Total 3 Marks]**

10 A football player kicked a ball downfield. The ball left the boot at $30°$ to the ground and with a velocity of $20\,\text{m s}^{-1}$. Calculate:

    a) the vertical and horizontal components of the initial velocity of the ball

    b) the time taken for the ball to strike the ground

    c) the horizontal distance travelled by the ball before it bounced   **[Total 3 Marks]**

11 Describe the motion of an object represented by each of the lines A, B, C and D on the graph (Figure 3.14) for:

    a) displacement–time graph   **[4]**

    b) a velocity–time graph.   **[4]**

    [Total 8 Marks]

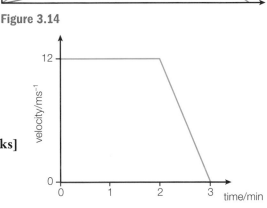

Figure 3.14

12 The graph in Figure 3.15 shows the variation of velocity with time for a body moving in a straight line.

    Calculate:

    a) the acceleration of the body during the final minute

    b) the total distance travelled in 3 minutes

    c) the average velocity over this time.   **[Total 5 Marks]**

13 a) Describe a method of determining the acceleration due to gravity in the laboratory. Include a labelled diagram of the equipment you would use and describe all the measurements you would take.   **[6]**

Figure 3.15

    b) State which of your readings is most likely to have the biggest effect on the uncertainty of the final answer and explain why.   **[3]**

    [Total 9 Marks]

14 An Olympic ski-jump run has a downhill slope with a take-off ramp and a landing slope.

    The ski-jumper jumps upwards to leave the ramp at an angle θ to the horizontal and lands at the 'K-point' 120 m down the landing slope.

    The landing slope is at an angle of $36°$ to the horizontal with its top point 3.0 m below the take-off ramp.

    The time taken between take-off and landing is 4.5 s.

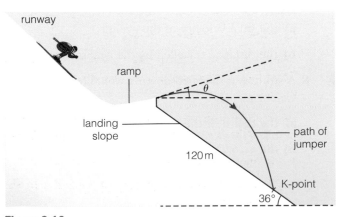

Figure 3.16

a) Calculate:

    i)   the horizontal distance travelled between take-off and landing **[1]**

    ii) the horizontal component of the take-off velocity   **[1]**

    iii) the vertical downward displacement of the jumper   **[1]**

    iv) the vertical component of the take-off velocity   **[1]**

    v)   the take-off angle, θ **[1]**

    vi) the speed of take-off.   **[1]**

b) Explain why the jumper would need a take-off speed much bigger than this to achieve this distance on this slope.   **[2]**

[Total 8 Marks]

15 A friction-driven toy car, with an ultrasound emitter attached, is energised and then released on a flat bench. An ultrasound receiver, placed at the starting point, is connected to a data logger that records the displacement from that point at regular time intervals. The results in the table were displayed on a computer loaded with suitable software.

t/ms	s/m
0	0
100	0.082
200	0.240
300	0.538
400	0.903
500	0.981
600	1.000

a) **i)**   Use the data in the table to plot a graph of displacement against time.   **[3]**

    ii) Use your graph to determine values of the velocity of the car at 100 ms intervals.   **[3]**

    iii) Plot a graph of velocity against time.   **[3]**

    iv) Use your second graph to describe the motion of the car over the complete journey.   **[3]**

b) The computer program is able to use the data to display instantaneous values of the velocity at the times used in part (a). Explain how this is achieved and why it is likely to give more accurate velocity values than the graphical method.   **[2]**

[Total 14 Marks]

Stretch and challenge

16 A tennis player returns a shot when the ball is 1.40 m from the net and 0.20 m above ground. She hits the ball at an angle of 30° to the horizontal and the ball leaves the racquet with a speed of 12 m s⁻¹.

a) Show that the ball will clear the net, which is 0.90 m high.   **[6]**

The distance from the net to the baseline on the other side is 11.9 m. The player's opponent allows the ball to bounce after it has crossed the net.

b) Perform suitable calculations to determine if the ball bounces in play (short of the baseline) or out of play (beyond the baseline).   **[4]**

Some players apply 'backspin' when playing such a shot.

c) Describe how the trajectory of the ball is likely to be affected by such spin.   **[2]**

[Total 12 Marks]

Momentum

4

Prior knowledge

In this chapter you will need to:
→ be able to apply the equations for rectilinear motion (Chapter 3)
→ have a basic knowledge of Newton's second and third laws of motion.

The key facts that will be useful are:
→ the difference between scalar and vector quantities
→ the meaning of the mass of a body
→ $F = ma$.

Test yourself on prior knowledge

1 Arrange the following into scalar or vector quantities: acceleration, displacement, distance, mass, speed, time and velocity.
2 Explain the difference between mass and weight.
3 Calculate the resultant force acting on a 2.0 kg mass falling through a liquid if the resistive force is 12 N.
4 What is the resultant force when forces of 3.0 N and 4.0 N act at right angles at a point?
5 Calculate the resultant force acting on a body of mass 5.6 kg that accelerates uniformly from rest to 14 m s⁻¹ in 12 s.
6 Calculate the acceleration of a trailer of mass 1.2 tonnes if the pull of the car is 800 N and the resistive forces on the trailer total 200 N.

Figure 4.1 On your marks, get set, go!

When you push on the ground, Newton's third law states that the ground pushes you back. The push on the sprinters in Figure 4.1 gives them forward momentum. An understanding of momentum in physics and its conservation is central to explaining how animals and vehicles move.

In this chapter you will develop your understanding of Newtonian mechanics, especially in the way bodies interact with one another. This will involve studying collisions, recoils and impulsive forces.

4.1 Linear momentum

When you push a loaded supermarket trolley to link with an empty stationary one, what happens depends on the speed with which you launch the loaded trolley but also on just how loaded it is. The product of a body's mass and velocity is useful in analysing such collisions: this product is called the body's momentum, or more carefully its **linear momentum**, p.

Key term

Momentum is given by the following formula:

momentum = mass × velocity

$p = mv$

The unit of momentum is $kg\,m\,s^{-1}$, but this can also be expressed as $N\,s$.

$$N\,s \equiv kg\,m\,s^{-1}$$

Because v stands for velocity and not speed in this definition, momentum is a vector.

Tip

Be careful when writing the unit for newton seconds to leave a gap between N and s, otherwise it looks a bit like newtons (plural). Similarly, when writing metres/second, leave a gap to avoid confusion with milliseconds.

Examples

1 How big is the momentum of a child of mass 25 kg walking at $0.38\,m\,s^{-1}$?

2 What is the momentum of a *Eurostar* train of mass 650 tonnes moving due South at a speed of $60\,m\,s^{-1}$?

Answers

1 Child's momentum = $25\,kg \times 0.38\,m\,s^{-1} = 9.5\,kg\,m\,s^{-1}$

2 *Eurostar*'s momentum is $650 \times 10^3\,kg \times 60\,m\,s^{-1} = 3.9 \times 10^7\,kg\,m\,s^{-1}$ due South.

Test yourself

1 What is the momentum of a body?

2 Is momentum a scalar or a vector quantity?

3 Write down the unit of momentum.

4 Calculate the momentum of a 640 g stone thrown with a velocity of $25\,m\,s^{-1}$.

5 Estimate the momentum of a charging rhinoceros.

Tip

Be careful to spot the difference between a question that asks for the size or magnitude of a body's momentum and one that asks for its (vector) magnitude *and* direction.

4.2 Collisions

The reason momentum is important is that it is *conserved*. In physics it is quantities like energy or charge or momentum, which are conserved, that help enormously in understanding how the world around us behaves. A study of the history of science supports this claim and we will see later, in the Year 2 Student's book, that it can even lead scientists to suggest the existence of undiscovered elementary particles.

Activity 4.1

A simple collision between trolleys on a friction-compensated slope

Trolley A of mass m_A is given a push so that, after its release, its interrupter card cuts through a light beam with the trolley moving at a constant velocity down the friction-compensated slope. The time the light beam is interrupted is recorded electronically and the (constant) velocity, u, of trolley A is calculated.

Trolley A has a cork attached with a pin sticking out of it that 'couples' to a cork attached to trolley B of mass m_B when A hits B. The two now move off together at a constant velocity, v, down the slope; a velocity that is calculated as the interrupter card on trolley A passes through the second light beam.

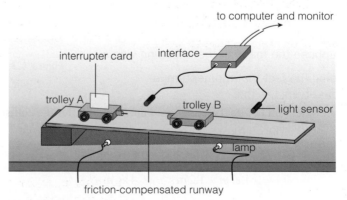

Figure 4.2

If momentum is conserved in this collision, then $m_A u$ should equal $(m_A + m_B)v$. The easiest case is when $m_A = m_B$, for which v should equal $\frac{1}{2}u$. By varying how fast trolley A is set moving, and also adding masses to either or both trolleys to alter their masses, the relation that conserves the trolleys' momentum

$$m_A u = (m_A + m_B)v$$

can be tested for a variety of different values in the expression.

A student used the above method and obtained the set of times given in Table 4.1.

Table 4.1 Experimental results.

t_1/s	t_2/s	u/m s^{-1}	v/m s^{-1}	Initial momentum/N s	Final momentum/N s
0.64	1.28				
0.73	1.45				
0.48	0.95				
0.60	1.20				

Mass of trolley A, m_A = 0.860 kg
Mass of trolley B, m_B = 0.840 kg
Length of interrupter card = 0.200 m

Questions

1 Copy and complete Table 4.1. (Use an Excel or Lotus spreadsheet if available.)
2 Comment on the values of the initial and final momentum.
3 **a)** What is meant by a 'friction-compensated' slope?
  b) Explain how the results would differ if the runway were horizontal.
4 State one method that you would use to ensure that the experiment was performed safely.

Instead of using light gates, the pin and cork can be got rid of and a transmitter can be attached to the back of the first trolley and a receiver fixed at the top of the slope. Using this technique, a displacement–time graph of the collision can be drawn, from which the software can make a speed–time graph as $v = \dfrac{\Delta s}{\Delta t}$. The experiment can, in principle, be tried with the supermarket trolleys mentioned at the beginning of this chapter, but measuring their speeds is quite difficult.

Example

A loaded supermarket trolley of unknown total mass is rolled into a stationary stack of two empty trolleys each of mass 8.0 kg. The speed of the loaded trolley before they link together is 2.2 m s^{-1} and the speed of the linked trolleys after the collision is 1.0 m s^{-1}. Calculate the mass of the shopping in the loaded trolley.

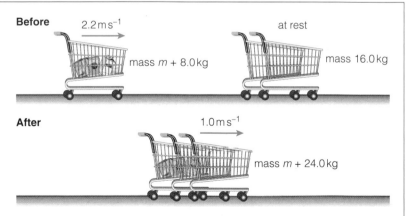

Figure 4.3

Answer

Suppose the mass of the shopping was m.

Using the fact that momentum is conserved, i.e. that

momentum before the collision = momentum after the collision

$$(m + 8.0\,\text{kg}) \times 2.2\,\text{m s}^{-1} = (m + 24.0\,\text{kg}) \times 1.0\,\text{m s}^{-1}$$

$$m \times (2.2\,\text{m s}^{-1} - 1.0\,\text{m s}^{-1}) = 24.0\,\text{kg m s}^{-1} - 17.6\,\text{kg m s}^{-1}$$

$$\text{i.e. } m \times 1.2\,\text{m s}^{-1} = 6.4\,\text{kg m s}^{-1}$$

$$m = 5.3\,\text{kg}$$

Tip

It really pays to sketch the situation before and after when solving problems about momentum conservation, but you can represent the colliding bodies, in this case trolleys, with simple blobs.

The experiment on page 45 can be developed to measure the speed of a small fast-moving object, for example, a rifle bullet. The bullet may be fired so as to lodge in a piece of wood attached to a free-running trolley or equivalent. If the mass of the bullet and wood + trolley are known, then measuring the speed of the trolley + wood + bullet after the collision enables the initial speed of the bullet to be found.

Safety note

If an airgun is going to be used, stringent safety precautions need to be followed. A model risk assessment should be consulted.

Example

An arrow of mass 120 g is fired into a block of wood with a mass of 0.60 kg resting on a wall. The arrow and wood fly off the wall at 8.0 m s^{-1}. Calculate the speed of the arrow.

Answer

We will use the concept that momentum is conserved. Let the speed of the arrow be v.

$$0.12\,\text{kg} \times v = (0.12\,\text{kg} + 0.60\,\text{kg}) \times 8.0\,\text{m s}^{-1}$$

$$v = 48\,\text{m s}^{-1}$$

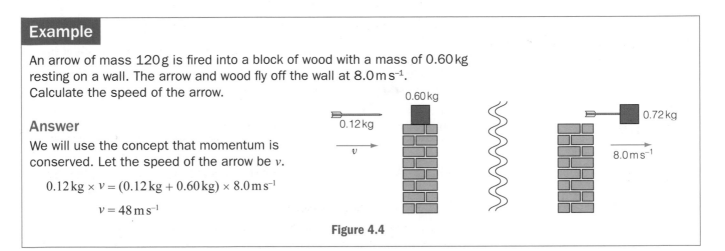

Figure 4.4

Collision experiments can also be simulated using gliders on linear air tracks where the friction is negligible. To get the gliders to stick together, a bit of sticky tack can be stuck to the gliders where they strike one another. More complex collisions can also be performed on the air track. Figure 4.5 shows a plan view of how gliders that bounce off each other when they collide can be demonstrated.

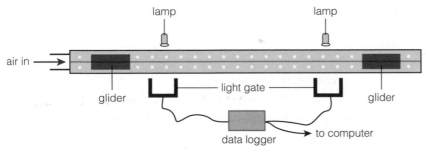

Figure 4.5 Testing momentum conservation using gliders on an air track.

The two gliders can be fitted with small magnetic buffers that are set to repel. The gliders are first set moving towards each other. When they collide they bounce apart and, provided the computer software can store the initial velocities and then record the velocities after the bounce, the **principle of conservation of linear momentum** can be tested. The experiment works just as well for other buffers such as pieces of cork or small pencil rubbers, showing that the principle works even when kinetic energy is lost, provided no external force acts on the gliders.

In practice, other forces such as friction often act, but forces that are perpendicular to the line in which momentum is being measured, for example, the weight of the gliders and the upward air force on them, can be ignored.

Key term

The **principle of conservation of linear momentum** states that in any interaction between bodies, linear momentum is conserved, provided that no resultant external force acts on the bodies.

Tip

Remember that momentum is a vector quantity. If the momentum is assigned a positive value in a left to right direction, then it must be negative when the body moves in the opposite direction from right to left.

Test yourself

6 What are the requirements for a system of interacting bodies if the law of conservation of momentum is to be tested in the laboratory?

7 A trolley of mass 400 g moving at 2.0 m s⁻¹ collides and sticks to a stationary trolley of mass 800 g. Calculate the velocity of the trolleys after the collision.

8 The trolleys in Question 7 are fitted with rubber bands so that the 400 g trolley bounces back from the larger trolley with a speed of 0.20 m s⁻¹. Calculate the velocity of the 800 g trolley after the collision.

9 An airgun pellet of mass 0.60 g is fired into a wooden block of mass 119.4 g resting on a smooth surface. If the initial velocity of the block and pellet after the impact is 0.30 m s⁻¹, calculate the speed of the pellet before the collision.

4.3 Momentum and Newton's laws

Key terms

Newton's second law of motion states that the resultant force exerted on a body is directly proportional to the rate of change of linear momentum of the body.

Impulse can be given by the following expression:

impulse = force × time
        = change in momentum

You may have used **Newton's second law of motion** for the case of a single force, *F*, acting on a body of fixed mass, *m*, namely $F = ma$. Newton's laws will be looked at in more detail in Chapter 5, but a link between the force applied to a single body and its momentum can be determined here. Using the equation for acceleration:

$$a = \frac{\Delta v}{\Delta t}$$

we can write the second law as

$$F = \frac{m\Delta v}{\Delta t}$$

or as

$$F = \frac{\Delta(mv)}{\Delta t}$$

In words,

$$\text{resultant force} = \frac{\text{change in momentum}}{\text{time interval}}$$

Newton's second law of motion can now be defined in terms of changes in momentum and can be applied to situations where the mass of a body may change, for example, a rocket burning fuel.

If the equation for Newton's second law is rearranged, we get

$$F\Delta t = \Delta(mv)$$

The product $F\Delta t$ is called the **impulse** of the force and has the unit N s. The equation shows that this is equal to the **change in momentum** of a mass when it is subjected to a resultant force.

Consider what happens in a situation where two people, such as the ice skaters in Figure 4.6, push each other apart as they move horizontally across the ice.

Figure 4.6 If the skaters push one another apart, the man's change of momentum will be equal and opposite to the woman's change of momentum.

As momentum must be conserved in the system, the change in momentum of the woman must be equal and opposite to that of the man. This means that the impulse on the woman must be equal and opposite to the impulse on the man.

As impulse equals $F\Delta t$ and the time interval, Δt, is the same for each skater, this can be shown mathematically as

$$F_w\Delta t = -F_m\Delta t$$

$$F_w = -F_m$$

This means that the force exerted by the man on the woman must be equal and opposite to that exerted by the woman on the man. This is an example of Newton's third law, which will be studied in detail in Chapter 5.

The case of the two skaters pushing one another apart is an example of **recoil**. The word recoil is often applied to a large gun after it fires a shell, but it can also be used to describe what happens when a nucleus emits an α-particle or when a rocket is fired in space.

Example

An α-particle is emitted from a polonium nucleus at a speed of $1.8 \times 10^7\,\text{m s}^{-1}$. The relative masses of the α-particle and the remaining nucleus are 4.002 and 212.0. Calculate the recoil velocity of the nucleus.

Answer

Let the masses be $4.002m$ and $212.0m$, respectively, and the recoil speed be v.

As momentum will be conserved, momentum before = 0, so momentum after = 0

$$4.002\,m \times 1.8 \times 10^7\,\text{m s}^{-1} - (4.002\,m + 212.0\,m)v = 0$$

The m cancels, giving $v = -3.4 \times 10^5\,\text{m s}^{-1}$; the – indicating that it is in the reverse direction to the α-particle.

at rest

Po   212.0m   4.002m

v $1.8 \times 10^7\,\text{m s}^{-1}$

Figure 4.7

Test yourself

10 State the unit of impulse.

11 Calculate the impulse needed to increase the velocity of a body of mass 2.5 kg from $3.2\,\text{m s}^{-1}$ to $4.8\,\text{m s}^{-1}$.

12 Calculate the average resultant force needed to increase the momentum of a cyclist from 400 N s to 600 N s in a time of 20 s.

13 A falling ball of mass 400 g hits the ground travelling at $6.0\,\text{m s}^{-1}$ and bounces up at $4.0\,\text{m s}^{-1}$. Calculate the change in momentum of the ball during the impact.

14 A cannon of mass 1200 kg fires a cannonball of mass 5.0 kg in a horizontal plane. If the recoil velocity of the cannon is $0.25\,\text{m s}^{-1}$, calculate the velocity of the cannonball as it leaves the barrel.

Tip

There is a change in direction when a ball bounces. This means that the momentum changes from a positive value to a negative one. The change in momentum will therefore be the sum of the magnitudes of the initial and final momentum.

4.4 Impulsive forces

Figure 4.8 shows the instant that a tennis racket hits a ball. Here the forces can be very large and the time during which they act very small. This is sometimes called an **impulsive force**.

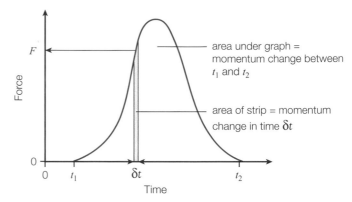

Figure 4.9 The variation of the force on the tennis ball with time.

labels on graph:
F — area under graph = momentum change between t_1 and t_2

area of strip = momentum change in time δt

Figure 4.8 An impulsive force.

The force on the tennis ball increases as the strings are stretched and then reduces to zero as the strings straighten and the ball leaves the racket. The variation of the force on the ball with time is shown in Figure 4.9.

The area of the strip represents the impulse, $F\delta t$ during an instant δt. The area under the curve equals the sum of all of the strips, $\sum F\delta t$, which is the same as the total change in momentum of the ball.

In general, the average size of a force can be estimated using the **impulse–momentum equation**

$$F\Delta t = m\Delta v$$

if the time of contact is known and the change in momentum is measured.

> ### Tip
>
> The units look different, but as the unit of $F\Delta t$ is the N s and the unit of $m\Delta v$ is the kg m s^{-1} they are equivalent because N $\equiv$ kg m s^{-2}.

Example

A gymnast first touches a trampoline moving downwards at $9.0\,\text{m s}^{-1}$ and leaves it moving upwards at the same speed. She has a mass of $55\,\text{kg}$ and is in contact with the trampoline surface for only $0.75\,\text{s}$ during a bounce.

What is the average resultant force acting on her during the bounce?

Answer
Her change of momentum is $55\,\text{kg} \times 18\,\text{m s}^{-1} = 990\,\text{kg m s}^{-1}$ upwards.

Therefore

$$F \times 0.75\,\text{s} = 990\,\text{kg m s}^{-1}$$

$$F = 1320\,\text{kg m s}^{-2}$$

So the average resultant force on her is $1300\,\text{N}$ to two significant figures.

Figure 4.10

Calculations of this kind can reveal just how large the forces involved in violent collisions can be – collisions like a car crashing into a wall. A crumple zone at the front of a car will increase the time taken for the car to come to a halt and so reduce the impulsive force thus reducing the chance of serious injury to the driver.

The impulse–momentum equation also helps us to analyse force versus time graphs, where $\Sigma F \Delta t$ represents the area under the graph. Such graphs are now routinely produced using modern force platforms in sports science laboratories.

It may seem difficult to apply the conservation of linear momentum to some situations. Consider a stone falling freely under gravity. The stone is clearly accelerating and so its momentum is continually changing. However, the law applies to **closed systems**, where no resultant external forces act. In this case, the system consists of the stone and the Earth. Using Newton's third law, the gravitational force applied by the Earth on the stone is equal and opposite to the gravitational force of the stone on the Earth. We can now see that the downward change in momentum of the stone is balanced by an upward change in momentum of the Earth. Because the mass of the Earth is so huge, however, there is no noticeable change in its momentum.

Test yourself

15 Give two SI units for impulse.

16 State Newton's second law of motion.

17 What is a closed system?

18 Calculate the average force needed to change the momentum of a body from 3.2 N s to 4.8 N s in a time of 0.50 s.

19 A tennis ball of mass 60 g is travelling at $10\,\mathrm{m\,s^{-1}}$ when struck by a tennis racket. It leaves the racket 0.20 s after the initial impact with a velocity of $30\,\mathrm{m\,s^{-1}}$ in the opposite direction. Calculate the average force of the racket on the ball.

Exam practice questions

1 a) How big is the momentum of a rifle bullet of mass 6.0 g moving at 450 m s⁻¹?

b) What is the momentum of a canal barge of mass 12 tonnes moving due East at a speed of 1.5 m s⁻¹?

c) Calculate the size of the momentum of a 8700 kg truck moving at 50 km h⁻¹. **[Total 3 Marks]**

2 The momentum of a male Olympic sprinter is about

    A 1800 kg m s⁻¹   **C** 300 kg m s⁻¹

    B 900 kg m s⁻¹   **D** 90 kg m s⁻¹   **[Total 1 Mark]**

3 Which of the following physical quantities is not conserved?

    A charge   **C** impulse

    B energy   **D** momentum   **[Total 1 Mark]**

4 The relationship $m_A u = (m_A + m_B)v$ might be applied to a collision where

    A a heavy object A strikes a stationary light object B

    B a heavy object A sticks to a stationary light object B

    C a heavy object B sticks to a stationary light object A

    D a heavy object B strikes a stationary light object A.   **[Total 1 Mark]**

5 A 'baddie' in a Western is shot in the chest. What would happen to him in real life if he were to be shot like this?

    A He would slump to the floor.

    B He would fall forward.

    C He would fall backwards.

    D He would not at first move.   **[Total 1 Mark]**

6 Which of these expressions does not have the units of energy?

    A $\frac{1}{2}mu$ **C** $\frac{2p^2}{m}$

    B $2mu$ **D** $\frac{1}{2}pu$ **[Total 1 Mark]**

7 A force of 250 N acts on a body of mass 80 kg moving at 6.0 m s⁻¹ for 12 s. The change of momentum of the body is

    A 480 N s   **C** 3000 N s

    B 1500 N s   **D** 5760 N s   **[Total 1 Mark]**

8 In a nuclear experiment an unknown particle moving at 390 km s⁻¹ makes a head-on collision with a carbon nucleus. After the collision a single particle continues at a speed of 30 km s⁻¹. Express the mass of the unknown particle as a fraction of the mass of a carbon nucleus.   **[Total 1 Mark]**

9 A man is stranded in the middle of a frozen pond. He is unable to grip the ice with his feet. To move to the edge of the pond he should

    A   hurl himself towards the edge

    B   fall towards the edge and repeat this movement

    C   lift one leg and lower it towards the edge

    D   throw his coat away from the edge.   **[Total 1 Mark]**

10 What is the change of momentum of a tennis ball of mass 55 g that bounces vertically on the ground (the speeds before and after impact are 6 m s^{-1} and 4 m s^{-1}, respectively)?   **[Total 3 Marks]**

11 a) Use a graphical method to add a momentum of 50 kg m s^{-1} North to one of 100 kg m s^{-1} West.

    b) Confirm your answer by calculating the result of the same addition.   **[Total 4 Marks]**

12 A rowing boat has a mass of 300 kg and is floating at rest in still water. A cox of mass 55 kg starts to walk from one end to the other at a steady speed of 1.2 m s^{-1}. What happens to the boat?   **[Total 2 Marks]**

13 The graph in Figure 4.11 shows the horizontal force exerted by a wall on a ball that is hit hard at it horizontally and rebounds. The mass of the ball is 57.5 g.

By estimating the area under the graph, deduce the change of velocity of the ball during this impact.   **[Total 4 Marks]**

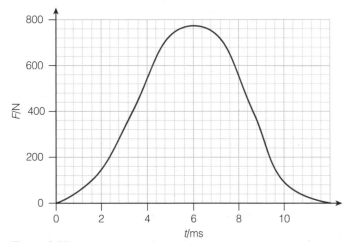

14 A rocket car can accelerate from 0 to 100 mph in a very short time. Outline the physics principles behind this method of propulsion.   **[Total 2 Marks]**

15 By suitable scale drawing deduce what must be added to a momentum of 64 kg m s^{-1} to the East to produce a resulting momentum of 64 kg m s^{-1} to the North–West.

Figure 4.11

    [Total 3 Marks]

16 A golf ball of mass 45 g is struck by a golf club with a force that varies with time as follows:

t/ms	0	0.5	1	1.5	2	2.5	3	3.5	4
F/N	0	700	1800	2500	2100	1500	800	300	0

Plot a graph of *F* up against *t* along and find the impulse exerted by the club on the ball. Hence find the initial speed of the ball after it is struck.   **[Total 3 Marks]**

17 One billion people in India are organised to jump up from the ground at noon local time on a certain day. The average mass of each person is 60 kg and the average speed at which they leave the ground is 2 m s^{-1}. The Earth has a mass of 6×10^{24} kg. Does the Earth recoil? Explain your answer.   **[Total 5 Marks]**

18 A stationary helicopter of mass 1200 kg hovers over the scene of a road accident. The whirling helicopter blades force air downwards at a speed of $20\,\text{m}\,\text{s}^{-1}$. Calculate the rate, in $\text{kg}\,\text{s}^{-1}$, at which air is projected downwards.

[Total 4 Marks]

19 A vertical wall has a surface area of $120\,\text{m}^2$. It lies perpendicular to a wind that blows at $25\,\text{m}\,\text{s}^{-1}$. The density of air is $1.3\,\text{kg}\,\text{m}^{-3}$. What force does the air exert on the wall, assuming that the air striking the wall is brought momentarily to rest.

[Total 4 Marks]

Stretch and challenge

20 Figure 4.12 shows how the horizontal force, F, on a jogger's foot varies with time, t, during a single stride. F is taken to be positive when it is in the direction of the jogger's motion.

  a) Describe how F varies during this stride.   **[4]**

  b) Estimate the average backward force on the jogger after his foot touches the ground.   **[2]**

  c) The area above the axis is the same as the area below the axis. Explain what this tells you about the motion of the jogger.   **[2]**

  d) Estimate the area under the graph during the period when the horizontal force of the jogger is forward.   **[3]**

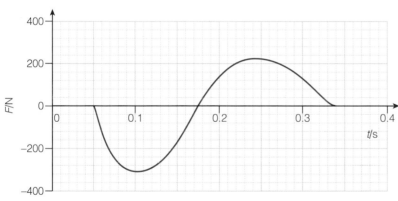

Figure 4.12

[Total 11 Marks]

5 Forces

Prior knowledge

In this chapter you will need to:
→ recall the ideas of momentum and Newton's second law of motion that were covered in Chapter 4
→ use the addition of vectors and the equations of motion covered in Chapters 1 and 3.

The key facts that will be useful are:
→ forces are vector quantities that can be considered as two components at right angles to each other
→ when two or more forces act at a point they can behave like a single, resultant force acting at that point
→ a force acting upon an object can cause a change in its momentum
→ for a fixed mass, m, the resultant force, $\Sigma F = ma$.

Test yourself on prior knowledge

1 A man applies a force of 50 N at 30° to the horizontal to a rope attached to a large stone. Calculate the vertical and horizontal components of the force applied to the stone.

2 A roller skater experiences a forward driving force of 480 N from the ground and a backward, resistive force of 80 N. If she also feels a vertical upward force of 500 N from the floor, calculate the magnitude and direction of the resultant force acting between the floor and the skates.

3 a) A ball of mass 120 g strikes a wall at right angles to the surface and bounces straight back again. The ball struck the wall moving with a velocity of $8.0\,\mathrm{m\,s^{-1}}$ and rebounded at $6.0\,\mathrm{m\,s^{-1}}$. Calculate the change in momentum of the ball during the collision.

    b) The collision occurred over a time of 40 ms. Calculate the average force exerted by the wall on the ball.

4 Calculate the acceleration of a 12 kg mass when a resultant force of 100 N acts upon it.

5.1 Nature and types of force

Forces push or pull and squeeze or stretch. There are several different types of force, most of which will be studied in more detail later. Forces fall into two categories: distant and contact forces.

Forces at a distance

Gravitational forces act over very large distances. The planets are kept in orbit by the gravitational pull of the Sun. We all experience a gravitational attraction to the Earth: this is called our weight and it pulls us down to Earth if we jump out of an aeroplane or fall off a chair.

The gravitational force between two objects depends on the mass of the objects and their separation. Gravitational forces are very small unless one or both of the masses is extremely large. The attractive force between two elephants standing close together would be very difficult to detect, but the force between a mouse and the Earth is noticeable.

The gravitational force exerted by the Earth on a mass of one kilogram is known as the **gravitational field strength**, g. It is defined by the equation $g = \dfrac{F}{m}$ and has a value of $9.8\,\text{N}\,\text{kg}^{-1}$ on the surface of the Earth. It follows that an object of mass, m, will have a weight, W, given by:

$$W = mg$$

The concept of a gravitational field will be considered in detail in the Year 2 Student's book.

Key term

Gravitational field strength is the force per unit mass on an object placed in a gravitational field.

Example

Estimate the weight of an elephant, a mouse, yourself and a bag of sugar.

Answer
For estimates, g can be taken as $10\,\text{N}\,\text{kg}^{-1}$.

Elephant: mass = 3–7 tonnes, so $W = 5000\,\text{kg} \times 10\,\text{N}\,\text{kg}^{-1} = 50\,\text{kN}$

Mouse: mass = 50 grams, so $W = 0.050\,\text{kg} \times 10\,\text{N}\,\text{kg}^{-1} = 0.5\,\text{N}$

Average adult: mass = 70 kg, so $W = 70\,\text{kg} \times 10\,\text{N}\,\text{kg}^{-1} = 700\,\text{N}$

Sugar: mass = 1 kg, so $W = 1\,\text{kg} \times 10\,\text{N}\,\text{kg}^{-1} = 10\,\text{N}$

Electrostatic and electromagnetic forces are also examples of forces that act over a distance.

Insulating rods can be given positive or negative charges by rubbing with woollen dusters or cotton rags. The electrostatic forces produced can be investigated by suspending one rod from a thread and holding another rod close to it (Figure 5.1). Unlike gravitational forces, both pulling and pushing effects are observed. Similar charges repel and opposite charges attract. A similar effect is observed when magnets are used.

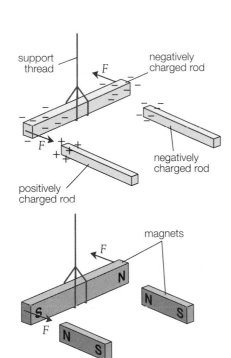

Figure 5.1 Electrostatic and electromagnetic forces.

Contact forces

When you stand on the floor, your weight pushes the floor. The floor and the soles of your shoes are in contact with each other and become slightly compressed. The electrons in the atoms are displaced and short-range forces result in an upward push of the floor on you. Other examples of contact forces include friction between moving surfaces, viscous forces in liquids and air resistance.

Tension

When a rubber band is stretched, the molecular separation increases. This leads to short-range attractive forces between the molecules. The band is in a state of tension, with the molecular forces trying to restore the band to its original length. All objects subjected to a stretching force are in a state of tension. Examples are the cables of a suspension bridge and tow ropes.

5.2 Forces in equilibrium

Newton's first law of motion

Imagine a stationary rock deep in space, where the gravitational fields of distant stars are negligibly small. The rock experiences no forces and does not move. If the rock were moving it would continue with constant velocity until it felt the gravitational effect of another body in space.

The above ideas seem obvious and trivial nowadays when space missions are commonplace, but they were the bedrock of the laws relating to forces and motion that were formulated by Sir Isaac Newton in the seventeenth century. At that time, laws were based on earthly experiences – for example, a constant force needs to be applied to a cart to make it continue to move at constant speed in a straight line. Newton was able to envisage a world without the hidden forces of friction, drag and weight.

Newton's first law of motion states that an object will remain in a state of rest or continue to move with a constant velocity unless acted upon by a resultant external force.

Newton's first law explains the dilemma of the moving cart: although a constant force is acting on the cart, it is opposed by an equal set of frictional forces, so there is no *resultant* force acting on it.

Equilibrium

Newton's first law also relates to the 'thought experiment' about the rock in space. If two or more forces were acting on the rock in space, the rock might still remain in a state of rest or uniform motion. Newton's first law adds the provision that no *resultant* force acts on the object. A pair of equal and opposite forces acting on the rock would not affect its motion.

When a number of forces act on a body, and the vector sum of these forces is zero, the body is said to be in **equilibrium**.

Free-body force diagrams

Objects may be subjected to a range of different forces and may themselves exert forces on other bodies.

The rock climber in Figures 5.2 and 5.3 feels his weight and the weight of his rucksack pulling down and the push of the rock on his feet, as well as the pull of the rope on his arms. It is clear that there must be frictional forces between the rope and the hands and the rock face and the climber's boots, and that the climber's companion and the rock will experience forces from the climber.

> **Key term**
>
> **Newton's first law of motion** states that an object will remain in a state of rest or continue to move with a constant velocity unless acted upon by a resultant external force.

> **Key term**
>
> A body is in **equilibrium** if, when acted on by a number of forces, the resultant force is zero.

Key term

The **centre of gravity** of an object is the point at which the weight of the object can be said to act.

It is very useful to isolate the forces acting on a single object by drawing a **free-body force diagram**. The motion of the climber can be analysed using the free-body force diagram in Figure 5.4. If the climber is in equilibrium, the line of action of all three forces will pass through a single point. In this case, it is simpler to represent the object as a point mass (known as the **centre of gravity**), with all of the forces acting at this point.

Figure 5.2 Rock climber.

Figure 5.3 Forces acting on a rock climber.

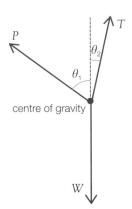

Figure 5.4 Free-body force diagram of a rock climber.

Tip

By representing a body as a point mass it becomes much easier to find the components of the forces.

For equilibrium, the forces can be represented on a scale drawing as a closed triangle that gives a vector sum of zero (see Chapter 1). Alternatively, the sum of the components in the horizontal and vertical planes must both be zero. So, using Figure 5.4:

In the vertical plane:

$$(P \cos \theta_1 + T \cos \theta_2) - W = 0$$

In the horizontal plane:

$$T \sin \theta_2 - P \sin \theta_1 = 0$$

Activity 5.1

Investigating the equilibrium of three vertical forces

Two pulley wheels are attached to a vertically clamped board. Three weights are connected by thread so that they are in equilibrium, as shown in Figure 5.5.

The direction of the upward forces is found by marking the position of the thread on to a sheet of paper fixed to the board (Figure 5.6). The angles that forces W_1 and W_2 make with the vertical line drawn between them are measured. For equilibrium:

$$W_1 \cos \theta_1 + W_2 \cos \theta_2 = W_3$$

The experiment is repeated using different values of W_1, W_2 and W_3.

An alternative approach is to find the resultant of W_1 and W_2. This is achieved with the help of a scale drawing that uses the parallelogram of forces method of vector addition. The directions of W_1 and W_2 are marked as before, and the lines

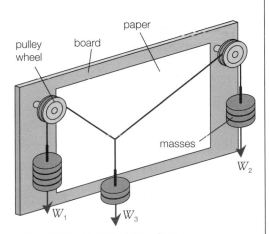

Figure 5.5 Equilibrium of three forces.

are drawn to scale to show the magnitudes of W_1 and W_2. A parallelogram is constructed as shown in Figure 5.7.

The parallelogram rule states that the sum of the forces represented in size and direction by adjacent sides of a parallelogram is represented in size and direction by the diagonal of the parallelogram. For equilibrium, the resultant, R, of W_1 and W_2 must be equal in magnitude to W_3 and must be acting vertically upwards.

A student set up the arrangement shown in Figure 5.5. She used a mass of 400 g to give W_1, 500 g for W_2 and 600 g for W_3. The angles θ_1 and θ_2 were measured as 35 ° and 65 °, respectively.

Questions

1 Draw a large-scale diagram representing the forces W_1 and W_2, and use the law of parallelogram of forces to determine the magnitude and direction of the resultant of these forces.
2 For the masses to be in equilibrium, the resultant of W_1 and W_2 should equal the weight W_3, and act vertically upwards. Suggest a reason why this is not quite true for the above experiment.

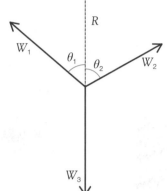

Figure 5.6

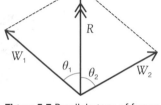

Figure 5.7 Parallelogram of forces.

Example

A 10 kg mass is suspended from a beam using a length of rope. The mass is pulled to one side so that the rope makes an angle of 40 ° to the vertical, as shown in Figure 5.8.

1 Write expressions for the vertical and horizontal components of the tension, T, in the rope.

2 Use the conditions for equilibrium in the vertical plane to show that the value of T is about 130 N.

3 Use the horizontal component to determine F.

Answers

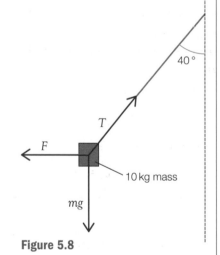

Figure 5.8

1 Vertical component of $T = T\cos 40°$; horizontal component of $T = T\sin 40°$ (if you wish to use cosines for all components use $T\cos 50°$ here).

2 For equilibrium in the vertical plane:

$$T\cos 40° = mg$$

$$T = \frac{10 \text{ kg} \times 9.8 \text{ N kg}^{-1}}{\cos 40°} = 128 \text{ N} = 130 \text{ N}$$

3 In the horizontal plane:

$$F = T\sin 40°$$

$$= 128 \text{ N} \times \sin 40° = 82 \text{ N}$$

Tip

For 'Show that…' questions, always calculate the answer to one more significant figure than asked for in the question. For example, in Question 2 of the example, the value of T is shown as 128 N.

Test yourself

1 Give two examples of forces that act at a distance and two examples of contact forces.

2 State Newton's first law of motion.

3 A ball is dropped inside a railway carriage. Use Newton's first law to explain whether the ball will hit the floor in front of, level with or behind the point of release if

   a) the train is stationary

   b) the train is moving forward at constant speed

   c) the train is moving backward at constant speed

   d) the train is accelerating in a forward direction.

4 Explain why a bicycle can move at a constant speed even when a driving force is applied.

5 What is meant by equilibrium?

6 A hanging sign of mass 24 kg is blown by the wind so that the chains holding it are at an angle of 30° to the vertical.

   a) Draw a free-body force diagram of the forces acting on the sign.

   b) Calculate the magnitude of the force of the wind on the sign.

5.3 Newton's second law of motion applied to fixed masses

Newton's first law states that a body will remain at rest or move with constant velocity if no resultant force acts on it. What will happen to the body if a resultant force does act on the body? Consider the rock in space described on page 57. Give the rock a push and its momentum will change. A static rock will move or the velocity of a moving rock will increase, decrease or change direction. We saw in Chapter 4 that such a change in momentum is called an **impulse** and that the impulsive force is given by the expression:

$$\Sigma F = \frac{\Delta(mv)}{\Delta t}$$

This equation represents a general version of **Newton's second law of motion**.

In most cases you will be considering the effects of forces on bodies of fixed mass so the equation becomes

$$\Sigma F = \frac{m\Delta v}{\Delta t} = ma$$

The relationship between the resultant force applied to a body and this acceleration is given in **Newton's second law of motion for fixed masses**, which states that the acceleration of a body of constant mass is proportional to the resultant force applied to it and in the direction of the resultant force. This definition can be represented by the equation:

$$\Sigma F = ma$$

provided that the unit of force is in newtons (N), the mass is in kilograms (kg) and the acceleration is in metres per second squared (ms^{-2}). It should be noted that this expression applies only for forces acting on fixed masses.

Key term

Newton's second law of motion for fixed masses states that the acceleration of a body of constant mass is proportional to the resultant force applied to it and in the direction of the resultant force.

Investigating factors that affect the acceleration of an object

The acceleration of the trolley in Figure 5.9 is found by measuring the time taken for the card to cut the light beams and the separation of the gates.

If l is the length of the card, t_1 and t_2 are the times to break the light beams and s is the separation of the gates:

initial velocity, $u = \dfrac{l}{t_1}$

final velocity, $v = \dfrac{l}{t_2}$

The acceleration can be calculated using the equation of motion (see Chapter 3):

$$v^2 = u^2 + 2as$$

The resultant force is provided by the weight on the hanger.

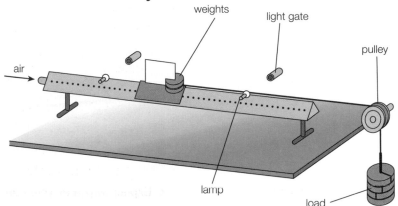

Figure 5.9 Investigating the acceleration of an object.

Effect of the force acting on a fixed mass

It is important to be aware that the gravitational force pulling down on the weights and hanger is acting on the total mass of the system – that is, the mass of the trolley plus the weights and hanger. If extra weights were put on to the hanger, the mass of the system would increase. To make the resultant force increase without changing the mass of the system, four (or more) 10 g weights are initially fixed on the trolley. The force can be increased, without altering the mass, by taking a weight off the trolley and placing it on to the hanger.

A range of forces is used and the corresponding accelerations are measured. A graph of acceleration against force is then plotted.

Effect of mass on the acceleration produced by a fixed force

The weight on the hanger is kept constant. Weights are added to or taken from the trolley to increase or decrease the mass of the system. The acceleration is found for each mass, as before, and a graph of acceleration against the inverse mass is plotted. If both graphs are straight lines through the origin, it follows that:

$a \propto F$ and $a \propto \dfrac{1}{m}$ $\Rightarrow a \propto \dfrac{F}{m}$

A typical set of results for an experiment to find the effect of the force acting on a fixed mass is as follows:

$l = 0.200 \, \text{m}$ $s = 0.400 \, \text{m}$

Questions

1 Complete Table 5.1 (use an Excel spreadsheet with appropriate equations in the column headings if you wish) and plot a graph of $a/\text{m s}^{-2}$ against F/N.
2 Use the graph to show that the mass of the trolley is about 0.21 kg.
3 The air track minimises the effects of friction on the above experiments. What other precaution must be taken to ensure that no further external forces act on the trolley?

Table 5.1 Experimental results.

F/N	t_1/s	t_2/s	$u/\text{m s}^{-1}$	$v/\text{m s}^{-1}$	$a/\text{m s}^{-2}$
0.10	0.48	0.29			
0.20	0.32	0.20			
0.30	0.25	0.16			
0.40	0.16	0.12			
0.50	0.15	0.11			

Tip

In many examples, a body may be acted upon by a number of forces. You must ensure that the *resultant* force is calculated before applying the equation $\Sigma F = ma$.

Example

The tension in the rope pulling a water skier is 500N and the resistive force of the water is 400N. The total mass of the water skier and skis is 78kg.

Figure 5.10 Water skier.

1 Draw a free-body force diagram for the water skier.

2 Calculate:

  a) the resultant force

  b) the acceleration of the skier.

Answers

1

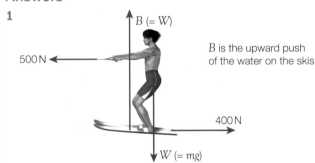

B is the upward push of the water on the skis

Figure 5.11 Free-body force diagram for the water skier.

2 a) $F_R = 500\,\text{N} - 400\,\text{N} = 100\,\text{N}$

  b) $a = \dfrac{F}{m} = \dfrac{100\,\text{N}}{78\,\text{kg}} - 1.3\,\text{m}\,\text{s}^{-2}$

Multi-body systems

An engine pulls three carriages along a track (Figure 5.12). The forward (frictional) force of the track on the wheels of the engine drives it forward. If the driving force exceeds the resistive forces, the train will accelerate.

D driving force of tracks on wheels   total resistive force F

Figure 5.12 Forces on a train.

The resultant force is required to accelerate not only the engine but also the carriages:

$$D - F = (m_e + 3m_c)a$$

The engine and the carriages accelerate at the same rate, but the carriages have no driving force. The free-body force diagram in Figure 5.13 is for a single carriage of the train.

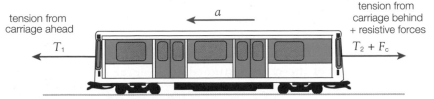

tension from carriage ahead
T_1

a

tension from carriage behind + resistive forces
$T_2 + F_c$

Figure 5.13 Free-body force diagram of carriage.

The carriage is accelerating like the rest of the train, so it must experience a resultant force. We can apply Newton's second law to the carriage as follows:

$$T_1 - (T_2 + F_c) = m_c a$$

Example

A lift and its load are raised by a cable, as shown in Figure 5.14.

1 If the total mass of the lift and its contents is 1.5 tonnes and the tension in the cable is 22.2 kN, calculate the acceleration of the lift.

2 a) Draw a free-body force diagram for a woman of mass 50 kg standing in the lift.

   b) Calculate the normal contact force of the floor of the lift on her feet.

Answers

1 Resultant force = $22\,200\,\text{N} - (1500\,\text{kg} \times 9.8\,\text{m s}^{-2}) =$ 7500 N

   $7500\,\text{N} = 1500\,\text{kg} \times a$

   $a = 5.0\,\text{m s}^{-2}$ upwards

2 a) See Figure 5.15.

   b) Resultant force = $N - (50\,\text{kg} \times 9.8\,\text{m s}^{-2}) = (N - 490\,\text{N})$

   $(N - 490\,\text{N}) = 50\,\text{kg} \times 5.0\,\text{m s}^{-2}$

   $N = 490\,\text{N} + 250\,\text{N} = 740\,\text{N}$

Figure 5.14 Lift.

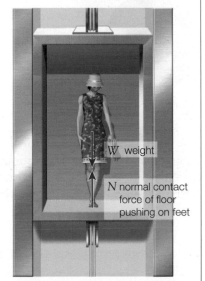

W weight

N normal contact force of floor pushing on feet

Figure 5.15 Free-body force diagram of woman in lift.

Test yourself

7 State Newton's second law of motion.

8 Show that 1 newton (N) is equivalent to $1\,\text{kg m s}^{-2}$.

9 Explain the difference between mass and weight.

10 Calculate the average braking force needed to bring a truck of mass 12 tonnes to rest from a speed of $20\,\text{m s}^{-1}$ in a distance of 40 m.

11 A 50 kg mass is acted upon by forces of 30 N and 40 N which are at right angles to each other.

   a) Calculate the resultant force acting on the mass.

   b) Calculate the acceleration of the mass.

Systems of interacting bodies

In order to examine the motion of a single body, we isolated the forces acting on the body and ignored the effect of the forces on surrounding objects. When more than one body is considered, the interacting objects make up a **system**. On a large scale, we have the Solar System, in which the Sun and planets interact by gravitational forces; on a smaller scale, two colliding snooker balls can be considered as a system in which contact forces prevail. Most of the systems studied at AS level will involve only two bodies.

Figure 5.16 Who is pushing?

Key term

Newton's third law of motion states that if body A exerts a force on body B, body B will exert an equal and opposite force on body A.

5.4 Newton's third law of motion

If you push against a wall, you will feel the wall pushing back on your hands. Imagine you are in space close to the stationary rock described on page 57. The rock and you will remain at rest unless either is subjected to a resultant force. If you push the rock, it will move away from you, but you will also 'feel' a force pushing back on your hands. This force will change your state of rest and you will move away from the rock.

The following experiment can be used to test this principle on Earth. Two students on roller skates face each other on a level surface (Figure 5.16).

First, student A pushes student B (the rock) and then student B pushes student A. In each case, both students will move away from the other.

Sir Isaac Newton used the same 'thought experiment' to formulate his third law of motion: if body A exerts a force on body B, body B will exert an equal and opposite force on body A.

Newton's third law pairs

Newton's third law of motion always applies to a pair of objects. Consider the Moon orbiting the Earth.

Figure 5.17 The Earth and the Moon.

The Earth exerts a gravitational force on the Moon, and the Moon exerts an equal and opposite gravitational force on the Earth. The single force on the Moon is needed to maintain its orbit around the Earth, while the force of the Moon on the Earth gives rise to the tides.

Activity 5.3

Illustrating Newton's third law

A spring is attached to trolley A. An identical trolley, trolley B, is connected to trolley A using a fine thread. The spring is compressed as shown in Figure 5.18. The trolleys are placed on a smooth, horizontal surface.

The thread is burned and the trolleys move apart. Assuming that the surfaces are uniform and the trolleys roll in a similar manner, the distances travelled will be proportional to the applied force.

The experiment is repeated with the spring attached to trolley B. In both cases, the trolleys should move the same distance.

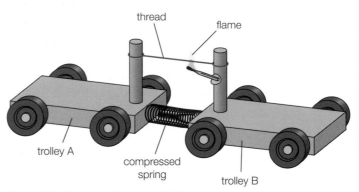

Figure 5.18 Illustrating Newton's third law.

In the case of the trolleys, trolley A exerts a contact (compression) force on trolley B and trolley B exerts an equal and opposite contact force on trolley A.

Newton's third law is sometimes defined by stating that every **action** has an equal and opposite **reaction**.

This definition can be misunderstood if the terms action and reaction are not clearly defined. In the definition given, action refers to the force exerted by one body on another and reaction is the resulting force of the second body on the first. Because of the ambiguity of the term, the word reaction is best avoided when describing forces.

Newton's third law pairs of forces must always:

- act on two separate bodies
- be of the same type
- act along the same line
- be equal in magnitude
- act in opposite directions.

In some cases there are two pairs of forces acting between two bodies. Consider the case of a woman standing on the Earth shown in Figure 5.19.

The woman and the Earth experience two pairs of forces. The Earth pulls the woman down with a gravitational force and the woman pulls the Earth up with an equal gravitational force. But there is also a pair of contact forces at the ground. The woman pushes down on to the surface of the Earth and the ground pushes up on her feet with an equal and opposite force.

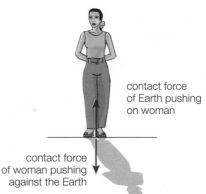

Figure 5.19 Newton's third law pairs.

Newton's third law and equilibrium

Let us consider forces in equilibrium. It is apparent that a woman at rest on the ground is subjected to a downward gravitational force (her weight) and an equal upward contact force (the normal force). The woman is in equilibrium as the resultant force is clearly zero. If the woman jumps in the air, there will still be a gravitational force on her by the Earth and an equal and opposite force on the Earth, but there is only one force acting on the woman, so she is not in equilibrium.

Tip

It is good practice to state what is providing the force and the object upon which it is acting in all cases. For example: force of car tyre on the road, force of the road on the car tyres, etc.

Never try to apply Newton's third law to a single body.

5.5 Turning forces

When a force is applied to an object that is not a point mass it may rotate. For example, when a frisbee is thrown, the force is applied at the edge of the disc making it spin as it flies through the air. The turning effect of a force is called its moment.

The moment of a force depends on the size of the force and the perpendicular distance of its line of action from the axis of rotation.

The moment of a force is calculated using the equation

moment = force × perpendicular distance between the line of action of the force and the axis of rotation

$$= F \times x$$

The unit of a moment is N m. Although this appears to be the same as the joule, it differs because it is a type of vector; it can cause either a clockwise or anticlockwise rotation.

Key term

The **moment** of a force is the product of the force and the perpendicular distance between the axis of rotation and the line of action of the force.

The principle of moments

Earlier in this chapter, we introduced the idea that a body is in equilibrium when the resultant force acting upon it is zero. However, it is possible to have equal and opposite forces acting on an object with lines of action at different directions to the axis of rotation so that, although it will not accelerate, it will rotate.

In order for an object to be in rotational equilibrium, the sum of the clockwise moments acting upon it about any point must equal the sum of the anticlockwise moments about the same point. This is known as the principle of moments.

Key term

The **principle of moments** states that if a system is in equilibrium, the sum of the clockwise moments must equal the sum of the anticlockwise moments.

Example

1) Figure 5.20 shows two children on a seesaw. The weight of the seesaw beam is 500 N and child A has a weight of 400 N.

   a) Draw a free-body force diagram of the system.

   b) What is the moment of the 400 N pull of the Earth on the beam?

   c) Calculate the weight of child B.

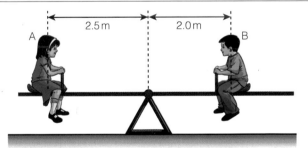

Figure 5.20

2) A delivery woman holds a wardrobe as shown in Figure 5.21, prior to pushing it into position by applying a force, P, at right angles to the wardrobe door. The mass of the wardrobe is 40 kg and she applies the force at a point 1.60 m from the bottom of the wardrobe. In this equilibrium position, the centre of gravity, G, is vertically above a point 12 cm to the left of the pivot.

   a) Use the principle of moments to calculate the size of the force that she applies to keep the wardrobe in this position

   b) Determine the value of the upward force, N, of the floor on the corner of the base

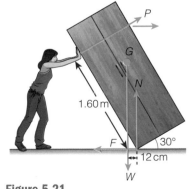

Figure 5.21

c) Explain why the wardrobe will topple into a stable position when the delivery woman pushes it past the point where the centre of gravity is vertically above the corner of the base.

Answers

1 a)

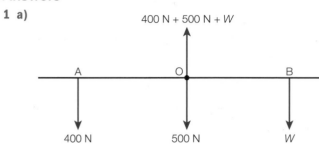

Figure 5.22

b) The weight of the beam acts through the axis of rotation and so the moment is zero.

c) Using the principle of moments:

sum of clockwise moments = sum of anticlockwise moments

$$400\,N \times 2.5\,m = W \times 2.0\,m$$
$$W = 500\,N$$

Child B has a weight of 500 N.

2 a) Taking moments about the bottom corner (the "pivot") sum of the clockwise moments = sum of the anticlockwise moments

$$P \times 1.60\,m = (40\,kg \times 9.8\,m\,s^{-2}) \times 0.12\,m$$
$$P = 29\,N$$

b) For equilibrium: the sum of the vertical (or horizontal) components of the forces must be zero

$$N + P \sin 30° = W \rightarrow N = 392\,N - 15\,N = 380\,N$$

c) When the line of action of the weight is to the right of the pivot, there will be a resultant clockwise moment acting on the wardrobe and so it will rotate until the base is flat on the floor.

Activity 5.4

Determining the position of the centre of gravity of an irregular object

This method uses the principle of moments and the concept that the weight acting on an object can be thought to act at the centre of gravity.

A piece of stiff card or thin plywood is cut into an irregular shape with three holes cut or drilled at positions close to the edges. The card is pivoted on a rod or nail at one of the holes as shown in Figure 5.23. If the card is moved to one side (Figure 5.23a), there will be an unbalanced anticlockwise moment ($W \times x$) acting upon it that will cause it to rotate. When the centre of gravity, G, is vertically below the pivot, the moment of the force is zero and the card is in equilibrium (Figure 5.23b).

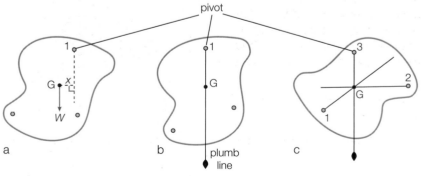

Figure 5.23

A piece of thread attached to a metal bob is suspended from the rod so that it acts as a plumb line. A vertical line is drawn alongside the plumb line onto the card from the point of suspension. The centre of gravity must lie along this line. The procedure is repeated with the card suspended from the other two holes. The centre of gravity is the point where the lines cross (Figure 5.23c).

Question
How would you check that the point of intersection of the lines is at the centre of gravity of the card?

Couples and torque

When a tap is turned on or off, it is usual to apply a pair of equal forces in opposite directions at each end of the tap head (Figure 5.24). The resultant moment is called a **couple**.

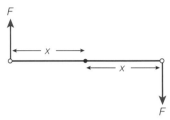

Couple $= F \times x + F \times x = 2(F \times x)$

= magnitude of each force $\times$ perpendicular separation of their lines of action

Figure 5.24 A couple.

In engineering, the turning effect due to two or more forces is usually referred to as the **torque**. The torque of the pistons connected to the crankshaft is often stated when the powers produced by cars are compared.

The unit of torque is N m (newton metre).

Test yourself

12 State Newton's third law.

13 Write down the conditions that apply to Newton's third law pairs.

14 Give the Newton's third law pairs for the forces acting upon an object suspended by a spring. State the nature and direction of each force.

15 Explain why you tend to slip backwards when trying to run across an icy surface.

16 The rate of flow of water from a tap is 12 kg min^{-1}. The initial velocity of the water is 0.60 m s^{-1}. Calculate the force exerted by the water on the tap.

17 Calculate the resultant moment produced on a pivoted rod when a force of 10 N acts at a perpendicular distance of 1.2 m to the right of the axis of rotation and a force of 15 N acts at a perpendicular distance of 0.9 m to the left of the pivot.

18 Describe how you would make an estimate for the position of the centre of gravity of a soup ladle.

Exam practice questions

1 a) The weight of the man in Figure 5.25 makes a Newton's third law pair with which other force?

Figure 5.25

   A The downward contact force of the man on the table.

   B The upward contact force of the table on the man.

   C The upward gravitational force of the man on the Earth.

   D The upward gravitational force of the man on the table.

                     [Total 1 Mark]

b) The man is in equilibrium because his weight is equal to:

   A the downward contact force of the man on the table

   B the upward contact force of the table on the man

   C the upward gravitational force of the man on the Earth

   D the upward gravitational force of the man on the table.

                     [Total 1 Mark]

2 Explain the differences between forces at a distance and contact forces. Give one example of each. **[Total 4 Marks]**

3 Copy and complete the table by including one example of each type of force.
                                                     [Total 3 Marks]

Type of force	Example
gravitational	
electromagnetic	
nuclear	

4 A car of mass 1500 kg tows a trailer of mass 2500 kg. The horizontal driving force of the road on the wheels of the car is 7000 N and both the car and the trailer experience resistive forces of 1000 N.

a) The acceleration of the vehicles is:

   A $1.25\,\mathrm{m\,s^{-2}}$ **C** $2.00\,\mathrm{m\,s^{-2}}$

   B $1.75\,\mathrm{m\,s^{-2}}$ **D** $3.33\,\mathrm{m\,s^{-2}}$ **[Total 1 Mark]**

b) The tension in the coupling when the driving force is reduced so that the acceleration falls to $1.00\,\mathrm{m\,s^{-2}}$ is:

   A 2500 N   **C** 4000 N

   B 3500 N   **D** 6000 N   **[Total 1 Mark]**

5 a) What condition applies for a body to be in equilibrium under the action of several forces? **[1]**

b) Describe an experiment to show this condition for three vertical forces. **[6]**

                     [Total 7 Marks]

6 A boat of mass 800 kg is pulled horizontally along the sand at constant velocity by a cable attached to a winch. The cable is at an angle of 15° to the horizontal.

   a) Draw a free-body force diagram for the boat, labelling the forces tension, friction, normal reaction and weight. **[4]**

   b) State which of the forces acts over a distance. **[1]**

   c) If the tension in the cable is 4.0 kN, show that:

   i) the horizontal frictional force opposing the motion of the boat is about 3.9 kN

   ii) the normal contact force of the sand on the boat is about 7.0 kN. **[3]**

   [Total 8 Marks]

7 A simple accelerometer consists of a metal sphere of mass 50 g attached to the ceiling of a moving train. Figure 5.26 shows the position of the sphere when the train is accelerating.

   a) Draw diagrams to show the position of the sphere when the train is

   i) moving at constant speed

   ii) slowing down with a lower negative acceleration than its initial forward acceleration. **[2]**

   b) Draw a free-body force diagram for the sphere when the train is accelerating. **[1]**

   c) Draw a vector diagram for the forces acting on the sphere when the train is accelerating and use this to determine the magnitude of the acceleration. **[3]**

   [Total 6 Marks]

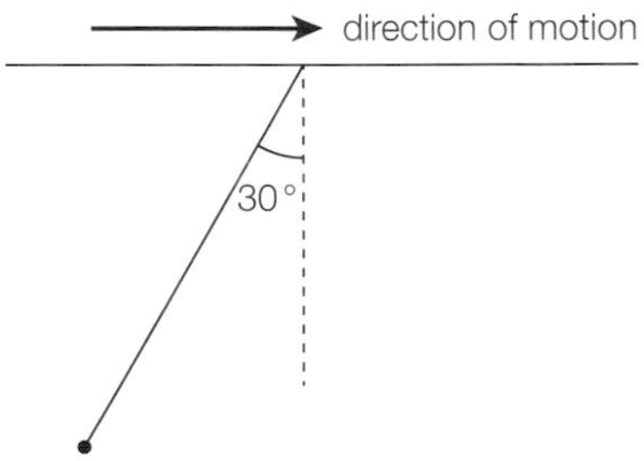

direction of motion

30°

Figure 5.26

8 a) State Newton's third law of motion. **[1]**

   b) Figure 5.27 shows a satellite in orbit around the Earth.

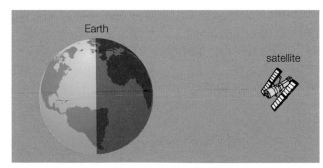

Earth

satellite

Figure 5.27

   i) Copy the diagram and show the forces acting on each body. **[2]**

   ii) State three properties of these forces that show them to be a Newton's third law pair. **[3]**

   [Total 6 Marks]

9 A jet-ski uses a pump to propel a stream of water from a nozzle at high speeds from the rear of the craft.

   a) Use Newton's third law to explain how this drives the jet-ski through the water. **[2]**

b) A particular jet-ski has a mass of 300 kg and, when carrying a rider of mass 80 kg at a constant speed of $15\,\text{m s}^{-1}$, its pump delivers 150 kg of water per second at a speed of $5.0\,\text{m s}^{-1}$ relative to the nozzle. Calculate the thrust developed by the pump. **[2]**

c) The power is increased so that $160\,\text{kg s}^{-1}$ is emitted at $6.0\,\text{m s}^{-1}$. Calculate the speed of the jet-ski if this power increase is maintained for 5.0 s. State any assumptions that you have made. **[4]**

[Total 8 Marks]

10 Figure 5.28 shows the forces acting on a suitcase held at rest by a vertical upward force, F, at the handle. The case and its contents have a total mass of 8.0 kg and the centre of gravity is labelled G.

a) Use Newton's first law to write an equation relating the size of the forces N, W and F. **[1]**

b) Use the principle of moments to find the magnitude of force F. **[2]**

c) Explain how the value of F would be affected if the centre of gravity was lower down in the case. **[3]**

[Total 6 Marks]

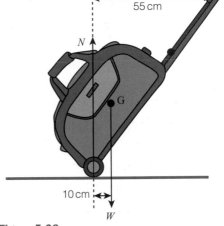

Figure 5.28

11 A tractor of mass 2500 kg starts to pull a plough of mass 1250 kg along a furrow. The horizontal driving force of the ground on the tractor's wheels is 4000 N and the resistive forces acting on the plough and tractor are 2250 N and 500 N, respectively.

a) Calculate the initial acceleration of the tractor and plough. **[3]**

b) Calculate the tension in the coupling between the tractor and the plough. **[2]**

c) Calculate the tension in the coupling when the tractor has reached a constant speed. (State what assumption you have made.) **[2]**

[Total 7 Marks]

12 Figure 5.29 shows how the momentum of two railway trucks A and B varies when they collide in a railway siding.

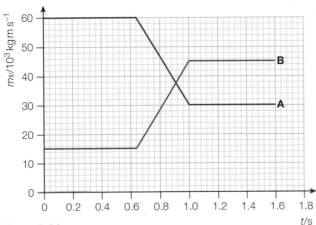

Figure 5.29

a) Calculate the force that truck A exerts on truck B. **[3]**

b) Calculate the force that truck B exerts on truck A. **[3]**

c) Comment on your answers to (a) and (b). **[2]**

13 A student is asked to explain why a book is at rest on a table. The student states that the book exerts a force on the table and, by Newton's third law, the table exerts an equal and opposite force on the book so that the book is in equilibrium. Explain why this is not a valid argument and give the true reason that the book is in equilibrium. **[Total 5 Marks]**

14 A teacher wishes to demonstrate Newton's second and third laws by performing an experiment in a lift (elevator). He stands on a bathroom scale balance and holds a suitcase of mass 10 kg suspended from a spring balance.

When stationary in the lift the bathroom scales read 800 N.

The lift accelerates upward at $2.0 \, \text{m s}^{-2}$ for a distance of 4.0 m, continues at constant velocity for a further 12.0 m and comes to rest (with uniform negative acceleration) 8.0 m higher up.

a) Describe the Newton's third law pairs that apply to the teacher plus suitcase and the suitcase alone. In each case give the magnitude, direction and nature of the forces. **[4]**

b) Calculate the readings on both balances for each of the three stages of the journey. **[8]**

[Total 12 Marks]

Stretch and challenge

15 A tall beaker containing water is placed onto a top-pan balance. The balance reading is 800 g. A hollow sphere of mass 20 g is dropped into the water from a height of 20 cm. The sphere momentarily comes to rest in the body of the liquid, rises back towards the surface and comes to rest, floating in the water.

a) Calculate the speed of the sphere as it hits the surface of the water. **[1]**

b) Determine the average force of the water on the ball if it brings it to rest in 0.50 s. **[2]**

c) Describe and explain the changes in the balance reading from the moment that the sphere enters the water until it is floating at rest within it. **[6]**

In practice the changes in the readings of the balance would be rapid and difficult to discern.

d) Suggest a method that could be employed to show these changes more clearly. **[1]**

[Total 10 Marks]

6 Work, energy and power

Prior knowledge

In this chapter you will need to:
→ be aware that work is done when a force is applied to an object in order to make it move and that energy is needed to do work
→ be able to apply the equations for rectilinear motion (Chapter 3) and Newton's laws of motion (Chapter 5) to systems of static and moving masses.

The key facts that will be useful are:
→ work = force × distance moved (measured in joules, J)
→ weight = force due to gravity = mg
→ energy is the ability to do work (measured in joules, J)
→ power is the rate of doing work or the rate of transfer of energy (measured in watts, W)
→ the law of conservation of energy tells us that energy cannot be created or destroyed.

Test yourself on prior knowledge

1 When is work done?
2 Calculate the work done when a force of 10 N is applied to an object that moves 25 m in the direction in which the force is acting.
3 Calculate the work done when a mass of 15 kg is raised 3.00 m from the surface of the Earth.
4 A mass of 5.0 kg accelerates from rest to 4.0 m s^{-1} in a time of 8.0 s.
   a) Calculate the acceleration of the mass.
   b) Calculate the resultant force acting on the mass.
   c) Calculate the distance travelled by the mass.
   d) Calculate the work done on the mass.
5 Describe the energy transformations that occur when a filament lamp is switched on.
6 Calculate the power output of a motor that can lift a load weighing 120 N through a height of 2.5 m in a time of 0.5 s.

6.1 Work and energy

We use energy continually in everyday life. Energy is needed to move around, to keep us warm and to manufacture things. But what is it and where does it go? In this section, you will investigate and apply the principle of conservation of energy, including the use of work done, gravitational potential energy and kinetic energy.

Work

Work is done when the point of application of a force is moved in the direction of the force.

One **joule** is the name for a newton metre; it is the work done when the point of application of a force of one newton is moved through one metre.

Lift this book about 50 cm above the table. You have done some work. Now lift the book about one metre above the table. You have done more work. If two books are used, even more work will be done. The amount of work you do depends on the force applied to lift the books and how far the books are moved.

Work is done when the point of application of a force is moved in the direction of the force:

work done = force × distance moved by the force in the direction of the force

Work is measured in joules (J).

Example

Calculate the work done in raising a book of mass 0.80 kg through a vertical height of 1.50 m.

Answer

Force applied to raise the book = $0.80 \, \text{kg} \times 9.8 \, \text{m s}^{-2} = 7.8 \, \text{N}$

Work done by the force = $7.8 \, \text{N} \times 1.50 \, \text{m} = 12 \, \text{J}$

A force often moves an object in a different direction to the direction of the applied force.

The work done by the man pulling the pram in Figure 6.1 is the product of the horizontal component of the force and the horizontal displacement, Δx, of the pram:

$$\text{work done on pram} = F \cos\theta \times \Delta x$$

This may also be expressed as the product of the force and the component of the displacement along the line of action of the force:

$$\text{work done on pram} = F \times \Delta x \cos\theta$$

In this example, the angle θ is between $0°$ and $90°$. The value of $\cos\theta$ will be between $+1$ and 0, and hence the work done will be positive. A value of θ between $90°$ and $180°$ indicates that the force is acting in the opposite direction to the movement of the pram. Work is done *by* the pram against the resistive forces, so $F \cos\theta$ will be negative.

In most cases, the forces acting on moving objects are not constant, and it is usual to express the work done in terms of the average force:

$$\Delta W = F_{av} \Delta s$$

Force and displacement are **vector quantities** and you might expect that their product should also be a vector. If a force is first moved in one direction, and then moved by the same distance in the opposite direction, the vector sum of the forces would equal zero. As work is done in both directions it follows that the total work done is the sum of these values and must therefore be a **scalar quantity**.

pram

F

θ

Figure 6.1

Relationship between work and energy

Work and energy are closely linked. Energy is transferred *to* an object when work is done on it. When energy is transferred *from* the object to another system we say that the object does work.

When you raised the book off the bench, you transferred energy to the book. To stop a moving bicycle by applying the brakes, energy is transferred from the bike to the braking system.

We have already defined work, so it is convenient to define energy in terms of work done: an object has **energy** when it has the ability to do work. This statement is useful, but it is not universal. Heating is energy transfer between regions of different temperatures. In heat engines, some but not all of the energy transferred can do work. Heat energy will be studied in more detail in the Year 2 Student's Book.

It follows that energy, like work, is measured in joules (J).

All forms of energy can be described in terms of potential energy and kinetic energy.

Key term

Energy is the ability to do work.

Potential energy

In Figure 6.2a, work is done on the box when it is lifted onto the bench. The upward force needed to raise the box has the same magnitude as its weight (*mg*). An object that can do work by virtue of its position or because it is stretched or twisted, is said to have **potential energy**.

Work done to lift the box = force × distance moved

$$\Delta W = mg \times \Delta h$$

Key term

Potential energy is the ability of an object to do work by virtue of its position or state.

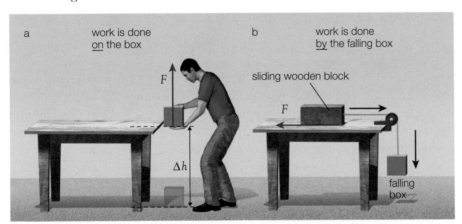

Figure 6.2

Imagine the box is attached to a block of wood by a length of string, as shown in Figure 6.2b. If the box falls off the bench, the pull of the string on the block will do work as the block moves along the bench. The box placed in Figure 6.2a has the *potential* to do work by virtue of its position. Because the work is done by the displacement of a gravitational force, the box is said to have **gravitational potential energy** (GPE or E_{grav}).

Whenever the box is raised by a height, Δh, its GPE is increased. The gain in GPE (ΔGPE or ΔE_{grav}) is equal to the work done on the box:

$$\Delta E_{grav} = mg\,\Delta h$$

Key term

Gravitational potential energy is the energy an object possesses by virtue of its position in a gravitational field.

If the box is raised more than a few kilometres above the surface of the Earth, the value of the gravitational field strength, g, will become noticeably smaller. The value will continue to fall as the distance from the Earth increases. The expression above therefore relates to variations in height close to the Earth's surface, where g can be assumed to be constant ($9.8\,\text{N}\,\text{kg}^{-1}$).

In Figure 6.3, work is done stretching the rubber band. When the band is released, the weight slides across the surface doing work against the frictional force. Energy has been stored in the rubber band, which enables it to do work. This is called **elastic potential energy** (EPE).

Key term

Elastic potential energy (EPE) is the ability of an object to do work by virtue of a change in its shape.

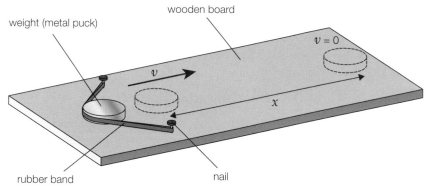

Figure 6.3

The EPE of a deformed object can be represented in terms of the work done to change its shape. For a spring or wire being stretched using an average force, F_{av}, so that it extends by Δx

$$\Delta \text{EPE} = F_{av}\Delta x$$

The elastic potential energy of stretched bands and wires is usually referred to as the **elastic strain energy** and will be further investigated in Chapter 13.

Kinetic energy

Key term

Kinetic energy (KE) is the ability of an object to do work by virtue of its motion.

Anything that moves is able to do work. Imagine a hammer driving a nail into a wall. Just before impact, the hammerhead is moving quickly. When the nail is struck, it moves into the wall and the hammer stops. The hammer has transferred energy to do work driving the nail into the wall. The energy to do this work is called **kinetic energy** (KE).

An expression for the kinetic energy of a moving object can be found using Newton's laws (Chapter 5) and the equations of uniformly accelerated motion (Chapter 3).

In Figure 6.4, a constant force, F, is applied to the car over distance s. The car accelerates from rest ($u = 0$) until it reaches velocity v.

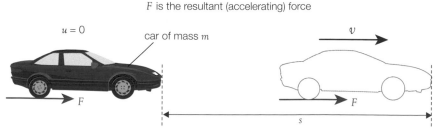

Figure 6.4 Accelerating car.

Gain in KE = work done on the car

$$= F \times s = ma \times s = m \times as$$

Using the equation:

$$v^2 = u^2 + 2as = 0 + 2as$$

$$\Rightarrow as = \frac{v^2}{2}$$

Gain in KE $= m \times \dfrac{v^2}{2} = \dfrac{1}{2}mv^2$

The kinetic energy (KE or sometimes E_k) of an object of mass m moving with a speed v is $\dfrac{1}{2}mv^2$.

Example

Estimate the kinetic energy of an Olympic sprinter crossing the finish line.

Answer

When an estimate is required, a wide range of answers will be accepted. You will be required to make sensible approximations of the mass and speed of the athlete.

Mass of male athlete = 80 kg (180 lb)

Mass of female athlete = 60 kg (130 lb)

Time for 100 m is around 10 s (11 s for women), so a reasonable estimate of speed = 10 m s^{-1}.

Kinetic energy $= \dfrac{1}{2}mv^2$

$$= \frac{1}{2} \times 60\,\text{kg} \times (10\,\text{m s}^{-1})^2$$

$$= 3000\,\text{J}$$

$$= 3\,\text{kJ}$$

Tip

As such a wide range of values is possible with estimates of this type, estimates should be written to one significant figure. If you had chosen m = 65 kg and v = 11 m s^{-1}, the kinetic energy would be 3932.5 mJ but should be given as 4000 J or 4 kJ.

If a mass changes speed from u to v:

change in kinetic energy,

$$\Delta \text{KE} = \frac{1}{2}mv^2 - \frac{1}{2}mu^2$$

Example

A car of mass 1500 kg travelling at 20 m s^{-1} is slowed down to 10 m s^{-1} by applying the brakes. The car travels 30 m during braking.

1 Determine the kinetic energy transferred from the car.

2 Calculate the average horizontal resistive force acting on the car.

Answers

1 $\Delta KE = \dfrac{1}{2}mv^2 - \dfrac{1}{2}mu^2$

KE transferred $= \dfrac{1}{2} \times 1500\,\text{kg} \times (20\,\text{m s}^{-1})^2$

$\qquad\qquad\quad - \dfrac{1}{2} \times 1500\,\text{kg} \times (10\,\text{m s}^{-1})^2$

$$= 2.25 \times 10^5\,\text{J}$$

$$= 2.3 \times 10^5\,\text{J}$$

2 Work done = force × distance

$$= F \times 30\,\text{m}$$

$$= 2.25 \times 10^5\,\text{J}$$

Average force, $F_{av} = 7.5 \times 10^3\,\text{N}$

Tip

When all the quantities are given to two significant figures, the final answer should also be given to two significant figures. However, intermediate calculations should not be rounded. In the previous example, $2.25 \times 10^5\,\text{J}$ should be used in the calculation of average force.

Gravitational potential energy

When the pendulum bob in Figure 6.5 is displaced to one extreme, it is raised by a height Δh. The GPE of the bob increases by $mg\Delta h$. When the bob is released, it begins to fall and accelerates towards the midpoint. The moving bob gains kinetic energy. The GPE of the bob decreases to a minimum value at the lowest point of the swing where the KE has a maximum value. The KE is then transferred back to GPE as the bob moves up to the other extreme.

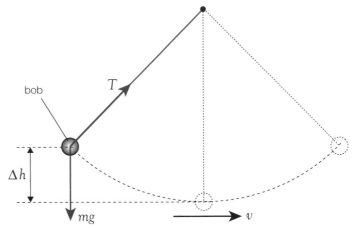

Figure 6.5 Pendulum.

The motion of the pendulum can be described as a continuous variation of GPE and KE:

$$\Delta GPE \rightarrow \Delta KE \rightarrow \Delta GPE$$

$$mg\Delta h \rightarrow \frac{1}{2}mv^2 \rightarrow mg\Delta h$$

All oscillators show a similar variation between PE and KE. For masses on springs or rubber bands, elastic potential energy stored at the ends of the oscillation is transferred to kinetic energy at the midpoint.

Test yourself

1 a) If the average, horizontal resistive force acting on the pram on page 74 is 14 N and the angle θ is 30°, calculate the work done by the man pulling the pram at constant speed over a distance of 40 m.

b) Why would more work be needed to push the pram?

2 Calculate the gravitational potential energy of a car of mass 2 tonnes on a ramp of height 80 cm.

3 A rubber band has a length of 10 cm when it is clamped at one end and supporting a load of 1.0 N. When a further 3.0 N is added to the load, the length becomes 16 cm. How much additional elastic potential energy has been transferred to the band?

4 Calculate the kinetic energy of a lorry of mass 7.5 tonnes travelling at 40 m s^{-1}.

Internal energy

In many transfers, energy seems to be lost to the surroundings. A pendulum swings with decreasing amplitude as work is done against air resistance and friction at the support. The brakes of a bicycle get hot when *KE* is transferred to them. Both of these conversions lead to an increase in the kinetic and potential energies of the particles (atoms or molecules) in the air and the brakes. This is often, wrongly, described as heat energy.

In a gas of fixed volume, the internal energy is the sum of the *KE* of all the molecules. In a solid, the atoms vibrate with continuously varying *PE* and *KE*.

An increase in internal energy usually results in a rise in temperature. Thermal energy and internal energy are studied in greater detail in the Year 2 Student's book.

Activity 6.1

Investigating the transfer of *GPE* to *KE*

A trolley is held at the top of a track, as shown in Figure 6.6. The difference in the height, Δh, of the trolley when it is at the top of the ramp and when it is on the horizontal board is measured.

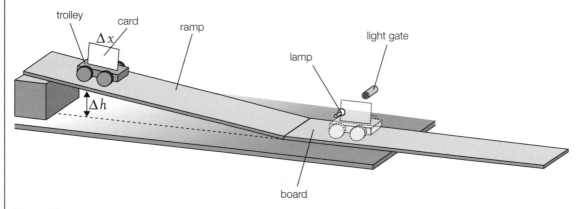

Figure 6.6

The trolley is allowed to run down the ramp onto the horizontal board, and it cuts through the light gate. The speed, v, of the trolley and masses is found by measuring the time, Δt, for the card (width Δx) to cross the light gate:

$$v = \frac{\Delta x}{\Delta t}$$

The experiment is repeated several times and an average value of v is found. If the mass of the trolley is known, the loss in potential energy and the gain in kinetic energy can be calculated.

ΔGPE of the trolley $= mg\,\Delta h$

ΔKE of the trolley $= \dfrac{1}{2}mv^2$

Ideally, the values of ΔGPE and ΔKE should be equal, but experimental uncertainties and work done against air resistance and the friction in the bearings are likely to affect the readings.

The energy transfers can be investigated further using a range of values of Δh and calculating the average velocities. A graph of ΔKE against ΔGPE is then drawn. The percentage of the original *GPE* converted to *KE* is found from the gradient of the graph.

A set of results taken from the experiment is as follows:

mass of trolley, $M = 0.400\,\text{kg}$

length of card, $x = 0.200\,\text{m}$

Questions

1 Copy and complete Table 6.1. (Use an Excel or Lotus spreadsheet if available.)

   a) Plot a graph of ΔKE/J against ΔGPE/J.

   b) Is there a systematic difference between ΔKE and ΔGPE?

   c) Find the gradient of the graph; this will give you the fraction of the GPE that is converted into KE. Why do you expect the value to be less than one?

The times given in the table are average values. The three readings taken for the first result are 0.24 s, 0.20 s and 0.22 s. The heights are measured with a rule with 1 mm divisions.

2 Calculate the percentage uncertainties of

   a) the first time reading

   b) the final value of Δh.

Table 6.1

Δh/m	ΔGPE/J	t/s	v/m s^{-1}	ΔKE/J
0.05		0.22		
0.10		0.16		
0.15		0.13		
0.20		0.11		
0.25		0.10		

An alternative method to observe the conversion of *GPE* to *KE* is to allow an object to fall from a range of heights and measure the time taken for the object to pass through a light beam. The ΔGPE is then compared with ΔKE as for the previous experimental set-up.

Example

The car on the funfair ride in Figure 6.7 has mass m and is released from rest at the top of the track, which is at height h above the lowest level of the ride. The car reaches speed v at the bottom of the dip.

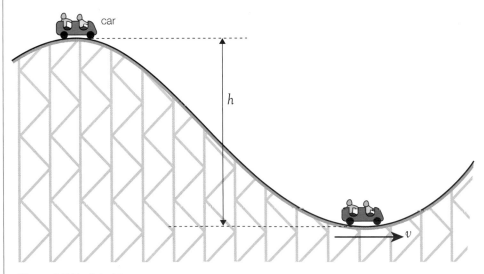

Figure 6.7 Funfair ride.

1 Write down expressions for:

   a) the gravitational potential energy transferred from the car as it runs from top to bottom

   b) the kinetic energy of the car at the lowest point.

2 If h is 20 m, show that the speed of the car should be about 20 m s^{-1}.

3 The speed of the car is measured as 19 m s^{-1}. Explain why this value is lower than the value calculated in Question 2.

4 Do you expect that the car will pass the next peak, which is at a height of 15 m above the dip? Explain your reasoning.

2 $mgh = \frac{1}{2}mv^2 \Rightarrow v = \sqrt{(2gh)}$

$v = \sqrt{(2 \times [9.81\,\mathrm{m\,s^{-2}}] \times [20\,\mathrm{m}])}$

$v = 19.8\,\mathrm{m\,s^{-1}} \approx 20\,\mathrm{m\,s^{-1}}$

3 The car does work against frictional forces and air resistance. Some of the *GPE* is transferred to increase the internal energy of the surroundings.

4 The car needs to transfer 75% of its original *GPE* to reach a height of 15 m. The energy transferred from the car to the surroundings on the downward journey is less than 10% of the *GPE*, so it is likely that the car will reach the top of the next peak.

Answers

1 a) $GPE = mgh$

 b) $KE = \frac{1}{2}mv^2$

Test yourself

5 An ice hockey puck of mass 170 g is struck at 80 m s^{-1}. After travelling 20 m it is saved by the goal minder. If the average resistive force between the puck and the ice is 2.0 N,

   a) calculate the initial kinetic energy of the puck

   b) calculate the work done by the resistive force on the puck

   c) determine a value for the speed of the puck as it strikes the goal minder's gloves.

6 A pendulum bob of mass 50 g is displaced to one side so that it is raised by 10 cm from the equilibrium position.

   a) Describe the energy changes of the pendulum bob during one complete oscillation.

   b) Calculate the maximum speed of the bob after it is released.

   c) Explain why this speed, in theory, will be the same irrespective of the mass of the pendulum.

   d) Suggest why this may not be the case in practice.

Tip

It is a common error for students to use the equation of motion $v^2 = u^2 + 2as$ to determine v. The so-called 'suvat' equations only apply to uniform motion in a straight line. The motion of the roller coaster car is not rectilinear and its acceleration is continually varying. You must clearly show that the conservation of energy has been correctly applied.

Other forms of energy

This section so far has been concerned mainly with **mechanical energy**. The potential and kinetic energy of objects have been explained in terms of mechanical work. Other forms of energy include:

- **chemical energy** – the ability to do work using chemical reactions (fuels and electric cells store chemical energy)
- **nuclear energy** – the ability to do work by changes in the constitution of nuclei (nuclear fission and fusion are examples, where energy stored in the nucleus can be transferred to other forms)
- **electrostatic potential energy** – the energy of a charged particle by virtue of its position in an electric field (charged capacitors store electrostatic energy)
- **radiant energy** – energy transferred by electromagnetic waves (radio waves, light and X-rays are examples of radiant energy).

6.2 Principle of conservation of energy

We have seen that energy is transferred readily from one form to another by work. Energy can also be transferred by heating. Often it seems that energy has been lost from a system. The examples of the pendulum and bicycle brakes on pages 79 and 80 illustrate that although some forms of energy have gone from the systems, they have simply been transferred to the surroundings in a different form.

Energy is never created or destroyed, but it can be transferred from one form into another. This is the principle of conservation of energy and it is universal. The Sun and other stars transfer nuclear energy to radiant energy in the form of electromagnetic waves. The Earth receives some of this energy (noticeably as light and infrared radiation) and also radiates energy back into space.

In nuclear reactions, such as alpha and beta decay, some mass is transferred to the particles in the form of kinetic energy.

A thermal power station illustrates the conservation of energy. The aim is to convert the chemical energy in the fuel (gas, oil or coal) into electrical energy. A simplified version of the transfers is:

chemical energy in the fuel → internal energy of the compressed steam → electrical energy

A large amount of internal energy is transferred from the low pressure steam to the cooling water in the turbine condenser. This is usually dissipated to the atmosphere, which results in an increase in the internal energy of the environment.

Energy transfers can also be represented in Sankey diagrams like that in Figure 6.8. The width of the arrow at the input represents the chemical energy in the fuel. The widths of the outgoing arrows represent the energy transfers. Note that the sum of the widths of the outputs equals that of the input. This indicates that the energy is conserved.

> **Key term**
>
> The **principle of conservation of energy** states that energy cannot be created or destroyed, it can only be transferred from one form to another.

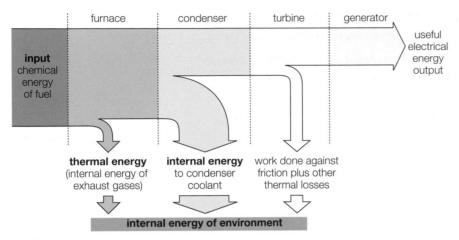

Figure 6.8 A Sankey diagram for a thermal power station.

Efficiency

The Sankey diagram shows that only a fraction of the chemical energy is converted into electrical energy. The ability to transfer the input energy to desired (useful) energy is the **efficiency** of the system.

Key term

$$\text{Efficiency} = \frac{\text{useful energy/power output}}{\text{energy/power input}} \times 100\%$$

> **Example**
>
> A pulley system is used to lift a 100 kg mass to a height of 1.5 m. The operator applies a force of 200 N and pulls the rope a total distance of 8.0 m.
>
> Calculate:
>
> **1** the gain in *GPE* of the load (useful energy output)
>
> **2** the work done by the operator (energy input)
>
> **3** the efficiency of the system.
>
> **Answers**
>
> **1** $\Delta GPE = mg\,\Delta h$
> $= 100\,\text{kg} \times 9.8\,\text{m s}^{-2} \times 1.5\,\text{m}$
> $\approx 1470\,\text{J}$
>
> **2** $\Delta W = F \times x$
> $= 200\,\text{N} \times 8.0\,\text{m}$
> $= 1600\,\text{J}$
>
> **3** $\text{Efficiency} = \dfrac{1470\,\text{J}}{1600\,\text{J}} \times 100\% = 92\%$

6.3 Power

A car engine transfers the chemical energy from the fuel to the work needed to move the car. If two cars use the same amount of fuel, an equal amount of chemical energy will be transferred to work by the engines. The more powerful car will use the fuel more quickly.

An electric lamp converts electrical energy into radiant (light) energy. A low-power lamp is much dimmer than a high-power one, but if it is left

Key term

$$\text{Power} = \frac{\text{Work done or energy transferred}}{\text{time taken}}$$

Key term

One **watt** is a rate of conversion of energy of one joule per second.

$$1\,W = 1\,J\,s^{-1}$$

on for a longer time, it can emit more energy. Both examples indicate that **power** depends on the rate at which energy is transferred:

$$\text{power} = \frac{\text{work done (energy transferred)}}{\text{time taken}}$$

Power is measured in **watts** (W).

Example

Estimate the minimum power you would need to generate to climb a hill of height 80 m in 5 minutes.

Answer

$$\Delta W = \Delta GPE$$
$$= mg\,\Delta h$$
$$= (\text{your mass})\,\text{kg} \times 9.8\,\text{m s}^{-2} \times 80\,\text{m}$$

For $m = 70\,\text{kg}$:

$$\Delta W = 70\,\text{kg} \times 9.8\,\text{m s}^{-2} \times 80\,\text{m}$$
$$\approx 55\,000\,\text{J}$$

$$\text{Power} = \frac{\Delta W}{\Delta t} \approx \frac{55\,000\,\text{J}}{5 \times 60\,\text{s}} \approx 200\,\text{W}$$

Activity 6.2

Measuring the output power of an electric motor

The motor in Figure 6.9 is switched on until the load is raised almost up to the pulley wheel and is then switched off. A stopclock is used to measure the time, Δt, when the motor is working, and a metre rule is used to find the distance, Δh, through which the load is lifted. The experiment is repeated for a range of masses to investigate the effect of the load on the power output of the motor using the equation:

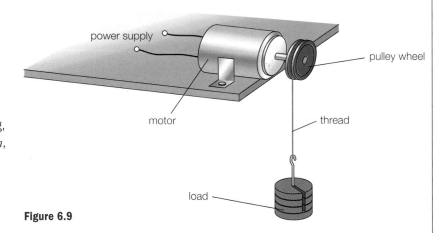

Figure 6.9

$$\text{power output of motor} = \text{power} = \frac{\Delta GPE}{\Delta t} = \frac{mg\,\Delta h}{\Delta t}$$

Safety note: Place a tray containing sand or foam beneath the weights to prevent them falling on your feet should they fall.

In Chapter 8, power in electrical circuits will be introduced. A similar experiment will be performed to measure the input power to the motor and hence the efficiency of the motor.

A motor used in the above experiment is rated as 3.0 W. The average time taken for a 100 g load to be raised through a height of 1.2 m was 0.8 s.

Questions

1 Show that the efficiency of the motor used for this purpose is about 50%.

2 Discuss what has happened to the other 50% of the energy.

3 Explain why it is necessary to take several readings of the time and take an average.

Power and motion

If, for example, a canoeist paddles at speed v for distance Δx against a uniform retarding force F, she will be working at a constant rate:

$$\Delta W = F \times \Delta x$$

$$\text{Power} = \frac{\Delta W}{\Delta t} = \frac{F \times \Delta x}{\Delta t} = F \times \frac{\Delta x}{\Delta t} = F \times v$$

Power (W) = retarding force (N) $\times$ speed (m s^{-1})

$$P = Fv$$

Example

A motorised wheelchair is driven at 2.5 m s^{-1} against an average resistive force of 80 N. Calculate the efficiency of the wheelchair if its motor has a power rating of 250 W.

Answer

Useful power output, P_o = 80 N $\times$ 2.5 m s^{-1} = 200 W

$$\text{Efficiency} = \frac{\text{useful power output}}{\text{power input}} \times 100\%$$

$$= \frac{200\,\text{W}}{250\,\text{W}} \times 100\% = 80\%$$

Test yourself

7 By taking measurements from the Sankey diagram in Figure 6.8, make an estimate of the efficiency of the power station.

8 a) Calculate the efficiency of a pulley system if the operator pulls on the rope with a force of 40 N over a distance of 2.0 m in order to raise a 100 kg mass by a height of 6 cm.

   b) Give a reason why such a system is not 100% efficient.

9 Calculate the average power generated by a cyclist travelling at a constant speed of 12 m s^{-1} against an average resistive force of 15 N.

10 Calculate the average power generated by a horse pulling a plough if the horizontal, resistive forces are 1200 N and the horse ploughs a furrow of length 186 m in 5 minutes.

It is interesting to note that James Watt (after whom the unit of power is now named) invented a unit of power called a 'horsepower' (hp). He did this as a marketing ploy to give some idea of the power of his steam engines compared with the power that could be obtained from a horse. Thus a 5 hp engine would supposedly generate the same power as five horses. Until about 50 years ago, cars in Britain were taxed according to a calculated horsepower. For example, the iconic Morris Minor 1000 was rated at 10 hp (although it actually developed 48 hp). In today's terms, 1 hp = 746 watts, so the Morris Minor developed a power of 48 $\times$ 746 W = 36 kW.

Exam practice questions

1 A barge is pulled along a canal by two men who each apply a force of 100 N to a rope at 60° to the banks.

The total work done by the men in moving the barge 100 m along the canal is:

A 5000 J   **C** 10 000 J

B 8660 J   **D** 17 400 J **[Total 1 Mark]**

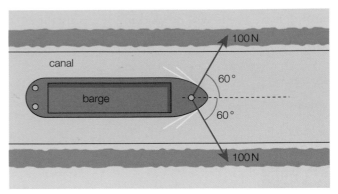

Figure 6.10

2 A woman performs a 'bungee jump', falling from a bridge with a length of elastic cord tied around her ankles. The gravitational potential energy lost by the woman at the lowest point has been transferred as:

A elastic strain energy in the cord

B elastic strain energy and internal energy in the cord

C kinetic energy of the jumper

D kinetic energy of the jumper and internal energy in the cord. **[Total 1 Mark]**

3 A forklift truck raises ten 100 kg sacks of rice off the floor onto a storage shelf 2.4 m high in 8.0 s. The average power generated by the truck is about:

A 300 W   **C** 3000 W

B 2400 W   **D** 24 000 W   **[Total 1 Mark]**

4 A toy tractor with a 120 W electric motor is driven at a constant speed of $2.0\,\text{m s}^{-1}$. If the motor is 75% efficient, the average resistive force on the tractor is:

A 45 N   **C** 180 N

B 60 N   **D** 240 N   **[Total 1 Mark]**

5 Define work and state the unit of work. **[Total 2 Marks]**

6 a) What is potential energy?   **[1]**

b) Explain the difference between gravitational potential energy and elastic potential energy.   **[2]**

     [Total 3 Marks]

7 a) What is kinetic energy?   **[1]**

b) Calculate the kinetic energy of a golf ball of mass 50 g travelling at $20\,\text{m s}^{-1}$.   **[2]**

     [Total 3 Marks]

8 State the meaning of the efficiency of a system.   **[Total 3 Marks]**

9 a) A sleigh is pulled along a level stretch of snow by a husky dog for a distance of 100 m. If the average force exerted on the sleigh is 60 N, how much work is done by the dog? **[2]**

b) The sleigh now moves across a slope so that the husky pulls with an average force of 80 N at an angle of 30° up the slope from the direction of travel of the sleigh. How much work does the dog do to move the sleigh 50 m along this path? **[2]**

[Total 4 Marks]

10 A ramp is set up in the laboratory so that a trolley is able to run from the top to the bottom. Describe an experiment you could perform to find out what percentage of the gravitational potential energy lost is converted to kinetic energy. Your answer should include:

- a diagram including any additional apparatus required **[3]**
- a description of how the apparatus is used **[3]**
- a description of how the results are analysed. **[3]**

[Total 9 Marks]

11 a) Define power and state the unit of power. **[2]**

b) A student of mass 70 kg runs up a flight of stairs in 5.0 s. If there are 20 steps each of height 25 cm, calculate the average power needed to climb the stairs in this time. **[2]**

c) Explain why the power generated by the student is likely to be greater than this value. **[1]**

[Total 5 Marks]

12 A cyclist pedals along a level road at 5.0 m s⁻¹. The rider stops pedalling and 'free wheels' until he comes to rest 20 m further down the road.

a) Show that the acceleration of the bike is approximately −0.6 m s⁻². **[2]**

b) If the mass of the bike plus rider is 100 kg, calculate:

   i) the average frictional force opposing the motion

   ii) the power needed for the cyclist to maintain a steady speed of 5.0 m s⁻¹. **[2]**

[Total 4 Marks]

13 The car on a fairground ride is launched by a compressed air piston gun capable of delivering an average force of 200 kN over a distance of 2.0 m. The car travels along 10 m of horizontal track before climbing 25 m to the top of the first peak of the ride, 18 m above the horizontal track.

a) Show that the minimum velocity of the car at the foot of the incline must be about 20 m s⁻¹ for the car to reach the peak. **[2]**

The car has a mass of 800 kg and, for a particular ride, the total mass of the passengers is 960 kg.

b) Calculate the maximum speed of the car at launch. **[3]**

c) Calculate the maximum speed of the car as it reaches the first peak. **[2]**

In practice, there are resistive forces acting on the car that will reduce these values.

d) The average resistive force acting on the car is 300 N. Calculate the maximum mass of passengers that the car could carry in order to reach the first peak. [3]

e) A student answering part **a** of this question used the equation $v^2 = u^2 + 2as$ and calculated a value for v of the correct magnitude. Explain why this method is not valid. [3]

[Total 13 Marks]

14 Nitrogen molecules in air at 20 °C have an average speed of about 500 m s^{-1}.

a) Calculate the average kinetic energy of the molecules at this temperature. [2]

b) Explain why the speed of a molecule is largely unaffected by its position in a room of height 2.5 m at this temperature. [2]

[Total 4 Marks]

Stretch and challenge

15 The Centre for Alternative Technology (CAT) at Machynlleth in Wales uses a water-driven funicular railway to carry passengers along an inclined track between the car park and the visitor centre. The track is 53 m long and the centre is at an altitude 30 m above the car park. Two cars, one initially at the top and the other at the bottom, are connected by a steel cable.

The cars have large water tanks so that when the top one is filled it becomes heavier than the lower one and so is able to move down the slope pulling the other car up.

A schematic diagram of the system is shown in Figure 6.11.

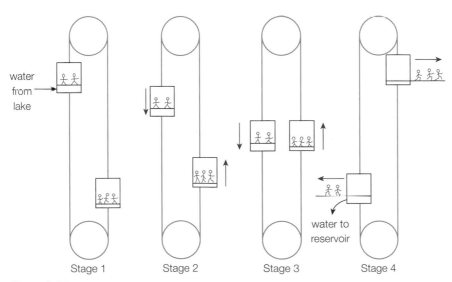

Figure 6.11

Stage 1: The cars are locked in position while the passengers enter. Water is channelled from the lake into the tank of the upper car until its total mass is greater than that of the bottom car.

Stage 2: The brakes are released and the cars begin to accelerate along the tracks.

Stage 3: When the optimum speed of $0.7\,\mathrm{m\,s^{-1}}$ is reached, a hydraulic braking system is applied to maintain this speed.

Stage 4: Towards the end of the journey, further braking is applied to bring the cars to rest. The passengers disembark and the water is released from the tank into a reservoir at the bottom station.

Each car has a mass of 3.50 tonnes and a tank of capacity 1600 litres. For a particular trip the lower car has ten passengers of total mass 700 kg, and the upper car has six people with a combined mass of 400 kg.

When the brakes are released, the cars are accelerated uniformly for the first 5.0 m of the trip when the speed reaches $0.7\,\mathrm{m\,s^{-1}}$. The hydraulic brakes are applied so that this speed is kept constant until 5.0 m from the end when extra braking brings the cars to a halt.

a) By considering changes in the gravitational potential energy and the kinetic energy of the cars, show that the minimum volume of water added to the upper tank should be about 400 litres. (The density of water is $1000\,\mathrm{kg\,m^{-3}}$.)   **[4]**

b) In practice, the volume of water added will be significantly greater than this value. Suggest a reason why this is so.   **[1]**

The hydraulic braking system operates by using the rotation of the wheels to drive a pump that pressurises oil. The high-pressure oil pushes the brake pads against the wheel discs, and also compresses a bag of nitrogen gas. The compressed gas is used to operate the inlet and outlet ports of the tanks and also to pump water from the lower reservoir back into the lake.

c) Calculate the work done by the braking system during the time the cars are moving at constant speed and the actual volume of water added is 500 litres.   **[3]**

d) Describe the energy conversions that occur during the journey.   **[4]**

[Total 12 Marks]

Charge and current

7

Prior knowledge

In this chapter you will need to:
→ be familiar with the concept of a simple model of the atom in which protons have a positive electric charge and electrons have a negative charge
→ understand that the movement of charge causes an electric current and that the current is equal to the rate of flow of charge.

The key facts that will be useful are:
→ charge (symbol Q) is measured in coulombs, C
→ current (symbol I) is measured in amperes, A
→ $I = \dfrac{Q}{t}$ so $A = \dfrac{C}{s} = C\,s^{-1}$

Test yourself on prior knowledge

1 The charge on an electron is $1.6 \times 10^{-19}\,$C. How many electrons are there in a charge of $10\,\mu$C?

2 A kettle has a current rating of $8.0\,$A and takes 3.0 minutes to boil some water. How much charge passes through the element during this time?

3 A fully charged car battery stores $1.8 \times 10^{5}\,$C of charge. Due to an electrical fault, it discharges at a steady $500\,$mA. How many hours will it take to discharge?

7.1 Electric charge

The Greek philosopher Thales (c. 600BC) discovered that rubbing the gemstone amber with a cloth caused it to attract small pieces of dry leaf. The Greek word for amber is *elektron*, which is the origin of our words 'electron' and 'electricity'.

In this section you will learn that all matter contains electric charges and that if these charges are made to move, an electric current is created. The difference between metallic conductors, semiconductors and insulators is then discussed in terms of the mobility of charges.

In simple terms, we can think of all matter as consisting of atoms. These may be considered to be made up of protons and neutrons, which form a nucleus that is surrounded by a 'cloud' of electrons. Protons and electrons have the property of **charge**, which gives rise to electrical forces. Historically, the charge on protons was called **positive** and that on electrons was designated **negative**.

Under normal circumstances we do not observe any effects due to these charges, because most of the time objects have equal numbers of protons and electrons so that the charges cancel out. Indeed, the idea of the **conservation of charge** (that is, equal quantities of positive and negative charge) is a fundamental concept in physics – rather like the conservation of energy. Only when charges move in some way does their effect become apparent.

Figure 7.1 confirms that there are two types of charge and that **like charges repel** while **unlike charges attract**. The strips become 'charged' by the transfer of electrons when rubbed with a duster.

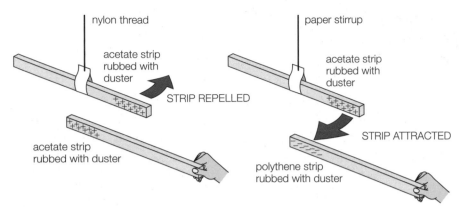

Figure 7.1 Like charges repel and unlike charges attract.

This is possible because the electrons, which are on the outside, can be detached fairly easily from their atoms. The process is as shown in Figure 7.2.

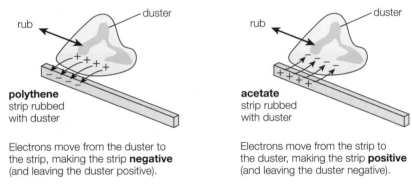

polythene strip rubbed with duster

Electrons move from the duster to the strip, making the strip **negative** (and leaving the duster positive).

acetate strip rubbed with duster

Electrons move from the strip to the duster, making the strip **positive** (and leaving the duster negative).

Figure 7.2 The charging process.

7.2 Electric current

Electric current is defined as the rate of flow of charge or, in other words, the amount of charge per second flowing past a given point. We can express this mathematically as:

$$I = \frac{\Delta Q}{\Delta t}$$

where I is the current and ΔQ is the amount of charge flowing in time Δt.

Current is measured in **amperes** (symbol A). The ampere is the base unit of electric current and is defined in terms of the force between two parallel wires each carrying a current of 1 A. In laboratory terms, an ampere is quite a large current, so we often use the units mA (milliampere $= 10^{-3}$ A) and μA (microampere $= 10^{-6}$ A).

We can arrange

$$I = \frac{\Delta Q}{\Delta t}$$

to give:

$$\Delta Q = I \Delta t$$

This tells us that the amount of charge flowing in a certain time is given by multiplying the current by the time. If a current of 1 A flows for 1 s, the quantity of charge flowing is said to be 1 **coulomb** (symbol C). The coulomb is a relatively large amount of charge, so we often use μC or even nC ('nano' $= 10^{-9}$); indeed, the charge on a single electron is only 1.6×10^{-19} C!

Key term

Electric current is the rate of flow of charge.

Tip

Remember to convert mA and μA to A when working out numerical problems.

Tip

When doing calculations, you must always remember to put the current in amperes, the charge in coulombs and the time in seconds.

Example

1 How much charge flows through the filament of an electric lamp in 1 hour when the current in it is 250 mA?

2 The electron charge is 1.6×10^{-19} C. How many electrons flow through the filament during this time?

Answers

1 $\Delta Q = I\Delta t$

$\qquad = 250 \times 10^{-3}\,\text{A} \times (60 \times 60\,\text{s})$

$\qquad = 900\,\text{C}$

2 Number of electrons $= \dfrac{\text{total charge}}{\text{charge on an electron}}$

$\qquad\qquad\qquad = \dfrac{900\,\text{C}}{1.6 \times 10^{-19}\,\text{C electron}^{-1}}$

$\qquad\qquad\qquad = 5.6 \times 10^{21}\,\text{electrons}$

Test yourself

1 A car battery is charged at a rate of 500 mA for 24 hours.

   a) How much charge is stored in the battery?

   b) The battery is then discharged at a steady current of 3.0 A. How long will it take to discharge?

2 An article states that 'a typical lightning bolt may transfer 10^{20} electrons in a fraction of a second, giving rise to a current of 10 kiloamperes'. If the electron charge is 1.6×10^{-19} C:

   a) how much charge is contained in the lightning bolt (note: 10^{20} is 1×10^{20} on your calculator)

   b) how many milliseconds is the 'fraction of a second'?

Activity 7.1

Investigation of current in a series circuit

A series circuit is set up as shown in Figure 7.3 with the meter set on the 200 mA DC range.

The current will be approximately 10 mA, depending on the exact values of the cell and the resistors. You should always remember that the values of resistors are only **nominal values** and that there is a manufacturing tolerance of probably 2% or possibly as much as 5% on the value stated.

The ammeter is now moved to point X and then point Y in the circuit. It is observed that (within experimental error) the current is the same wherever the ammeter is in the circuit.

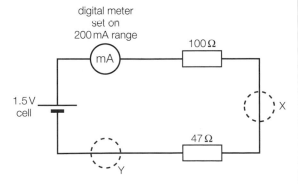

Figure 7.3 Current in a series circuit.

Questions

1 By considering current as the rate of flow of charge, suggest why the current in each resistor in a series circuit such as this must be the same.

2 The total circuit resistance is the sum of the two resistors, i.e. nominally 147 Ω. If each resistor has a tolerance of 5%, how would you record the value of the circuit resistance, with its uncertainty?

Activity 7.2

Investigation of current at a junction

The parallel circuit shown in Figure 7.4 is set up. The variable power supply is adjusted so that the current I_z at point Z is approximately 5 mA. Without altering the power supply, the current is measured, in turn, at points X and Y. This procedure is then repeated by adjusting the power supply to give four further values of I_z, tabulating I_z, I_X, I_Y and $I_X + I_Y$.

It is found that, allowing for experimental error, $I_z = I_X + I_Y$ in all cases. This shows that the current flowing out of a junction is always equal to the current flowing into the junction. This is another example of the conservation of charge – the rate of charge flowing out of a junction must always equal the rate at which it enters the junction because charge cannot be lost or gained.

If a graph of I_Y against I_X is plotted, it is found that the gradient is equal to the **inverse** of the ratio of the resistor values (Figure 7.5).

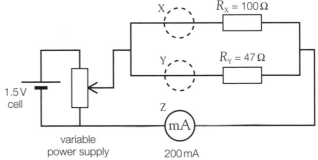

Figure 7.4 Current at a junction.

Questions

A typical set of results from such an experiment is recorded in Table 7.1.

Table 7.1

I_z/mA	I_X/mA	I_Y/mA	$I_X + I_Y$/mA
5.0	1.6	3.4	
15.0	4.9	10.3	
25.0	8.1	16.8	
35.0	11.2	23.7	
45.0	14.5	30.6	

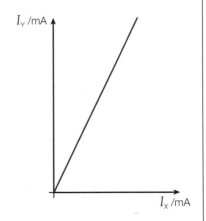

Figure 7.5

1 Copy and complete Table 7.1 by adding values for $I_X + I_Y$.
2 Plot a graph of I_Y against I_X.
3 Compare the gradient with the ratio of R_X to R_Y and comment on your finding.

7.3 Current in series and parallel circuits

The current in a component can be measured by connecting an **ammeter** in **series** with the component. Ammeters have a low resistance so that they do not affect the current that they are measuring.

In a series circuit, the current is the same in each component. This is because the rate at which electrons leave any component must be the same as the rate at which they enter the component – if this were not the case, electrons would be lost from the circuit, which would contravene the principle of conservation of charge (see Section 7.1).

> **Tip**
> When setting up a series circuit, the ammeter can be placed at any position in the circuit.

> **Tip**
> Remember that charge (and therefore current) must always be conserved at a junction.

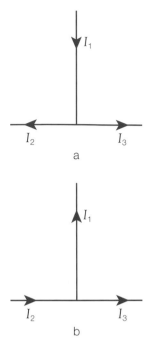

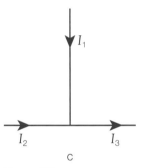

Figure 7.7

Key term

A **charge carrier** is either an electron detached from its atom or an ion in a conducting material that is free to flow in order to create a current.

3 In the circuit shown in Figure 7.6, ammeter A_1 reads 180 mA and ammeter A_2 reads 60 mA. What is the current in

   a) the 5 Ω resistor

   b) the 10 Ω resistor

   c) the 2 Ω resistor

   d) the 3 Ω resistor?

4 For each of the three situations **a**, **b** and **c** shown in Figure 7.7, write down an expression for I_1 in terms of I_2 and I_3.

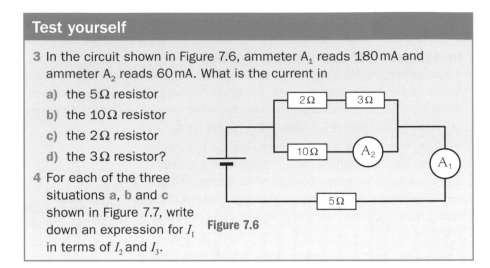

Figure 7.6

7.4 Drift velocity

In order for a current to flow in a material, suitable **charge carriers** must be present within the material – for example, loosely bound electrons in metals or ions in electrolytes and gases.

The idea of a charge carrier is demonstrated in the experiment in Figure 7.8. When the high voltage is switched on, the metallised sphere swings back and forth between the metal plates, transferring charge as it does so. Figure 7.8b shows that positive charge is carried from left to right and vice versa. This is equivalent to a current flowing clockwise around the circuit, which is recorded by the nanoammeter (nA). In this experiment, the sphere can be considered to be the 'charge carrier'.

Safety note: Ensure that the high voltage supply unit has a specified short-circuit current that does not exceed 5 mA.

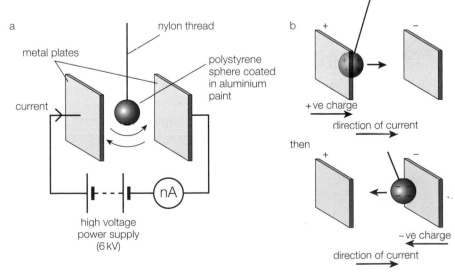

Figure 7.8 Metallised polystyrene sphere acting as charge carrier.

It is conventional to take the direction of an electric current to be the direction in which a positive charge would move. If the charge carriers are negative – for example, the electrons in the connecting wires – the charge carriers actually move in the opposite direction to the current.

Example

In the experiment in Figure 7.8, the average current is 30 nA when the sphere takes 16 s to travel across the plates and back a total of 40 times. How much charge is carried on the sphere each time?

Answer

Total charge $\Delta Q = I \Delta t = 30\,\text{nA} \times 16\,\text{s} = 480\,\text{nC}$

In this time, the sphere has travelled across and back 40 times – that is, a total of 80 transits. The charge carried on the sphere each time is therefore $\dfrac{480\,\text{nC}}{80} = 6\,\text{nC}$

Tip

In the example you will see that the current has been left in nA, which means that the charge will be given directly in nC. The use of 'quantity algebra' – that is, including the units with each numerical value – will help ensure that errors are not made when doing a calculation in this way.

In a metallic conductor, the charge carriers are loosely bound outer electrons – so-called 'free' or 'delocalised' electrons. On average, there is approximately one free electron per atom. These electrons move with **random thermal motion**, to and fro within the crystal lattice of the metal at speeds approaching one-thousandth the speed of light. They are also mainly responsible for metals being good conductors of heat. When a potential difference is applied to a circuit (see Section 8.1), an electric field is created and exerts a force on the free electrons, which causes them to 'drift' in the direction of the force. In accordance with Newton's second law, the electrons would accelerate continuously if it were not for the fact that they collide with the regularly spaced atoms in what is called the crystal lattice. The 'atoms' are, in effect, positive ions because the free electrons are detached from the atoms, leaving the atoms with a positive charge. These collisions cause an equal and opposite force to be exerted on the electrons, which, by Newton's first law, continue with a constant '**drift velocity**', giving rise to a constant current.

Tip

Remember that the direction of flow of the electrons in a metallic conductor is always in the opposite direction to that of the conventional current.

As the charge carriers are electrons, which have a negative charge, they drift towards the positive terminal of the cell. The electrons therefore drift anti-clockwise round the circuit shown in Figure 7.9. Before scientists knew it was the drift of electrons that caused a current in a metallic conductor, they defined the direction of the current to be from positive to negative. The conventional current direction in Figure 7.9 is therefore clockwise – in the opposite direction to the flow of electrons!

For a conductor, the current I is given by:

$$I = nAvq$$

where A = area of cross-section of conductor, n = number of charge carriers per cubic metre, q = charge on each charge carrier and v = drift velocity of charge carriers.

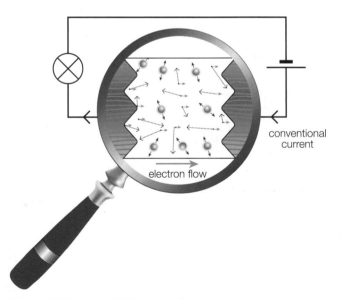

Figure 7.9 Electron drift in a metallic conductor.

This equation may be deduced as follows, with reference to Figure 7.10:

- Consider a section of wire of length Δx
- The volume of wire in this section will be

$$\Delta V = A \Delta x$$

- If there are n charge carriers per unit volume in the wire, the number in volume ΔV will be

$$\Delta N = n\Delta V = nA\Delta x$$

- If the charge on each charge carrier is q, the quantity of charge within the section will be

$$\Delta Q = nAq\Delta x$$

- Suppose each charge carrier takes a time Δt to travel the distance Δx.
- Dividing both sides of the equation by Δt gives us

$$\frac{\Delta Q}{\Delta t} = nAq\frac{\Delta x}{\Delta t}$$

- But $\dfrac{\Delta Q}{\Delta t} = I$ and $\dfrac{\Delta x}{\Delta t} = v$

- Hence $I = nAvq$

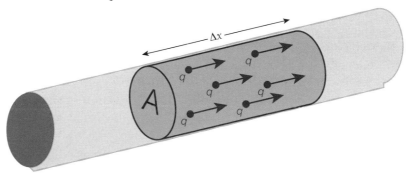

Figure 7.10

Example

Consider the circuit in Figure 7.11.

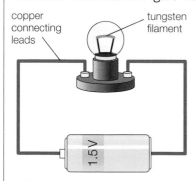

Figure 7.11

The diameter of the tungsten filament is 0.025 mm and the diameter of the copper connecting leads is 0.72 mm.

Calculate the drift velocity of the electrons in:

a) the filament

b) the connecting leads

when the circuit current is 160 mA.

(You may take the number of charge carriers per cubic metre to be $4.0 \times 10^{28}\,\text{m}^{-3}$ for tungsten and $8.0 \times 10^{28}\,\text{m}^{-3}$ for copper.)

Answers

a) For the filament: $I = nAvq$

Rearranging:

$$v = \frac{I}{nAq}$$

$$= \frac{160 \times 10^{-3}\,\text{A}}{4.0 \times 10^{28}\,\text{m}^{-3} \times \pi \times (0.5 \times 0.025 \times 10^{-3}\,\text{m})^2 \times 1.6 \times 10^{-19}\,\text{C}}$$

$$= 0.051\,\text{m s}^{-1}$$

$$= 51\,\text{mm s}^{-1}$$

b) For the leads:

$$v = \frac{I}{nAq}$$

$$= \frac{160 \times 10^{-3}\,\text{A}}{8.0 \times 10^{28}\,\text{m}^{-3} \times \pi \times (0.5 \times 0.72 \times 10^{-3}\,\text{m})^2 \times 1.6 \times 10^{-19}\,\text{C}}$$

$$= 3.1 \times 10^{-5}\,\text{m s}^{-1}$$

$$= 0.031\,\text{m s}^{-1}$$

The previous example shows two rather surprising but very important facts. Firstly, it shows how slow the drift velocity is – particularly in the copper connecting leads. In fact, an electron would probably not get around the circuit before the cell ran out! Secondly, the drift velocity is much faster in the very thin tungsten filament than in the thicker connecting leads. This is analogous to the flow of water in a pipe speeding up when it meets a constriction – just think of water squirting out at high speed when you put your finger over the end of a tap. (This is not recommended as a practical exercise in the laboratory!) As we saw in Section 7.3, the current in a series circuit is the same at every point in the circuit. So, in order for this to be the case, the electrons in the filament have to speed up as there are far fewer of them.

Continuing the water analogy, there is a large water pressure difference across the constriction; in electrical terms, a large **potential difference** is developed across the filament (virtually all of the 1.5 V or so provided by the cell), which applies the necessary force to speed up the electrons. This is dealt with more fully in Section 8.1.

Consider switching on an electric light. How is it that the light seems to come on instantly and there is a current in the order of 0.25 A when the drift velocity of the electrons in the wires is only a fraction of a millimetre per second? The answer to the first question lies in the fact that although the electrons themselves are travelling so slowly, the electric field that causes them to move travels at nearly the speed of light ($\approx 3 \times 10^8 \, \text{m s}^{-1}$). All of the electrons therefore start to move almost instantly. Secondly, although the individual electrons are moving along the wire very slowly, there is simply an enormous number of them (remember, $n \approx 10^{29} \, \text{m}^{-3}$) and therefore the charge flowing per second equates to a significant current.

Activity 7.3

Experiment to observe and estimate drift speeds
The arrangement shown in Figure 7.12 is set up.

The filter paper is soaked with 1 M ammonium hydroxide solution. A small crystal of copper sulfate and a small crystal of potassium permanganate are carefully placed near the centre of the slide as shown. The power supply is then switched on.

The movement of the blue copper ions and the purple permanganate ions can be observed and an estimate of their drift speeds can be made. It is found that the drift speed is very slow and the experiment needs to be left to run for an hour or so.

Questions
1 What can you deduce about the signs of the respective charges on the copper and permanganate ions?
2 Describe how you could estimate the drift speed of the ions.
3 State any safety precautions you would take while doing this experiment.

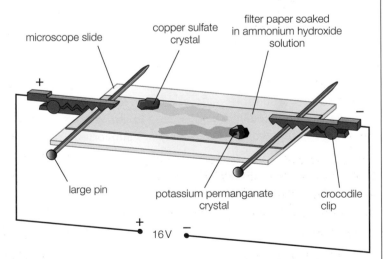

Figure 7.12 Ion drift.

7.5 Metals, semiconductors and insulators

Theories of conduction in different types of material are complex and go well beyond the scope of the A-level course. However, we can give a simple explanation of the electrical conductivity of metals, semiconductors and insulators in terms of the by now familiar equation $I = nAvq$.

The charge q is usually the electron charge, e (or possibly $2e$ for a doubly charged ion) and A is not a property of the material itself. This means that the value of n really determines the ability of a given material to conduct an electric current.

Metals typically have a value of n in the order of $10^{28}\,\text{m}^{-3} - 10^{29}\,\text{m}^{-3}$, while insulators such as glass and polystyrene have virtually no charge carriers at room temperature. In between, semiconductors such as germanium ($n \approx 10^{19}\,\text{m}^{-3}$) and silicon ($n \approx 10^{17}\,\text{m}^{-3}$) are able to conduct, but not as well as metals. The ability of semiconductors to conduct is greatly enhanced by increasing their temperature or adding 'impurity atoms'. This is discussed more fully in Section 10.5.

Test yourself

5 A copper wire of area of cross-section A has a current I in it. Copper contains n charge carriers per m^3 (electrons, each with a charge q). The drift velocity v of the electrons is proportional to:

   A A **B** I **C** n **D** q

6 Two copper wires of different diameter are connected in series. Which of the following will NOT be the same for each wire?

   A I **B** n **C** q **D** v

7 a) State the units of each of the quantities in the equation $I = nAvq$.

  b) With reference to the equation, explain why metals, such as copper, are much better conductors of electricity than semiconductors, such as silicon.

  c) Suggest a suitable value of n for an insulator, such as porcelain, which is used to insulate high-voltage cables where they are supported by pylons (towers).

  d) State another material that is a good insulator of electricity and give an example of how it is used.

8 The 'lead' in a pencil is actually made of graphite, which is a form of carbon. About 1 cm of the wood round the lead is carefully cut off at each end of a pencil. The pencil lead and the copper connecting wires each have a diameter of 0.50 mm.

In order to test whether the lead conducts electricity, the pencil is connected with crocodile clips into the circuit shown in Figure 7.13.

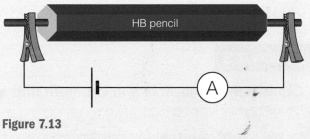

Figure 7.13

  a) The ammeter shows that there is a circuit current of 220 mA. If copper contains 8.5×10^{28} electron charge carriers per m^3, each having a charge of 1.6×10^{-19} C, what is the drift velocity of these electrons through the copper connecting wires?

  b) Would you expect the drift velocity of the charge carriers in the pencil lead to be faster, slower or the same as in the connecting wires? Explain your answer with reference to the equation $I = nAvq$.

  c) Describe how you would check that the diameter of the pencil lead was 0.50 mm.

Exam practice questions

This information is for use in Questions 1 and 5.

The beam current in a cathode ray tube is 32 mA. The electrons travel at a speed of $4.2 \times 10^7\,m\,s^{-1}$ through a distance of 21 cm.

1 The number of electrons striking the screen in 1 hour is:

    A 2.0×10^{17} **C** 1.2×10^{22}

    B 7.2×10^{20} **D** 7.2×10^{23} **[Total 1 Mark]**

2 The fact that copper is a much better electrical conductor than silicon can be explained with reference to the equation $I = nAvq$. Which of the quantities in the equation is significantly larger for copper than for silicon?

    A n **C** v

    B A **D** q **[Total 1 Mark]**

3 A typical flash of lightning lasts for about 0.5 ms, during which time it discharges approximately 8 C of charge.

    a) What is the average current in such a strike?   **[2]**

    b) The current is caused by positive ions having a charge of $1.6 \times 10^{-19}\,C$ travelling from the Earth to the thundercloud. How many such ions are there in the strike?   **[2]**

                                               [Total 4 Marks]

4 **a)** State what physical quantity is represented by each term in the equation $I = nAvq$.   **[3]**

    b) A semiconducting strip 6.0 mm wide and 0.50 mm thick carries a current of 10 mA, as shown in Figure 7.14.

Figure 7.14

        If the value of n for the semiconducting material is $7.0 \times 10^{22}\,m^{-3}$, show that the drift speed of the charge carriers, which carry a charge of $1.6 \times 10^{-19}\,C$, is about $0.3\,m\,s^{-1}$.   **[3]**

    c) The drift speed for electrons in a copper strip of the same dimensions and carrying the same current would be about $10^{-7}\,m\,s^{-1}$. Use the equation to explain why this value is very different from that of the semiconductor.   **[2]**

                                               [Total 8 Marks]

5 The number of electrons in the beam in Question 1 at any instant is:

A 1.0×10^9 **C** 1.0×10^{12}

B 1.0×10^{11} **D** 1.0×10^{14} **[Total 1 Mark]**

6 Figure 7.15 shows a copper cylinder that can roll along model railway lines. The diameter of section X is half that of section Y. When there is a current I in the cylinder, in the direction shown, the ratio of the drift velocity of the electrons $v_X : v_Y$ is

A $1:4$ **C** $2:1$

B $1:2$ **D** $4:1$ **[Total 1 Mark]**

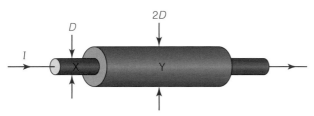

Figure 7.15

7 A manufacturer states that tinned copper fuse wire rated at 15 A has an area of cross-section of $0.20 \, \text{mm}^2$.

a) Show that the diameter of the wire is about 0.5 mm.   **[2]**

b) You are asked to find an accurate value for the diameter of the wire.

   i) State what instrument you would use and explain why it would be suitable.

   ii) Describe how you would try to make your determination of the diameter as accurate as possible.

   iii) Estimate the percentage uncertainty in your value for the diameter.

   iv) What would be the percentage uncertainty in the area of cross-section of the wire determined by using the measured value of the diameter?   **[6]**

[Total 8 Marks]

8 a) The current in the tungsten filament of a torch bulb is 25 mA.

   i) How much charge will conduct through the filament in 8.0 minutes?

   ii) How many electrons will flow through the filament during this time?   **[4]**

b) The bulb is connected to the cell in the torch by two copper leads. Discuss the factors that would cause a difference between the drift speed of the electrons in the copper leads and that of the electrons in the filament.   **[4]**

[Total 8 Marks]

9 An electric light is 3.0 m away from the switch. It is connected to the switch by copper wire of cross-section $0.50 \, \text{mm}^2$. The carrier density for copper is $8.5 \times 10^{28} \, \text{m}^{-3}$ and the electron charge is $1.6 \times 10^{-19} \, \text{C}$.

    a) If the light takes a current of 0.25 A, show that an electron in the wire near the switch will take almost a day to reach the lamp. **[5]**

    b) Explain, referring to your answer to part **a**, why the light appears to come on instantly. **[2]**

                                                                                [Total 7 Marks]

Stretch and challenge

10 The relative atomic mass ('atomic weight') of copper is 63.5. This means that there are 6.0×10^{23} atoms in 63.5 g of copper. The density of copper is $8.9 \times 10^3 \, \text{kg m}^{-3}$.

    a) Use these data to show that the charge density n for copper is about $8 \times 10^{28} \, \text{m}^{-3}$. What simplifying assumption do you have to make in obtaining this value? **[4]**

    b) Calculate the maximum drift velocity of conduction electrons in a 5 A copper fuse wire of diameter 0.20 mm (electron charge = $1.6 \times 10^{-19} \, \text{C}$). **[3]**

    c) The copper connecting wire in the circuit containing the fuse has a diameter of 1.0 mm. Without doing any calculation, explain how the drift velocity in the connecting wire compares with that in the fuse. **[4]**

    d) Use the data given to estimate the atomic spacing in copper. State any assumptions that you make. **[4]**

                                                                           [Total 15 Marks]

Prior knowledge

In this chapter you will need to:
→ be familiar with the term potential difference (p.d.), V, across a component in a circuit (you may well have called it 'voltage' for short)
→ recall from Chapter 7 that electric current, I, is the rate of flow of charge
→ recall from Chapter 6 the terms work, energy and power in relation to mechanics
→ be familiar with electrical power being given by the product of p.d. and current.

The key facts that will be useful are:
→ potential difference is measured in volts, V, i.e. joules per coulomb, JC^{-1}
→ work = force × distance moved (measured in joules, J)
→ energy is the ability to do work (measured in joules, J)
→ power is the rate of doing work or the rate of transfer of energy (measured in watts, W)
→ power developed electrically is given by $P = VI$
→ energy (in J) is given by power (in W) multiplied by time (in s).

Test yourself on prior knowledge

1 Calculate the energy given out, in kJ, by a 60 W lamp in one hour.
2 If a 60 W lamp runs off a 240 V supply, calculate the current in the filament of the lamp.
3 A car of mass 840 kg is driven up a hill of vertical height 15 m in 3.1 s.
   a) Calculate the gravitational potential energy gained by the car.
   b) Hence show that the power needed do this is about 40 kW.
   c) Suggest why, in practice, the car would have to generate more power than this.

Tip

In this chapter you will also relate electrical power to mechanical power, for example an electric motor lifting a weight. This will require you to combine your knowledge from Chapter 6 with that acquired in this chapter. This emphasises the importance of keeping on top of earlier work, as you will often need to build on this later in the course.

8.1 Potential difference

In Chapter 7 we saw that a moving electric charge gives rise to an electric current. We will now look at how we can make charges move. We will see that a cell produces an electromotive force, which creates a potential difference (or 'voltage') in a circuit. The charges in the circuit components and connecting wires experience a force, causing the charges to move round the circuit. The potential difference therefore does work on the charges, thus developing power in the circuit.

In order to understand how electrons can be made to flow around a circuit and thereby create an electric current, comparison with the flow of water in a pipe is often helpful. Such an analogy is illustrated by Figure 8.1.

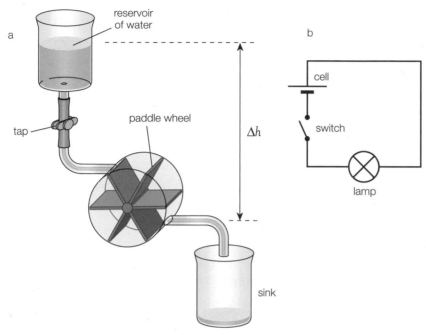

Figure 8.1 Water analogy of electric current.

In Figure 8.1a, there is a **pressure difference** between the water level in the reservoir and that in the pipe. When the tap is opened, this pressure difference causes the water to flow through the pipe and turn the paddle wheel. The **gravitational potential energy** stored by the water in the reservoir is converted into **kinetic energy** of rotation of the paddle wheel.

In Figure 8.1b, a chemical reaction in the cell creates an electromotive force (see Section 8.3 on page 106) which gives rise to an **electrical potential difference** (symbol V) across the cell terminals. The electrons in the circuit experience a force. When the switch is closed, this force causes the electrons to move around the circuit and through the lamp, i.e. work is done on the electrons. The **chemical potential energy** stored in the cell is converted into vibrational **kinetic energy** of the atoms in the filament of the lamp, thereby causing its temperature to rise.

In an electric circuit, electrical energy is converted into other forms of energy – for example in the circuit of Figure 8.1b, the lamp converts electrical energy into thermal energy and light. Conversely, an electric motor converts electrical energy into mechanical energy. In our analogy, if the paddle wheel in Figure 8.1a were to be connected to an electric motor, the water could be pumped up again from the sink to the reservoir.

These energy conversions in electrical circuits form the basis of the definition of potential difference (often abbreviated to p.d.).

Tip

'Heating' is defined as the energy transfer between regions at different temperatures. It is therefore strictly incorrect to say that the lamp converts electrical energy into 'heat' (although we invariably do!); you should instead use the term 'thermal energy'.

Tip

Remember: $V = \dfrac{W}{Q}$ or $W = QV$

Key term

The **potential difference** between two points in a circuit is the electrical energy per unit charge converted into other forms of energy.

As the potential difference between two points in a circuit is the electrical energy per unit charge converted into other forms of energy (thermal, light, mechanical, etc.) it can be expressed as:

$$\text{potential difference} = \frac{\text{electrical energy converted into other forms}}{\text{charge passing}}$$

or, as energy is the result of doing work, W:

$$\text{potential difference} = \frac{\text{work done}}{\text{charge passing}}$$

From $V = \dfrac{W}{Q}$ it follows that the unit of p.d. is joules per coulomb ($J\,C^{-1}$).

This unit is called the volt.

Because potential difference is measured in volts, potential difference is often called the **voltage** between two points in a circuit.

> ### Key term
>
> A **volt** is a joule per coulomb. For example, there would be a p.d. of 1 V between two points in a circuit if 1 J of energy is converted when 1 C of charge passes between the points.

Examples

1 The element of an electric kettle that takes a current of 12.5 A produces 540 kJ of thermal energy in 3 minutes.

   a) How much charge passes through the element in these 3 minutes?

   b) What is the potential difference across the ends of the element?

2 A 12 V pump for a fountain in a garden pond can pump water up to a height of 0.80 m at a rate of 4.8 litres per minute.

   a) How much work does the pump do per minute when raising the water to a height of 0.80 m? (You may assume that one litre of water has a mass of 1 kg.)

   b) If the pump is 75% efficient, how much charge passes through the pump motor in one minute?

   c) What current does the motor take when operating under these conditions?

Answers

1 a) $\Delta Q = I\Delta t = 12.5\,A \times (3.0 \times 60\,s) = 2250\,C$

   b) $V = \dfrac{W}{Q}$

   $= \dfrac{540 \times 10^3\,J}{2250\,C} = 240\,V$

2 a) $W = mg\Delta h = 4.8\,kg\,min^{-1} \times 9.8\,m\,s^{-2} \times 0.80\,m = 37.6\,J\,min^{-1} = 38\,J\,min^{-1}$

   b) $V = \dfrac{W}{Q}$

   So:

   $Q = \dfrac{W}{V} = \dfrac{37.6\,J}{12\,V} = 3.1(3)\,C$ (if 100% efficient)

   As the pump is only 75% efficient:

   $Q = \dfrac{100}{75} \times 3.1(3)\,C = 4.2\,C$

   c) $I = \dfrac{Q}{t} = \dfrac{4.2\,C}{60\,s} = 70\,mA$

Test yourself

1 When a mobile phone charger is charging a phone, it uses energy at the rate of 3.0 W. If the charger is left on 'standby' (i.e. plugged in but with no phone being charged) it still uses 0.13 W. Calculate:

   a) the energy (in kJ) used when charging the phone for one hour

   b) the energy used (in kJ) if the charger is left on standby for 23 hours.

   Comment on your answers.

2 A manufacturer advertises an LED torch as having a power of 1.0 W and requiring three AAA (1.5 V) batteries. If such a torch were to be left on for 15 minutes, calculate:

   a) the current in the LED

   b) the charge passing through it in this time

   c) the energy taken from the batteries.

8.2 Using a voltmeter

The potential difference between two points in a circuit is measured by connecting a **voltmeter** between the points. We talk about connecting a voltmeter *across*, or in *parallel* with, a component to measure the p.d. between its ends.

In the circuit shown in Figure 8.2, the voltmeter is measuring the p.d. across the lamp. To find the p.d. across the resistor, the voltmeter would have to be connected between A and B, and to measure the p.d. across the cell, it would have to be connected between A and C.

Tip

When setting up a circuit, always set up the series part of the circuit first and check that it works. Then connect the voltmeter in the required position.

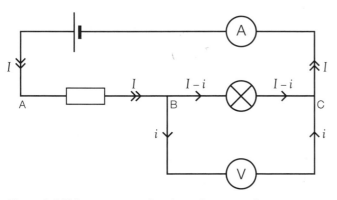

Figure 8.2 Voltmeter measuring the p.d. across a lamp.

A voltmeter must take some current in order to operate. In the circuit in Figure 8.2, the ammeter records the circuit current, I, but the current through the lamp is only $I - i$, where i is the current taken by the voltmeter. In order to keep i as small as possible, voltmeters should have a very **high resistance**. Typically, a 20 V digital voltmeter might have a resistance of 10 MΩ; analogue meters need more current for their operation and are likely to have resistances in the order of kΩ.

Activity 8.1

Investigating potential differences in a circuit

The circuit shown in Figure 8.3 is set up and Table 8.1 is prepared.

The potential difference V across the four-cell power supply is recorded in the table.

The voltmeter is now connected across each of the resistors R_1, R_2 and R_3 in turn and the corresponding potential differences recorded. The sum of V_1, V_2 and V_3 is then calculated.

The experiment is then repeated using three cells instead of four. Within experimental error it is found that $V = (V_1 + V_2 + V_3)$.

This shows that the sum of the energy per unit charge converted in each resistor is equal to the energy per unit charge produced by the cells. As the resistors are in *series* with the cells, the current, or rate of flow of charge, is the same in each component (see Section 7.3 on page 93). The amount of energy converted in the resistors is therefore equal to the amount of energy produced by the cells. This is an example of the fundamental law of conservation of energy (see Section 6.2 on page 82).

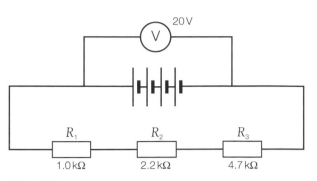

Figure 8.3

Table 8.1

V/V	V_1/V	V_2/V	V_3/V	$(V_1 + V_2 + V_3)$/V

Questions

A student wants to operate a filament lamp rated at 6 V, 300 mW from a 9 V power supply. She sets up the circuit shown in Figure 8.4.

She sets the variable resistor to its maximum resistance and then reduces the resistance until the lamp shines brightly. She notices that the variable resistor also gets slightly warm.

1 What, approximately, will be the potential difference across
   a) the lamp
   b) the resistor?
2 **a)** Calculate the current in the filament.
   b) Explain why the current in the resistor is the same as that in the lamp.
3 Explain, with the aid of a calculation, why the variable resistor gets slightly warm.
4 Explain why it was essential to start with the resistor at maximum resistance. What could have happened otherwise?

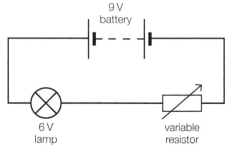

Figure 8.4

8.3 Electromotive force

Key term

The **electromotive force** of an electrical source is defined as the energy per unit charge converted into electrical energy by the source.

A cell or generator (dynamo or alternator) does work on charges just as a pump does work on water (see the Example on page 104). Chemical energy is converted into electrical energy in a cell, while mechanical energy (for example, from the engine of a car) is converted into electrical energy in an alternator. The cell or alternator is said to produce an **electromotive force**. This is rather misleading as it is not a force at all but a form of energy transfer! Electromotive force is usually abbreviated to e.m.f. and is given the symbol ε.

As the e.m.f. of an electrical source is defined as the energy (chemical, mechanical, etc.) per unit charge converted into electrical energy, it can be expressed as:

$$\text{e.m.f.} = \frac{\text{energy converted into electrical energy}}{\text{charge passing}}$$

$$\varepsilon = \frac{W}{Q} \quad \text{or} \quad W = \varepsilon Q$$

As $\frac{W}{Q}$ has units of $J\,C^{-1}$, the units of e.m.f. are the same as those of potential difference – that is, volts, V.

Tip

Remember:

- e.m.f. and p.d. are both measured in volts, or joules per coulomb ($1\,V = 1\,J\,C^{-1}$)
- e.m.f. is the creation of electrical energy *from* other forms of energy
- p.d. is the conversion of electrical energy *into* other forms of energy.

Think cause and effect: think of e.m.f. as being the *cause* and p.d. as the *effect*.

Test yourself

3 A battery of e.m.f. 12 V and two resistors, R_1 and R_2, are connected in series. The potential difference across R_1 is found to be 9.0 V.

   a) Draw a circuit diagram of this arrangement.

   b) Explain, with reference to energy, what is meant by an electromotive force of 12 V and a potential difference of 9 V.

   c) Add a voltmeter to your circuit diagram to show how you would measure the p.d. across R_2.

   d) What would you expect the p.d. across R_2 to be?

8.4 Power

In Chapter 6, we defined power as the *rate of doing work*:

$$P = \frac{\Delta W}{\Delta t}$$

This was based on doing work mechanically. What is the situation if the work is done electrically? From the definition of potential difference:

$$W = QV$$

If both sides of the equation are divided by time, t:

$$\frac{W}{t} = \frac{Q}{t} \times V$$

But, $\frac{W}{t} =$ power, P and $\frac{Q}{t} =$ current, I

so $P = IV$

In other words, the electrical power converted in a device is given by the product of the current in it and the voltage across it. The units of electrical power are watts, W– the same as mechanical power – as long as the current is in amperes and the p.d. is in volts.

Tip

Beware! The symbol 'W' can stand for work done (which is measured in joules) *or* watts (the units of power). In printed text, symbols for physical quantities are expressed in *italics* while units are written in normal text: so *W* stands for work and W means watts.

Examples

1 An X-ray tube operates at 50 kV. The maximum beam current is specified as 1.0 mA. What is the maximum safe power?

2 An electric iron is marked as 240 V, 1.8 kW. What is the current in the heater filament under normal operating conditions?

Answers

1 For safe power:

$$P = IV = 1.0 \times 10^{-3}\,\text{A} \times 50 \times 10^{3}\,\text{V} = 50\,\text{W}$$

2 From $P = IV$:

$$I = \frac{P}{V} = \frac{1800\,\text{W}}{240\,\text{V}} = 7.5\,\text{A}$$

Test yourself

4 The following are units you should be familiar with: A, C, J, s, V, W.

Unit name	Unit	Relationship
(i)	C	A × (ii)
electromotive force	(iii)	$\dfrac{J}{(iv)}$
(v)	(vi)	$\dfrac{J}{s}$
current	(vii)	$\dfrac{(viii)}{V}$

a) Copy out the table and fill in the missing names and symbols (i)–(viii).

b) The power of alternating current generators is sometimes quoted in units of volt-ampere, e.g. 2000 kVA. Using expressions for potential difference and current, show that a volt-ampere is equivalent to a watt.

Activity 8.2

Investigating how the efficiency of an electric motor varies for different loads

This is an extension of the experiment measuring the power of a motor in Chapter 6.

The arrangement shown in Figure 8.5 is set up.

The power supply is adjusted so that the motor can lift a mass of 100 g (that is, a force of 0.98 N) through a height of about 1 m in a few seconds (say 3–5 s).

The time, t, that it takes to lift the mass, m, through a measured height, h, is measured and the corresponding p.d., V, and current, I, are recorded in a table such as Table 8.2.

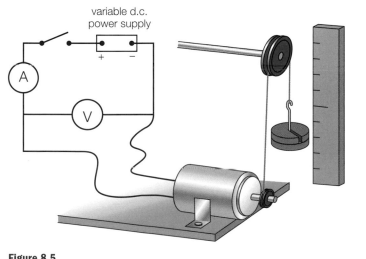

variable d.c. power supply

Figure 8.5

Questions

A typical set of results is shown in Table 8.2.

Table 8.2

V/V	I/A	Power in, IV/W	Force, mg/N	Height, h/m	Time, t/s	Power out, $\dfrac{mgh}{t}$/W	Efficiency, $\dfrac{P_o}{P_i} \times 100\%$
6.0	0.24	1.44	0.98	0.95	4.1	0.227	15.8
5.9	0.28		1.96	0.95	5.8		
5.8	0.31		2.94	0.95	7.5		
5.6	0.36		3.92	0.95	9.3		
5.4	0.40		4.90	0.95	11.6		

1 Copy and complete Table 8.2 (it would be a good exercise to do this as a spreadsheet).
2 Plot a graph of the efficiency of the motor for different loads (force).
3 What deductions can you make from this graph? (This graph may differ significantly from one you might have obtained if you tried this experimentally – different motors operating under varying conditions can produce very different results.)

8.5 Electrical energy

In the circuit shown in Figure 8.6, electrical energy is converted into thermal and light energy in the filament of the torch bulb.

In the previous section we deduced that:

$$P = IV$$

If both sides of this equation are multiplied by time Δt:

$$P\Delta t = IV\Delta t \quad \text{or} \quad \Delta W = IV\Delta t$$

where ΔW is the work done (or the energy converted) in time Δt.

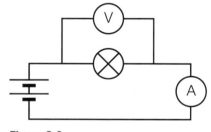

Figure 8.6

Example

In the circuit shown in Figure 8.6, the current in the filament is 250 mA and the p.d. across the bulb is 2.2 V. How much electrical energy is converted into thermal energy in one minute?

Answer

$$\Delta W = IV\Delta t$$
$$= 0.25\,\text{A} \times 2.2\,\text{V} \times 60\,\text{s}$$
$$= 33\,\text{J}$$

In practice, the thermal energy created in a filament lamp raises the temperature of the filament so that it radiates electromagnetic radiation. However, only about 10% is in the visible region and so only about 3 J of light is produced in the above example. The rest of the energy is mainly in the infrared part of the spectrum. A filament lamp is therefore a very inefficient way of generating light. This is why legislation is on its way to compel us to switch over to 'low-energy' compact fluorescent lamps (CFLs) and light emitting diode (LED) lamps.

Although a CFL costs more to manufacture than a filament lamp, it produces the same amount of light for less than 25% of the energy and also lasts about eight times longer (typically about 8000 hours compared with 1000 hours for an average filament lamp). It is estimated that if a CFL were to be on for as little as one hour per day, it would only take $2\frac{1}{2}$ years to recover the initial cost and that the savings over the lifetime of the lamp would amount to several times its original cost.

Furthermore, the benefit for the environment would be huge. Replacing a single filament lamp with a CFL would reduce emissions of carbon dioxide by 40 kg per year, and if each house in the UK were to install just three low-energy lamps, it would save enough energy to run the country's street lights for a year. This is why governments are encouraging the switch to low-energy lamps, and legislation may eventually lead to filament lamps becoming illegal.

The technology of LED lamps is advancing rapidly, which in turn reduces the cost of the lamps. These are even more energy efficient in use than CFL lamps. They are now being used to replace halogen lamps in ceiling 'down lighters' as they can be manufactured to exactly the same size and shape as the halogen lamps they replace. Typically, a 4 W LED lamp will give an equivalent light output to a 50 W halogen lamp. Many traffic lights and road signs are now LEDs and before long LED lamps will also provide all street lighting.

The significance of LED lamps is reflected in the award of the 2014 Nobel Prize for Physics to three Japanese scientists for, "*the invention of efficient blue light-emitting diodes which has enabled bright and energy-saving white light sources.*"

Figure 8.7 Comparison of different types of light bulb.

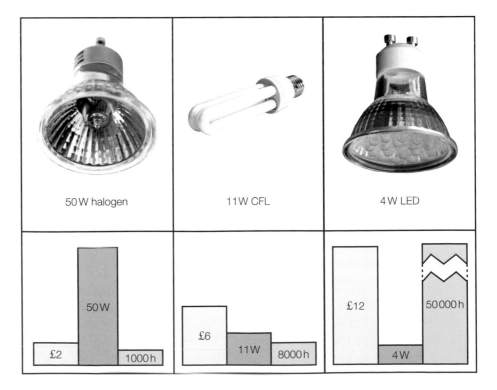

5 An iPad charger rated at 5.1 V, 2.1 A fully charges an iPad in 4.0 h. Calculate:

a) the power rating of the charger

b) the energy stored in the fully charged iPad.

An iPhone charger, which has an identical USB connection to that of the iPad charger, has a rating of 5 V, 5 W.

c) What is the current rating of the iPhone charger?

d) Suggest the main problem with using the iPhone charger to charge the iPad.

WARNING: You should note that although this will work, continuously charging an iPad in this way could have an adverse long-term effect on the life of the battery.

6 An electrically assisted cycle has a 36 V battery. The battery is fully charged after 4.5 hours at an average current of 2.0 A. Calculate:

a) the charge stored in the battery after this time

b) the energy stored in the battery in MJ.

c) The manufacturer states that the battery has a capacity of 324 Wh, i.e. it could deliver a power of 324 W for an hour. Comment on this in relation to your answer for part (b).

7 The table below gives some data for three different types of lamp, each giving out the same amount of light.

Type of lamp	Power/W	Voltage/V	Current/mA
filament	40	240	(i)
halogen	(ii)	240	150
LED	4.5	(iii)	500

a) Copy out and complete the table by calculating the missing data (i)–(iii).

b) Calculate the energy, in kJ, used

   i) by the filament lamp, and

   ii) by the LED in 5 hours.

c) Explain why the filament lamp uses much more energy than the LED.

Exam practice questions

1 The SI base units for the coulomb are

   A $A\,s^{-1}$ **C** $A^{-1}s^{-1}$

   B $A\,s$ **D** $A^{-1}s$ **[Total 1 Mark]**

2 An electric iron is rated at '230V, 1.5 kW'.

   a) What is the value of the current in the heating element?   **[2]**

   b) How much charge passes through the element during 20 minutes of use?   **[2]**

   c) How much thermal energy is produced during this time?   **[2]**

                             [Total 6 Marks]

3 The data in the table shows how the power consumed by typical plasma and LED televisions depends on screen size, i.e. the diagonal distance across the viewable area of the screen.

   a) Calculate the energy, in MJ, consumed by a 42-inch LED television if it is left on for 12 hours.   **[2]**

Screen size	Plasma	LED
42-inch (107 cm)	210 W	50 W
50-inch (127 cm)	300 W	71 W

   b) Energy companies measure energy in units of kW h. Show that 1 kW h is equivalent to 3.6 MJ.   **[2]**

   c) If electricity costs 15p per unit, show that the annual cost of running a 42-inch LED television for an average of 12 hours per day would be about £33.   **[3]**

                             [Total 7 Marks]

4 A manufacturer claims that replacing a 60 W filament lamp with an 18 W low-energy lamp, which provides the same amount of light, will save more than £40 over the lifetime of the lamp.

   a) Use the following data to test this claim:

       Cost of low-energy lamp: £2

       Estimated life of low-energy lamp: 8000 hours

       Cost of a 'unit' of electricity: 15p

       (A 'unit' of electricity is 1 kW h, which is the energy used when a device rated at 1 kW is used for one hour.)   **[5]**

   b) Discuss any other advantages of replacing filament lamps with low-energy lamps.   **[2]**

                             [Total 7 Marks]

5 The SI base units for the watt are

   A $kg\,m\,s^{-2}$ **C** $kg\,m\,s^{-3}$

   B $kg\,m^2\,s^{-2}$ **D** $kg\,m^2\,s^{-3}$ **[Total 1 Mark]**

6 The following quantities are often used in the study of electricity:
charge   current   energy   potential difference   power

Copy and complete the table by putting the appropriate quantity
or quantities in the second column.   **[Total 6 Marks]**

	Quantity or quantities
Which quantity is the product of two other quantities?	
Which quantity is one of the quantities divided by one of the other quantities? (3 possible answers)	
Which quantity, when divided by time, gives another quantity in the table? (2 possible answers)	

7 The safe dose for a patient being X-rayed is stated as being:

X-ray voltage: $800\,kV$

Tube current: $500\,mA$

Maximum exposure time: $1.0\,s$

a) Calculate the energy received by the patient in $1.0\,s$.   **[2]**

b) Estimate the time that the patient would have to be in front of
a $1.0\,kW$ electric fire to receive the same amount of energy.   **[2]**

c) State any assumptions that you made in determining your answer
to part **b**.   **[1]**

[Total 5 Marks]

8 This is a synoptic question, which requires
understanding of Chapters 6 and 8.

An experiment to investigate the efficiency of an
electric motor is set up as shown in Figure 8.8.

The mass of $600\,g$ was lifted from the floor to a
height of $0.94\,m$ in $7.7\,s$ and the meter readings
were $3.2\,V$ and $0.75\,A$.

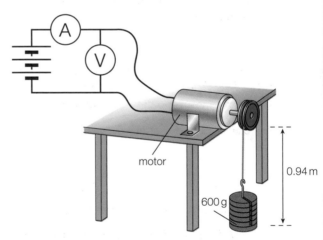

a) How much work did the motor do lifting
the mass?   **[2]**

b) What was the useful output power of the
motor?   **[2]**

c) How much electrical power was being supplied
by the cells?   **[2]**

Figure 8.8

d) Use your answers to parts **b** and **c** to show that the efficiency of the
electric motor was about 30%.   **[2]**

e) Explain what happened to the rest of the energy.   **[2]**

[Total 10 Marks]

9 In Question 7, the time, t, to lift the mass was the average of the following
four measurements:

t/s: 7.6, 7.9, 7.7, 7.5

a) Estimate the percentage uncertainty in the average value of *t*. **[2]**

b) Assuming that the height was measured to the precision recorded (i.e. ±1 cm), and that the uncertainty in the mass value is negligible, estimate the uncertainty in the value calculated for the useful output power of the motor. **[3]**

[Total 5 Marks]

10 The figures in the table are taken from the owner's manual of a car.

Alternator	14 V/70 A
Starter motor	12 V/1.5 kW
Battery	12 V/62 Ah
Maximum engine power	180 kW
Headlights (×2)	12 V/60 W
Sidelights (×4)	12 V/5 W
Spare fuses	8 A and 16 A
Mass	1740 kg (laden)

a) How much power is delivered by the alternator? **[2]**

b) How much current is taken by the starter motor? **[2]**

c) The headlamps are fused individually. Explain which value of spare fuse you would use for a headlight. **[3]**

d) How much energy can be produced by the battery (62 Ah means the battery can provide a current of 1 A for 62 hours)? **[2]**

e) Show that nearly one-third of the battery capacity is used up if the sidelights are accidentally left on for 12 hours overnight. **[3]**

[Total 12 Marks]

11 This synoptic question requires understanding of Chapters 6 and 8.

Referring to the data in Question 10, show that only about $\frac{1}{4}$ of the maximum power that can be developed by the engine is used doing work against gravity when the car is driven at 90 km per hour up a gradient of 10%. **[Total 5 Marks]**

12 A 1.0 m length of 32 swg copper wire is found to have a current of 1.8 A in it when the potential difference across it is 0.53 V.

a) Draw a circuit diagram to show how you would check these values. **[3]**

b) Calculate the rate at which the power supply does work on the electrons in the wire. **[2]**

c) If copper has 8.0×10^{28} free electrons per cubic metre, and 32 swg wire has a cross-sectional area of 0.059 mm², calculate the drift speed of the electrons in the wire. **[2]**

d) Use your answers to parts **b** and **c** to show that the total force exerted on the electrons in the wire is about 400 N. **[2]**

e) Hence find the force acting on each electron. **[4]**

[Total 13 Marks]

13 The SI base units for the volt are

A $kg\,m^2\,s^{-2}\,A^{-1}$ **C** $kg\,m^2\,s^{-3}\,A$

B $kg\,m^2\,s^{-3}\,A^{-1}$ **D** $kg\,m^2\,s^{-1}\,A$ **[Total 1 Mark]**

14 Figure 8.9 shows how the current varies with time when a 12 V car battery is being charged.

a) Calculate the charge stored in the battery after 4.0 h. **[2]**

b) Estimate the energy stored in the battery after 4.0 h. What do you have to assume? **[3]**

c) Show that the charge stored in the battery after 10 h is approximately 1×10^5 C. **[3]**

d) The battery is stated by the manufacturer to have a capacity of 25 A h, which means it could theoretically provide a current of 1.0 A for 25 hours. To what extent does your answer to part **c** confirm this? **[4]**

[Total 12 Marks]

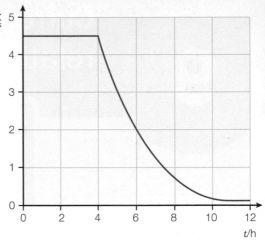

Figure 8.9

15 Use the data from Question 3 to answer this question.

a) If a 50-inch LED TV costs £950 to buy and a 50-inch plasma TV costs £650, how many years would it be before the saving in energy costs for the LED TV repays its initial higher cost, assuming that on average it is used for 12 hours per day? **[5]**

b) Show that the data in the table implies that the power used by a television is proportional to the square of its screen size. Suggest why this might be. **[6]**

[Total 11 Marks]

Stretch and challenge

16 A few years ago, a leading DIY store marketed a wind turbine that could be fitted to a house. The turbine had blades of diameter 1.8 m and was said to be capable of generating 1 kW when the wind speed was 12.5 m s⁻¹.

a) Show, by consideration of the loss of kinetic energy when the wind is stopped by the blades, that the theoretical power, P, developed is given by

$$P = \tfrac{1}{2}\rho A v^3$$

where ρ = density of air (1.3 kg m⁻³),
A = area swept by blades and v = wind speed. **[5]**

b) In practice, Betz law limits the actual power. A more rigorous analysis by Betz showed that the maximum theoretical power cannot be more than 0.59 of that given by the equation above. For the conditions under which this turbine was expected to operate, the reduction factor is about 0.3. Suggest what might cause the reduction factor to be considerably less than the Betz value of 0.59. **[2]**

c) Use the data given to determine the validity of the DIY store's claimed power. **[4]**

d) The turbine does not operate for wind speeds below about 4 m s⁻¹ and is very inefficient for wind speeds above 13 m s⁻¹. The average wind speed in the UK is about 6 m s⁻¹. Suggest why, after two years, following numerous customer complaints and a critical BBC television programme, the turbines were withdrawn. **[4]**

[Total 15 Marks]

9 Current–potential difference relationships

Prior knowledge

In this chapter you will need to:
→ be familiar with how resistance affects the current in a circuit
→ know how to use a variable resistor (rheostat) to vary the current in a circuit
→ know how to use ammeters and voltmeters to measure current and potential difference, respectively
→ have an understanding of how the current in a resistor, diode and filament lamp varies with potential difference.

The key facts that will be useful are:
→ potential difference across a component (V) = current (I) × resistance (R)
→ $R = \dfrac{V}{I}$
→ resistance is measured in ohms (Ω)
→ resistances connected in series add up, i.e. $R = R_1 + R_2$
→ unit prefixes:
  • μ = micro = 10^{-6} (e.g. $5\,\mu C = 5 \times 10^{-6}\,C$)
  • m = milli = 10^{-3} (e.g. $20\,mA = 20 \times 10^{-3}\,A$)
  • k = kilo = 10^{3} (e.g. $50\,kV = 50 \times 10^{3}\,V$)
  • M = mega = 10^{6} (e.g. $4.7\,M\Omega = 4.7 \times 10^{6}\,\Omega$).

Test yourself on prior knowledge

1 What is:

   a) the p.d., in mV, across a 220 kΩ resistor when the current in it is 200 μA

   b) the current, in μA, in a 10 MΩ resistor when the p.d. across it is 240 V

   c) the resistance, in MΩ, of a voltmeter if the current in it is 0.15 μA when the p.d. across it is 3.0 V?

2 A lamp rated at 12 V, 36 W is operating at the stated voltage. What is:

   a) the current in the lamp

   b) the resistance of the lamp?

9.1 Varying the potential difference and current in a circuit

In order to investigate how the current in an electrical component depends on the potential difference (p.d.) across it, we need to be able to vary the applied p.d. Some power supplies are continuously variable, but it is important that you understand how to set up a variable supply using a fixed power source and a variable resistor called a **rheostat**.

A rheostat can be used to either control the current in a circuit (Figure 9.1a) or give a continuously variable potential difference (Figure 9.1b). In the latter case, it is said to be a **potential divider**. Rheostats manufactured specifically for this purpose are called **potentiometers**; this term is rather misleading as 'meter' suggests that it is measuring something, which it is not! We will discuss potential dividers in more detail in Chapter 11.

Activity 9.1

Investigation of a rheostat as a variable resistor and as a potential divider

The circuit shown in Figure 9.1a is set up. Care must be taken to connect the rheostat as shown. The rheostat is used to vary the current in the circuit and the minimum and maximum values of current and p.d. that can be achieved are recorded.

The circuit in Figure 9.1b is now set up. Look carefully at the way in which the rheostat is connected to act as a potential divider. The rheostat is then used to vary the p.d. across the lamp and the minimum and maximum values of p.d. and current are recorded.

Questions

Some typical results are shown in Table 9.1

1 Explain which you think is the more useful circuit for providing a variable voltage.
2 In circuit **a**, why is it advisable to set the rheostat to its maximum resistance before connecting the circuit?
3 For safety, what extra component might you include in the circuits?

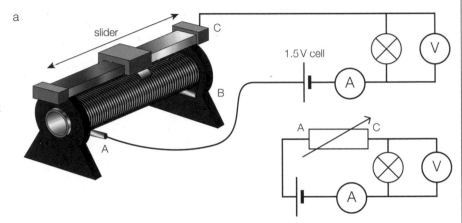
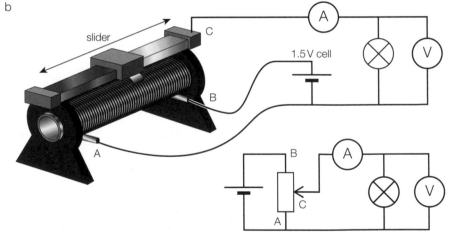

Figure 9.1 Rheostat as a variable resistor and as a potential divider.

Table 9.1

Circuit	V_{min}/V	V_{max}/V	I_{min}/mA	I_{max}/mA
a) variable resistor	0.50	1.50	10	30
b) potential divider	0.00	1.50	0.00	30

Activity 9.2

Investigation of the *I–V* characteristics for a length of resistance wire

Just over a metre length of 32 swg nichrome wire (or the 0.25 mm metric equivalent) is cut and taped to a metre rule as shown in Figure 9.2. A potential divider circuit is set up to provide a variable p.d. and a 1.000 m length of the wire is connected to the circuit by means of crocodile clips. It is important to **ensure that the clips make a good connection.** The p.d. is set to its minimum value (that is, with the rheostat slider at the end nearer the negative contact) and switched on. The p.d. is gradually increased and the values of p.d., *V*, and current, *I*, are recorded, up to the maximum p.d. possible. The values of *V* and *I* are tabulated.

The experiment is repeated with the terminals of the cell reversed. The polarity of the meters should not be changed – they will now indicate negative values. The p.d. across the wire and the current in the wire are now in the **opposite direction** to that in the first experiment and are therefore recorded as having **negative** values, as displayed on the meters.

The data are plotted on a set of axes as shown in Figure 9.3. Note that it is conventional to plot *I* (on the *y*-axis) against *V* (on the *x*-axis).

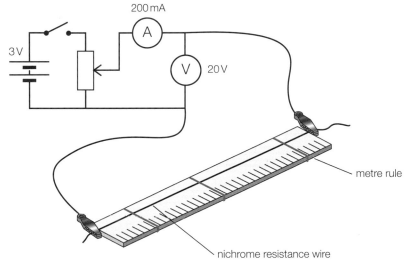

Figure 9.2

Questions

A typical set of results is shown in Table 9.2.

Table 9.2

V/V	0.00	0.50	1.00	1.50	2.00	2.50	3.00
I/mA	0	28	54	83	108	138	163

With cell terminals reversed

V/V	0.00	−0.50	−1.00	−1.50	−2.00	−2.50	−3.00
I/mA	0	−26	−56	−81	−110	−136	−165

1 Plot the data on a suitable set of axes.
2 What can be deduced about the relationship between *I* and *V* from your graph, which should look like Figure 9.3?

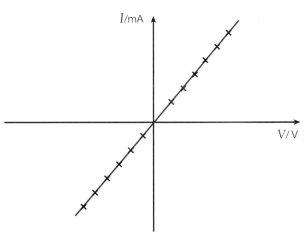

Figure 9.3

Tip

To show **proportionality**, a graph must be a straight line **through the origin.** If the line does not pass through the origin, it may indicate that there is a systematic error in the experiment or that there is a linear but **not** proportional relationship between the quantities, of the form $y = mx + c$.

9.3 Ohm's law

From experiments similar to that above, Ohm discovered that in metals the current was proportional to the potential difference provided the *temperature remained constant*.

This is now known as **Ohm's law** and any electrical component for which the current is proportional to the voltage is said to be '**ohmic**'.

9.4 Resistance

The **resistance** (symbol R) of an electrical component can be thought of as its opposition to an electric current flowing in it. This resistance is caused by collisions of the electrons with the vibrating lattice ions as the electrons 'drift' through the material of the conductor (see also $I = nAvq$ on page 95). As a result of these collisions, electrical energy is dissipated as thermal energy and the component heats up.

It is important that you understand that resistance is *defined* by the equation:

$$R = \frac{V}{I}$$

The relationship between V, I and R can be remembered by means of the diagram in Figure 9.4, i.e.

$$R = \frac{V}{I} \qquad V = IR \qquad I = \frac{V}{R}$$

The unit of resistance is the **ohm** (Ω).

From the equation $R = \frac{V}{I}$ it follows that

$$\text{ohms} = \frac{\text{volts}}{\text{amps}} \text{ or } \Omega = V\,A^{-1}$$

Key terms

Ohm's law states that for metals at a constant temperature, the current in the metal is proportional to the potential difference across it.

Resistance is *defined* by:

$$\text{resistance} = \frac{\text{potential difference}}{\text{current}}$$

The **ohm** is the name for a volt per ampere ($V\,A^{-1}$). For example, a conductor will have a resistance of $1\,\Omega$ if a potential difference of $1\,V$ across it produces a current of $1\,A$ in it.

Figure 9.4 Relationship between V, I and R.

Tip

Take care – resistances are often quoted in $k\Omega$ or $M\Omega$ and currents in mA or μA, so make sure you remember to convert to base units.

Example

What is the potential difference across a $22\,k\Omega$ resistor when the current in it is $50\,\mu A$?

Answer

$$V = IR = 50 \times 10^{-6}\,A \times 22 \times 10^{3}\,\Omega$$

$$= 1.1\,V$$

Be careful to distinguish between **Ohm's law** and **resistance**. It is a common mistake to state that Ohm's law is $R = \frac{V}{I}$, but this is *not correct*. It is **resistance** that is defined as $R = \frac{V}{I}$.

Ohm's law (Section 9.3) says that, under certain conditions, the current is proportional to the potential difference or, putting it another way, the ratio $\frac{V}{I}$ is *constant*.

It therefore follows that if a conductor obeys Ohm's law, *its resistance is constant* within the conditions specified (as in Figure 9.3 on page 118). Note that because this graph has been plotted as I against V (as opposed to V against I) the resistance is actually the *inverse* of the gradient.

1 Discuss whether the nichrome wire used in Activity 9.2 (page 118) obeys Ohm's law.

2 Use your graph to calculate the resistance of a 1.000 m length of nichrome wire.

Answers

1 As the graph is a straight line through the origin, the current is proportional to the potential difference, which indicates that the nichrome wire does obey Ohm's law.

2 The resistance of the wire is given by $R = \dfrac{V}{I}$.

As the graph is a straight line through the origin, its gradient is equal to $\dfrac{I}{V}$.

This means that the resistance can be found from the inverse of the gradient:

$$\text{resistance} = \frac{V}{I} = \frac{1}{\text{gradient}}$$
$$= \frac{1}{0.0547\,\text{A V}^{-1}}$$
$$= 18.3\,\Omega$$

Test yourself

1 The circuit shown in Figure 9.5 is set up. The ammeter registers a current of 500 µA and the voltmeter reads 2.35 V.

a) Calculate the resistance of resistor X.

b) Calculate the potential difference across each of the other two resistors.

c) Explain, in terms of conservation laws, why the potential differences across the three resistors add up to 9.0 V.

2 A physics student friend shows you his homework. He has written the following: 'Ohm's law states that resistance is equal to voltage divided by current'. Explain what is wrong with this statement by referring to the definitions of resistance and Ohm's law.

3 A student sets up the circuit shown in Figure 9.6 to investigate the I–V characteristics of a 47 Ω resistor.

She obtains the following data:

V_{min}/V	V_{max}/V	I_{min}/mA	I_{max}/mA
0.50	1.50	2.5	31.0

a) Use the maximum p.d. and current data to show that the value of the resistor is 47 Ω within a stated tolerance of ±5%.

b) Use the minimum p.d. and current, together with the value of the resistor found in a, to find the maximum resistance of the rheostat. (As the resistor and rheostat are connected in series, their combined resistance is the sum of their individual resistances – see page 154.)

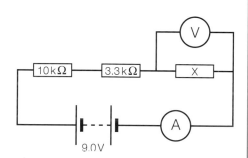

Figure 9.5

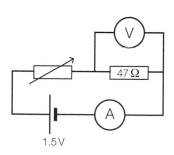

Figure 9.6

c) Her teacher suggests it would be better to use the rheostat as a potential divider. Using the same components as in Figure 9.6, draw a circuit diagram to show how this could be done.

d) For this new circuit, the student records another set of data:

V_{min}/V	V_{max}/V	I_{min}/mA	I_{max}/mA
0.00	1.50	0.0	31.0

Explain why the new circuit is a better arrangement than the first circuit for this investigation.

9.5 $I–V$ characteristics for a tungsten filament lamp

Activity 9.3

Investigation of the $I–V$ characteristics for a tungsten filament lamp

The circuit shown in Figure 9.7 is set up, with the voltage from the power supply set at its **minimum** value. The circuit is switched on and the p.d. is gradually increased. The p.d., V, and corresponding current, I, are recorded at regular intervals up to a maximum p.d. of 12 V.

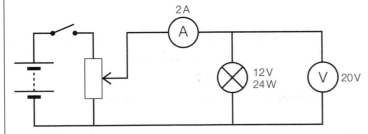

Figure 9.7

The results are tabulated and the circuit is **switched off**. A graph with I on the y-axis against V on the x-axis is then drawn.

After at least five minutes have elapsed, the power supply connections are reversed and the experiment is repeated, remembering to start with the minimum p.d. as before. These data are then added to the graph to obtain a curve as shown in Figure 9.8.

Questions

1 Why is it important to:
  a) always start with the power supply set at its minimum value
  b) allow at least 5 minutes to elapse before repeating the experiment with the power connections reversed?
2 What deductions can you make from the graph regarding the electrical characteristics of the filament? Explain your answer.

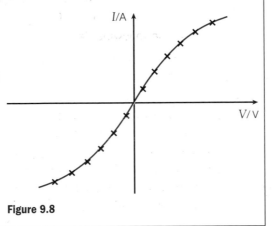

Figure 9.8

Example

A typical I–V graph for a tungsten filament lamp is shown in Figure 9.9.

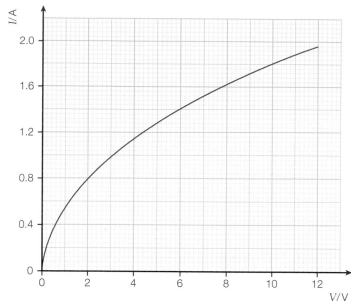

Figure 9.9

1 Discuss whether the lamp obeys Ohm's law.

2 The lamp from which the data were obtained was rated as 12 V, 24 W. What is the resistance of the filament when the lamp is operating under these conditions?

3 Use the graph to determine the resistance of the filament when:

   a) the p.d. across it is 8.0 V

   b) the current in it is 0.80 A.

Answers

1 As the graph of I against V is not a straight line through the origin, I is not proportional to V and so the lamp does not obey Ohm's law. This is because its temperature rises as the current increases.

2 Using $P = IV$:
$$I = \frac{P}{V} = \frac{24\,\text{W}}{12\,\text{V}} = 2.0\,\text{A}$$
$$R = \frac{V}{I} = \frac{12\,\text{V}}{2.0\,\text{A}} = 6.0\,\Omega$$

3 a) When the p.d. across the filament is 8.0 V, we can read off from the graph that the current is 1.62 A, so:
$$R = \frac{V}{I} = \frac{8.0\,\text{V}}{1.62\,\text{A}} = 4.9\,\Omega$$

   b) When the current in the filament is 0.80 A, we can read off from the graph that the p.d. is 2.0 V, so:
$$R = \frac{V}{I} = \frac{2.0\,\text{V}}{0.80\,\text{A}} = 2.5\,\Omega$$

Tip

Remember that resistance can be found from the gradient of an I–V graph only if the graph is a straight line through the origin – that is, if Ohm's law is obeyed. In all other cases, R must be calculated from $R = \frac{V}{I}$ using discrete values of I and V.

4 A student wishes to operate a lamp rated at 12V, 24W from a fixed 16V power supply. He proposes to set up the circuit shown in Figure 9.10.

a) Calculate the current in the lamp when it is operating at its stated rating.

b) The student has the following 10% tolerance resistors available:

1.0Ω, 2.2Ω, 3.3Ω, 4.7Ω and 6.8Ω.

Explain which value he should choose for resistor X.

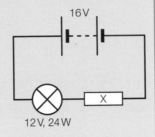

c) Calculate the resistance of the lamp when it is operating at its stated rating.

Figure 9.10

d) Assuming the lamp has an $I–V$ characteristic similar to that shown in Figure 9.9, estimate what the resistance of the lamp would be if it were operating at 6V instead of 12V.

5 A student writes the following statement for homework: 'A car bulb has a tungsten filament. As tungsten is a metal, the bulb will obey Ohm's law.' Explain what is wrong with this statement.

9.6 $I–V$ characteristics for a semiconductor diode

Activity 9.4

Investigation of the $I–V$ characteristics for a semiconductor diode

The circuit is set up as shown in Figure 9.11a. Note that it is *essential* to observe the **polarity** of the diode (see Figure 9.11b). The diode is said to be **forward biased** when connected this way around. The 10Ω resistor is included to limit the current in the diode and prevent it being damaged.

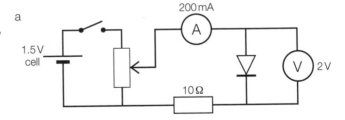

The circuit is switched on and the p.d. is very slowly increased. Values of I and V up to the maximum current obtainable are recorded.

The power supply connections are now reversed so that the diode is **reverse biased** and the experiment is repeated. All the data are plotted on a graph with I on the y-axis against V on the x-axis.

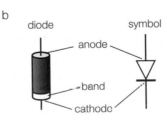

Figure 9.11

A typical graph is shown in Figure 9.12. You will see that the current in the diode when it is reverse biased seems to be zero. In reality, a very small current flows, but it is so small that it is not detected by the milliammeter. However, we normally think of a diode not conducting at all when it is reverse biased.

Questions

A typical set of results is shown in Table 9.3.

Table 9.3

V/V	0.00	0.40	0.56	0.64	0.67	0.71	0.73	0.74
I/mA	0.0	0.0	1.0	5.0	10.0	20.0	30.0	40.0

With power supply terminals reversed.

V/V	0.00	−0.20	−0.40	−0.60	−0.80	−1.00	−1.20	−1.40
I/mA	0.0	0.0	0.0	0.0	0.0	0.0	0.0	0.0

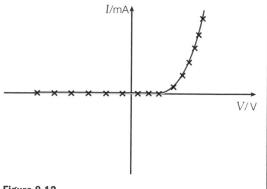

Figure 9.12

1 Plot the data on a set of axes as shown in Figure 9.12.
2 Use your graph to determine:
  a) the p.d. needed to make the diode conduct
  b) the resistance of the diode when the current in it is 6.5 mA
  c) the current in the diode when its resistance is 50 Ω. (Hint: construct a straight line on your graph corresponding to a resistance of 50 Ω.)
The answers to Question 2 should be about:

  a) 0.45 V   **b)** 100 Ω   **c)** 14 mA.

> **Tip**
>
> Remember that when a diode is reverse biased its resistance is *infinite* (well, actually very, very large) and *not* zero!

Test yourself

6 Figure 9.13 shows a computer printout of the *I–V* characteristics for a diode.

  a) Draw a diagram of the circuit you would use to obtain such data.

  b) Estimate the potential difference needed to make the diode conduct.

  c) What is the resistance of the diode when the p.d. across it is 0.20 V?

7 **a)** What is meant by 'forward bias' and 'reverse bias' for a diode? Illustrate your answer by drawing two simple circuits showing a diode connected in series with a cell and a small lamp, labelling clearly which is forward bias and which is reverse bias.

  b) If the diode was not marked, how would you be able to tell which way round it was connected?

  c) Sketch an *I–V* curve for a diode, labelling each region to show which is forward biased and which is reverse biased.

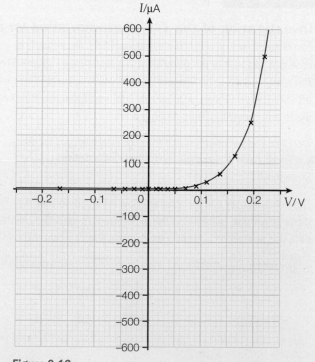

Figure 9.13

9.7 *I–V* characteristics for a thermistor

Activity 9.5

Investigation of the *I–V* characteristics for a thermistor

The circuit as shown in Figure 9.14 is set up, with the voltage from the power supply set at its **minimum** value. The circuit is switched on and the p.d. is gradually increased. The p.d., *V*, and corresponding current, *I*, are recorded at regular intervals up to a maximum current of 20 mA.

The results are tabulated and the circuit is **switched off**. A graph with *I* on the *y*-axis against *V* on the *x*-axis is then drawn.

After at least five minutes have elapsed, the power supply connections are reversed and the experiment is repeated, remembering to start with the minimum p.d. as before. These data are then added to the graph to obtain a curve as shown in Figure 9.15.

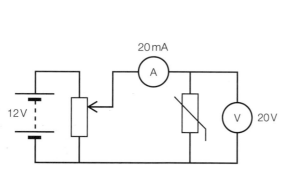

Figure 9.14

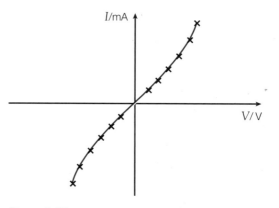

Figure 9.15

Questions

A typical set of results is shown in Table 9.3.

Table 9.3

V/V	0.00	1.00	2.00	3.00	4.00	5.00	6.00	7.00	8.00
I/mA	0.00	2.01	4.09	6.23	8.47	10.95	13.45	16.25	19.30

With power supply terminals reversed.

V/V	0.00	−1.00	−2.00	−3.00	−4.00	−5.00	−6.00	−7.00	−8.00
I/mA	0.00	−2.04	−4.10	−6.25	−8.50	−10.90	−13.50	−16.31	−19.25

1 Plot a graph of *I* against *V* for the data.
2 Calculate the resistance of the thermistor when the current in it is
  a) 3.00 mA
  b) 15.00 mA.
  Estimate the uncertainty in each of these resistance values.
3 Discuss the behaviour of the thermistor with reference to your graph and your answers to Question 2.
4 Explain why it is necessary to wait at least 5 minutes before taking the measurements with the current reversed.
5 Discuss why it is not advisable to use currents significantly larger than the 20 mA suggested.

Exam practice questions

Use Figure 9.16 to answer Questions 1, 2, and 3

1 Which of the graphs could be a plot of current against potential difference for a semiconductor diode? **[Total 1 Mark]**

2 Which of the graphs could be a plot of resistance against potential difference for a tungsten filament lamp? **[Total 1 Mark]**

3 Which of the graphs could be a plot of power against potential difference for a resistor that obeys Ohm's law? **[Total 1 Mark]**

Use Figure 9.17 to answer Questions 4 and 5.

4 Which of the graphs could be a plot of resistance against current for a fixed resistor that obeys Ohm's law? **[Total 1 Mark]**

5 Which of the graphs could be a plot of current against resistance for a variable resistor in a circuit in which the potential difference across the resistor is constant? **[Total 1 Mark]**

6 Graphs A and B in Figure 9.18 show the *I–V* characteristics for a carbon resistor of nominal value $33\,\Omega \pm 5\%$ and a tungsten filament lamp rated 12 V, 5 W.

  a) Explain the different shapes of the two graphs. **[4]**

  b) i) What value does graph A give for the resistance of the carbon resistor?

    ii) Does this fall within the stated tolerance? **[4]**

  c) i) What value does graph B give for the power of the lamp when the potential difference across it is 12.0 V?

    ii) By what percentage does this differ from the stated value? **[3]**

  d) What is the resistance of the lamp when the p.d. across it is:

    i) 12.0 V   **ii)** 1.0 V? **[3]**

        [Total 14 Marks]

7 The $33\,\Omega$ resistor and 12 V, 5 W lamp in Question 6 are now connected in series with a 12 V power supply and ammeter, as shown in Figure 9.19.

  a) A student mistakenly thinks that the current in the circuit should be about 0.18 A. Why do you think this is? **[4]**

  b) The current is actually found to be 0.24 A. Explain, with reference to the graph in Figure 9.18, why this is the case. **[4]**

  c) Explain quantitatively why the lamp glows dimly. **[4]**

  d) Discuss whether the resistor or the lamp consumes the greater power. **[3]**

        [Total 15 Marks]

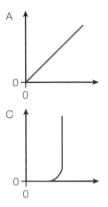
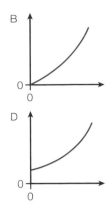

Figure 9.16

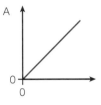
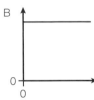
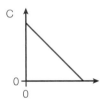
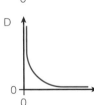

Figure 9.17

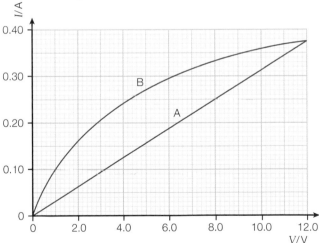

Figure 9.18

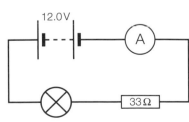

Figure 9.19

8 A student carried out an experiment on a 1.00 m length of nichrome ribbon taken from the heating element of an old toaster that was rated at 240 V, 1000 W. She found that the ribbon obeyed Ohm's law and had a resistance of 11.3 Ω when she applied potential differences of up to 12 V.

a) Define:

    i)   Ohm's law   **ii)**   resistance.   **[3]**

b) What would be the value of the current in the wire when the potential difference across it was 2.0 V?   **[2]**

c) Sketch a graph of current against p.d. for values of p.d. up to 2.0 V, labelling the axes with appropriate values.   **[3]**

d) Explain, with a circuit diagram, how you could obtain this graph using a data logger. State one advantage and one disadvantage of using a data logger for an experiment such as this.   **[5]**

e) Estimate the length of ribbon making up the heating element of the toaster. What assumption do you have to make?   **[5]**

[Total 18 Marks]

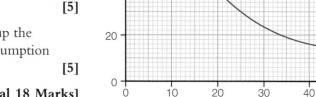

Figure 9.20

9 A graph of the resistance of a diode for different values of current is shown in Figure 9.20.

The circuit shown in Figure 9.21 is to be constructed so that the diode operates at a current as close to 25 mA as possible.

The following resistors are available: 10 Ω, 22 Ω, 33 Ω and 47 Ω. Which value would be the most suitable?

[Total 5 Marks]

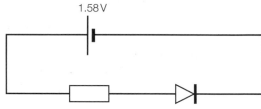

Figure 9.21

Stretch and challenge

10 a) Draw a circuit diagram to show how you would investigate the *I–V* characteristics of a forward biased silicon diode.   **[3]**

b) Sketch the curve you would expect. Label the potential difference axis with the voltage at which the diode starts conducting, which you may take as 0.7 V.   **[3]**

c) The circuit shown in Figure 9.22 is set up using a 5.0 V (peak) a.c. supply of frequency 50 Hz.

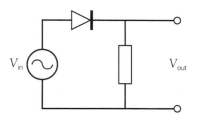

Figure 9.22

    i)   On a sheet of graph paper, sketch a large graph of two cycles of the input voltage. You should label your axes with appropriate values.

    ii)   Directly below your graph, using the same scale, draw a graph of the output voltage, referring to (b) above.

    iii)   Explain the shape of your graph in (ii).   **[9]**

[Total 15 Marks]

Resistance and resistivity

10

Prior knowledge

In this chapter you will need to:
→ recall that resistance is caused by collisions of the electrons with the vibrating lattice ions as the electrons 'drift' through the material of the conductor
→ know that as a result of these collisions, electrical energy is dissipated as thermal energy and the component heats up
→ understand that for a metallic conductor, such as a tungsten filament, its resistance increases as it gets hot while for a semiconductor, such as a thermistor, its resistance decreases with a temperature rise.

The key facts that will be useful are:

→ $\text{resistance} = \dfrac{\text{potential difference}}{\text{current}}$

→ $R = \dfrac{V}{I}$

→ $P = VI$

→ $I = nAvq.$

Test yourself on prior knowledge

1 Explain why copper is a much better conductor of electricity than carbon.

2 A resistor in a supplier's catalogue is rated as:
   Carbon/$33\,\Omega \pm 5\%/1\,\text{W}$.

   a) What is the smallest resistance that it should have?

   b) What is the maximum current that it should take?

   c) Discuss what may happen if the current exceeds this value.

   d) Discuss whether it would be safe to connect a 6 V battery across the resistor.

▌ 10.1 Resistance

It was convenient to introduce the concept of resistance in Chapter 9 when looking at the I–V characteristics for different electrical components. The resistance R of a component was defined as:

$$R = \frac{V}{I}$$

with V measured in volts, I in amperes and R, therefore, in ohms(Ω). The following example will act as a reminder.

Example

1 Explain why the graph in Figure 10.1a shows that the component obeys Ohm's law and calculate the resistance of the component.

2 Calculate the resistance of the component in Figure 10.1b when the current in it is 80 mA.

Answers

1 As the graph is a straight line *through the origin*, the current in the component is proportional to the potential difference across it, which is Ohm's law.

$$R = \frac{V}{I}$$
$$= \frac{6.0\,\text{V}}{120 \times 10^{-3}\,\text{A}}$$
$$= 50\,\Omega$$

2 Clearly the component in Figure 10.1b does not obey Ohm's law, and, as its resistance is not constant, we cannot use the gradient of the graph. We simply use the equation: $R = \frac{V}{I}$.

Putting in the values for $I = 80\,\text{mA}$:

$$R = \frac{V}{I}$$
$$= \frac{0.80\,\text{V}}{80 \times 10^{-3}\,\text{A}}$$
$$= 10\,\Omega$$

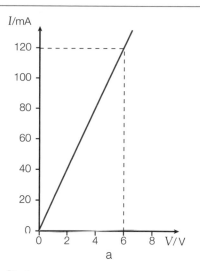

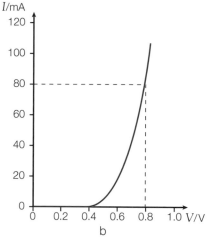

Figure 10.1

10.2 Power dissipation in a resistor

In Chapter 8, we derived that the power, P, transferred in an electrical component was given by:

$$P = IV$$

and that if I was in amperes and V in volts, then P would be in watts – the same unit as mechanical power.

Example

Show that the units of electrical power as defined by $P = IV$ are consistent with those for mechanical power, defined as the rate of doing work.

Answer

$$I = \frac{Q}{t} = \text{C}\,\text{s}^{-1} \text{ and } V = \frac{W}{Q} = \text{J}\,\text{C}^{-1}$$
$$\Rightarrow IV = \text{C}\,\text{s}^{-1} \times \text{J}\,\text{C}^{-1}$$
$$P = \text{J}\,\text{s}^{-1} = \text{W}$$

When a charge flows through a resistor, work is done on the resistor. It is sometimes convenient to express the power transferred in the resistor in terms of its resistance, R, and the current, I, in it.

We know that:

$$R = \frac{V}{I} \text{ or, rearranged, } V = IR, \text{ so:}$$

$$P = IV = I \times (IR) = I^2R$$

We can similarly show, by substituting $I = \frac{V}{R}$, that:

$$P = \frac{V^2}{R}$$

Tip

Learn that:
From the definition that $P = IV$ you need to be able to derive that

$$P = I^2R = \frac{V^2}{R}$$

Example

A manufacturer indicates that the maximum safe power for a particular range of resistors is 250 mW. What is the maximum safe current for a 47 kΩ resistor from this range?

Answer

$$P = I^2R$$

Rearranging $I^2 = \frac{P}{R} \Rightarrow I = \sqrt{\frac{P}{R}} = \sqrt{\frac{250 \times 10^{-3}\,\text{W}}{47 \times 10^3\,\Omega}}$

$$I = 2.3\,\text{mA}$$

Key term

The **internal energy** of a body is the sum of the potential energy and the kinetic energy of its molecules.

We often say that power is **dissipated** in a resistor, particularly if that resistor is in the form of the filament of a lamp or the element of an iron or kettle. The word 'dissipated' means 'scattered'. The electrical energy transferred in the resistor increases the potential energy and the random kinetic energy of the atoms of the material of the resistor – what we call the internal energy of the atoms.

Example

An electric filament lamp is rated at 240 V, 60 W.

1 When it is operating under these conditions, what is:

   a) the current in the filament

   b) the resistance of the filament?

2 At the instant the lamp is switched on, its resistance is 80 Ω. At this instant, what is:

   a) the current in the filament

   b) the power dissipated in the filament?

Answers

1 a) $P = IV$, so $I = \frac{P}{V} = \frac{60\,\text{W}}{240\,\text{V}} = 0.25\,\text{A}$ 2 a) $I = \frac{V}{R} = \frac{240\,\text{V}}{80\,\Omega} = 3.0\,\text{A}$

   b) $R = \frac{V}{I} = \frac{240\,\text{V}}{0.25\,\text{A}} = 960\,\Omega$ b) $P = IV = 3.0\,\text{A} \times 240\,\text{V} = 720\,\text{W}$

The previous example shows the large current (and power) 'surge' that occurs at the instant the lamp is switched on. This explains why the filament of a lamp heats up and reaches its operating temperature very quickly.

Activity 10.1

Investigation of current surge in a lamp

The current surge when a lamp is switched on can be investigated using a data logger, which has the advantage of being able to take a large number of readings very quickly.

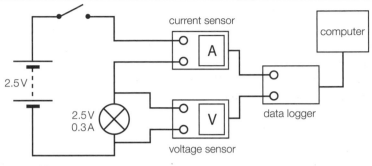

Figure 10.2

Using the circuit shown in Figure 10.2, the current and potential difference can be measured for the first 200 ms after the lamp has been switched on.

Questions

Figure 10.3 shows a typical computer print-out from such an experiment. The data logger is switched on at $t = 0$ and the lamp is switched on at $t = 100$ ms.

Use the print-out to determine:

1 the resistance of the filament when:
   a) it is operating normally
   b) the current in it is at its peak
2 the power developed in the filament when:
   a) it is operating normally
   b) the current in it is at its peak.

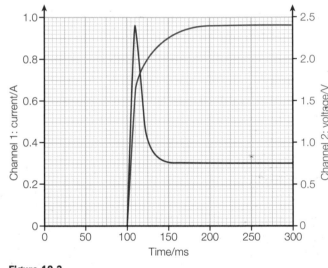

Figure 10.3

Test yourself

1 A carbon resistor has a stated value of $150\,\Omega \pm 5\%$. It has a power rating of 250 mW.

   a) The resistor is ohmic within the stated power rating. What does this mean? Suggest why it may no longer be ohmic if the power exceeds 250 mW.

   b) What should the minimum and maximum values of resistance be?

   c) What is the maximum safe current for the resistor?

   d) Discuss whether it would be safe to connect the resistor across the terminals of a 6V battery.

   e) What might happen if it were to be connected across the terminals of a 12V car battery?

2 An electric hair dryer is rated 1.2 kW, 220 V. You may assume that the blower motor uses 100 W.

   a) Calculate the current the heating element takes when operating at 220 V.

   b) Calculate the resistance of the heating element under these conditions.

   c) Discuss the effect of connecting the dryer to a 110V supply.

   d) Discuss what would happen if you connected a hair dryer rated at 1.1 kW, 110V to a 220V supply.

Activity 10.2

Investigation of the factors affecting the resistance of a wire

The factors that affect the resistance of a wire can be investigated using the same 1 m length of 32 swg (diameter 0.2743 mm) nichrome wire taped to a metre rule, as used in the activity in Section 9.2 on page 118.

For the purpose of this activity, it is more convenient to measure the resistance of the wire directly using an **ohmmeter** set on its 200 Ω range, as shown in Figure 10.4.

One crocodile clip is kept fixed at the zero end of the wire, while the other is pressed firmly on the wire to make contact at different lengths along the wire. The corresponding resistance, R, for each length, l, is recorded, and a graph of R against l is plotted.

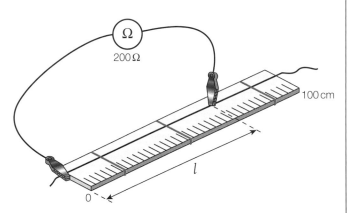

Questions

1 What check should you make before taking readings?
2 Why should you press firmly on the wire, but not too firmly?
3 Suggest, with your reasoning, at what intervals of length it would be appropriate to measure resistance values.

Figure 10.4

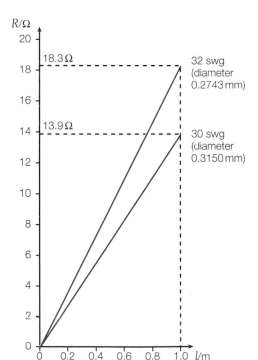

Figure 10.5

A typical set of results has been plotted in Figure 10.5, which also shows the graph for a length of 30 swg (diameter 0.3150 mm) nichrome wire.

What can we deduce from these graphs? Firstly, as they are both straight lines through the origin, it is obvious that:

$$R \propto l$$

The second deduction requires some calculation. The ratio of the areas of cross-section of the two wires is:

$$\frac{A_{30}}{A_{32}} = \frac{0.3150^2\,\text{mm}^2}{0.2743^2\,\text{mm}^2} = 1.32 \left(\text{since } A = \frac{\pi d^2}{4}\right)$$

The corresponding ratio of the resistances of a 1 m length of each wire is:

$$\frac{R_{30}}{R_{32}} = \frac{13.9\,\Omega}{18.3\,\Omega} = 0.76 = \frac{1}{1.32}$$

In other words:

$$R \propto \frac{1}{A}$$

From the experiment, we can see that for a conductor (for example, a wire) of length l and uniform area of cross-section A:

$$R \propto l \text{ and } R \propto \frac{1}{A}$$

If we think about what is happening in the wire, this makes sense: if the wire is longer, it will be more difficult for the electrons to drift from one end to the other. If the wire has a larger cross-sectional area, it will be easier for the electrons to flow (like water in a larger diameter pipe).

Combining the two relationships:

$$R \propto \frac{l}{A}$$

or

$$R = (constant) \times \frac{l}{A}$$

This constant is a property of the *material* of the wire, called its **resistivity**.

Resistivity has the symbol ρ ('rho'). The equation that defines resistivity is therefore:

$$R = \frac{\rho l}{A}$$

The units of resistivity can be derived by rearranging the equation:

$$\rho = \frac{RA}{l} = \frac{\Omega\,m^2}{m} = \Omega\,m$$

Key term

Resistivity (symbol ρ) is a property of a material and is defined by the equation:

$$R = \frac{\rho l}{A}$$

Example

Use the data for the 32 swg wire in Figure 10.5 to calculate a value for the resistivity of nichrome.

Answer

$$R = \frac{\rho l}{A}$$

where $A = \pi \left(\frac{d}{2}\right)^2$.

Rearranging:

$$\rho = \frac{RA}{l}$$

Substituting $R = 18.3\,\Omega$ when $l = 1.000\,m$ and $d = 0.2743\,mm$:

$$\rho = \frac{18.3\,\Omega \times \pi(0.5 \times 0.2743 \times 10^{-3}\,m)^2}{1.000\,m}$$

$$= 1.08 \times 10^{-6}\,\Omega\,m$$

Tip

Remember that:

- resistivity is a property of the *material*
- the unit of resistivity is ohms *times* metres and **not** ohms per metre
- in calculations, l must be in metres and A must be in m^2.

Example

A carbon chip of resistivity $3.0 \times 10^{-5}\,\Omega\,m$ has the dimensions shown in Figure 10.6. What resistance does the chip have for a current in the direction shown?

Answer

Using the formula:

$$R = \frac{\rho l}{A}$$

where $l = 10\,mm = 10 \times 10^{-3}\,m$

and $A = 5\,mm \times 1\,mm = 5 \times 10^{-3}\,m \times 1 \times 10^{-3}\,m$

$$= 5 \times 10^{-6}\,m^2$$

$$R = \frac{3.0 \times 10^{-5}\,\Omega\,m \times 10 \times 10^{-3}\,m}{5 \times 10^{-6}\,m^2}$$

$$= 0.060\,\Omega$$

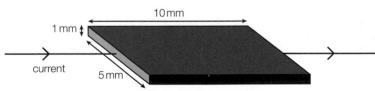

Figure 10.6

Measuring the resistivity of graphite

A standard technique for determining the resistivity of a material in the form of a wire is to calculate the cross-sectional area from measurements of the diameter and then get a value for $\frac{R}{l}$ by a graphical method.

An interesting experiment is to do this for graphite in the form of a pencil 'lead'. The diameter, d, of the lead is measured with a micrometer, or digital calipers, at four places along its length, at different orientations, to give a good average. The lead is then taped to a half-metre rule, with about 1 cm protruding beyond the zero end of the rule, and the circuit in Figure 10.7 is set up.

Firm contact is made at different lengths, l, along the lead. The current, I, and potential difference, V, are recorded, and the corresponding resistance, R, is calculated for each value of l. A graph of R against l is then plotted – it is sensible to plot l in metres at this stage (Figure 10.8).

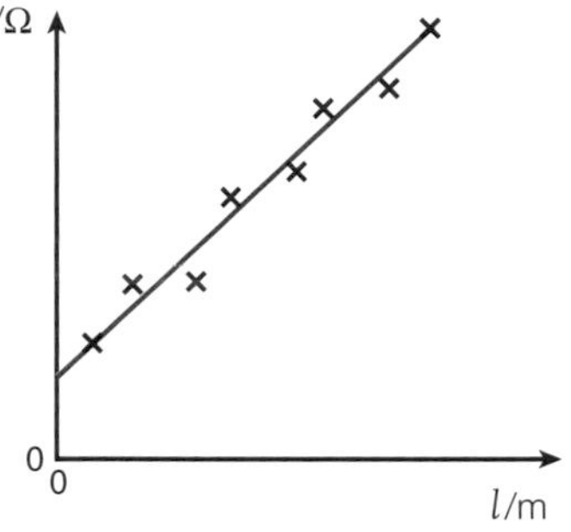

$$R = \frac{\rho l}{A}$$

where $A = \frac{\pi d^2}{4}$ and $\frac{R}{l}$ is the gradient of the graph.

Questions

A typical set of results is shown in Table 10.1.

Table 10.1

d/mm	2.24	2.23	2.25	2.24

1 Why is a micrometer, or digital calipers, a suitable choice of instrument for measuring the diameter of the pencil lead?
2 What check should you make before using this instrument?
3 Explain why it is a good technique to measure the diameter of the pencil lead at four places along its length, at different orientations.
4 Using the data in Table 10.1, plot a graph of resistance R against length l and hence find a value for the resistivity of graphite. (You should get a figure of about $7 \times 10^{-5}\,\Omega\,\text{m}$.)

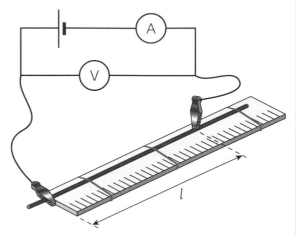

Figure 10.7

Figure 10.8

l/mm	V/V	I/A	R/Ω
20	1.48	0.925	
40	1.51	0.786	
60	1.53	0.648	
80	1.55	0.587	
100	1.58	0.527	
120	1.60	0.465	

The experiment shows the value of using a graphical technique for finding $\frac{R}{l}$.

The scatter of the points on the graph (Figure 10.8) indicates that there is considerable **random error** – probably caused by variation in the pressure applied when making contact with the pencil lead or possibly by inconsistencies in the composition of the graphite. The fact that the graph does not go through the origin shows that there is also a significant **systematic error** caused by **contact resistance** due to poor contact between the crocodile clip and the pencil lead. A graphical method minimises the random error by averaging out the values and virtually eliminates the systematic error when the gradient is taken.

Test yourself

3 The material in the table below is taken from a data book:

Material	Gauge no.	Diameter/ mm	Area/mm²	Ohms per metre	Resistivity/ Ω m
copper	26	0.4572	0.1642	a)	1.72×10^{-8}
constantan	30	b)	0.0779	6.29	c)
nichrome	34	0.2337	d)	25.2	e)

Calculate the missing data a)–e).

4 The 'lead' in a pencil is made of graphite, with clay added to make it harder. Data for two types of pencil lead are shown in the table.

Type of pencil	Length/cm	Diameter/mm	Resistance/Ω	Resistivity/Ω m
B (soft)	17.5	0.69	15.8	
2H (hard)	17.0	2.61	33.8	

a) Copy and complete the table by calculating the resistivity of the two types of pencil. (You might like to use a spreadsheet.)

b) You should find that the resistivity of the 2H pencil is about 30 times that of the B pencil. What can you deduce about the electrical nature of clay from these observations?

10.4 Effect of temperature on the resistivity of a metal

In Section 9.5, we looked at the $I–V$ characteristics for a tungsten filament lamp and produced the graph shown in Figure 10.9.

We said that the tungsten filament did not obey Ohm's law because the temperature of the filament rises as the current increases. We can infer from the graph that the resistance of the filament increases as its temperature rises.

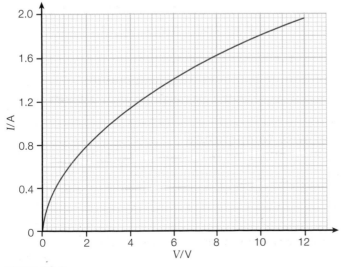

Figure 10.9

Example

Use the graph in Figure 10.9 to calculate the resistance of the tungsten filament when the potential difference across it is 0.20 V, 6.0 V and 12.0 V.

Answer

Remember, we *cannot* use the gradient. We must read off the current, I, at each value of p.d., V, and then calculate the resistance from $R = \dfrac{V}{I}$.

From the graph: at $V = 0.20\,V$, $I = 0.20\,A$

$$R = \frac{V}{I} = \frac{0.20\,V}{0.20\,A} = 1.0\,\Omega$$

You should check for yourself that the resistances at 6.0 V and 12.0 V are 4.3 Ω and 6.1 Ω, respectively.

The example shows that the resistance of the filament increases significantly as the temperature rises. The resistance of the filament at its normal operating temperature (about 3000 K) is some ten times greater than when it is 'off' – that is, at room temperature. This gives rise to a momentary 'surge' of current (and therefore power) when the lamp is first switched on, as shown in Activity 10.1 on page 131.

We can account qualitatively for the increase in the resistivity of metallic conductors with temperature by looking again at the 'free electron' model that we proposed for metals in Section 7.4. You will remember that we said that 'resistance' was caused by the vibrating positive ions in the crystal lattice of the metal impeding the flow of electrons. When the temperature of the metal is raised, the amplitude of vibration of the lattice ions increases and, as a result, there is an increased interaction between the lattice and the 'free' electrons.

A good analogy is to imagine 500 students standing one metre apart from each other in a sports hall. If they just moved a little from side to side, you could run through them fairly easily from one end of the sports hall to the other. If, on the other hand, they jumped around quite vigorously (equivalent to an increase in temperature) you would find it much more difficult to get from one end to the other (more resistance). In doing so, you would expend considerable energy and get hot!

In terms of the drift velocity equation, $I = nAvq$, A and q are constant for a given wire. For a metallic conductor, **n does not depend on the temperature** and so n is also constant. As the temperature rises, the increased vibrations of the lattice will reduce the drift velocity, v, of the electrons and so I will also decrease – that is, the **resistance increases with temperature**.

Activity 10.3

Investigation of the change in resistance of copper with temperature

This can be investigated with the arrangement shown in Figure 10.10.

A coil of copper is formed by winding about 2 metres of enamelled 34 swg copper wire around a short length of plastic tube. About 2 cm of each end of the wire is scraped clean to form a good electrical contact in the connector block.

As the resistance of the coil is less than 1 Ω, a voltmeter/ammeter method is needed to measure the resistance. The p.d., V, and current, I, are recorded for a range of temperatures, θ, from room temperature up to the boiling point of the water. The results are tabulated and a graph of R against θ is plotted as shown in Figure 10.11.

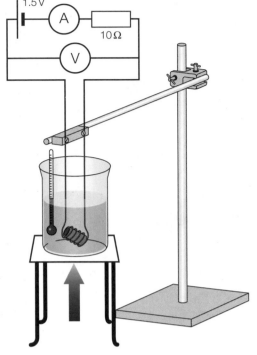

Note that the graph starts at 0 °C on the x-axis and a scale on the y-axis has been chosen such that a value for R_0 – the resistance at 0 °C – can be read off. The graph shows that there is a linear increase of resistance with temperature. However, it is *not* a *proportional* relationship, as the graph does not go through the origin but has an intercept of R_0.

Questions

1 Describe any experimental techniques that you would use to make your measurements as accurate as possible.
2 Some typical results are shown in Table 10.2.

Table 10.2

$\theta/°C$	20	40	60	80	100
V/mV	114	122	130	137	144
I/mA	143	142	141	140	139
R/Ω					

Copy and complete the table by adding values for R and then plot a graph of R against θ, choosing your scales carefully as above.

Figure 10.10

3 Determine values for the gradient and intercept of your graph.
4 There is a theoretical relationship (which you do *not* have to know for examination purposes) between the resistance and temperature of the form:

$$R = R_0(1 + \alpha\theta)$$

where α is a constant – called the **temperature coefficient of resistivity** of copper. The equation can be rearranged to:

$$R = R_0\alpha\theta + R_0$$

The gradient is therefore equal to $R_0\alpha$ and the intercept on the y-axis is R_0. Using the values for the gradient and intercept that you found above, show that α has a value of about $4 \times 10^{-3}\,K^{-1}$ for copper.

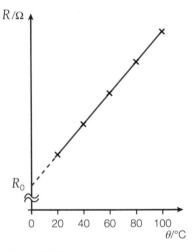

Figure 10.11

As the gradient of the graph obtained in Activity 10.3 is positive, α must also be positive. Copper, along with other metals, is therefore said to have a **positive temperature coefficient** (PTC).

Key term

A material has a **positive temperature coefficient** if its resistivity increases when its temperature increases.

10.5 Effect of temperature on the resistivity of a semiconductor

In order to demonstrate the effect of temperature on the resistance of a semiconductor, we can use a semiconducting device called a **thermistor**. The term comes from an abbreviation of 'thermal resistor' which, as its name suggests, is a resistor whose resistance is temperature dependent. Figure 10.12 shows two types of thermistor (a disc thermistor and a rod thermistor) and the circuit symbol for a thermistor.

Figure 10.12

Key term

A material has a **negative temperature coefficient** if its resistivity decreases when its temperature increases.

The dependence on temperature for a thermistor can be shown very simply by connecting a disc thermistor to an ohmmeter, holding it by its leads and noting its resistance at room temperature (typically about 20 °C). If it is then held firmly between the thumb and forefinger so that its temperature is your body temperature (about 37 °C), a very significant drop in the resistance of the thermistor should be observed. Such a thermistor, where the resistivity *decreases* with temperature, is said to have a **negative temperature coefficient** (NTC).

A negative temperature coefficient can also be explained using $I = nAvq$. In a semiconductor, an increase in temperature can provide extra energy to release more charge carriers. This means that n increases with temperature. To a good approximation, n increases **exponentially** with the absolute temperature, which means that n shows a rapid increase as the temperature rises. Applying $I = nAvq$, A and q are constant as before but n increases by much more than the relatively small decrease in v. The overall effect is that I increases, so the *resistivity decreases with temperature*.

If the temperature is high enough, even some materials that we normally think of as **insulators** (for which n is very small) can begin to conduct. This is because the energy associated with the very high temperature breaks down the atomic structure so that more charge carriers are released.

A spectacular demonstration of this for glass is shown in Figure 10.13.

Warning! In the interests of safety you MUST NOT attempt this experiment as accidental contact with the mains voltage could be extremely dangerous or even fatal!

Instead, you are strongly advised to watch good demonstrations of this experiment that can be found on *Youtube*, for example:

www.youtube.com/watch?v=xeFtfJHcigk or
www.youtube.com/watch?v= ee9bj4mhosY.

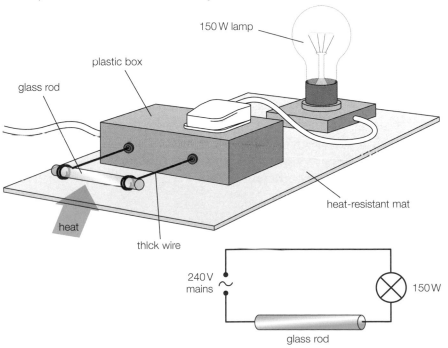

Figure 10.13 Electrically-conducting glass.

At room temperature, the lamp does not light at all as the glass rod acts as an insulator. When the rod is heated very strongly so that it glows red hot, the lamp begins to light, indicating a current – the glass has become a conductor! When the glass is really glowing, the source of heat can be removed and the power dissipated by the current in the rod ($P = I^2R$) is sufficient to maintain the temperature of the glass and keep it conducting.

<table>
<tr><td colspan="2">Tip</td></tr>
<tr><td>

It helps to remember diagrammatically what happens in metallic conductors and semiconductors when the temperature is increased (Table 10.3).

You should note that for AS Level you only need to understand that for an exponentially increasing quantity, the larger the quantity gets, the faster it increases. However, a more rigorous mathematical treatment is expected at A level, as shown in Question 4 of Activity 10.4 on page 140.

</td><td>

Table 10.3

Metallic conductor	Semiconductor
n constant	n ↑↑
A constant	A constant
v ↓	v ↓
q constant	q constant
⇒	⇒
I ↓	I ↑
R ↑	R ↓

</td></tr>
</table>

For the sake of completeness, it should be noted that semiconductors with a **positive temperature coefficient** (PTC) can also be constructed – their resistance actually increases with temperature. Indeed, it is even possible to make a semiconductor for which, over a reasonable temperature range, the increase in n is balanced by the decrease in v so that the resistance is constant! For example, the resistance of carbon resistors – of the type commonly found in school laboratories and in electronic circuits – remains more or less constant over a fairly wide range of temperatures.

Such a process is technically called '**doping**', whereby a pure semiconductor such as germanium or silicon has atoms of an 'impurity' (such as arsenic or boron) added to it. The properties of the semiconductor created are very sensitive to the type and quantity of the impurity atoms. Such materials are extensively used in electronic chips for devices such as mobile phones, notebooks and computers.

Investigation of the effect of temperature on the resistance of a thermistor

This can be investigated using the apparatus shown in Figure 10.14.

The investigation is started with the thermistor immersed in crushed ice. The lowest steady temperature, θ, and the corresponding resistance, R, of the thermistor are recorded. An ohmmeter range that gives the most precise reading is selected.

The beaker of ice is now heated up, ensuring that the leads are kept away from the flame and hot gauze. At intervals of about 10 °C, the Bunsen burner is taken away and the water is stirred thoroughly before the temperature and corresponding resistance are recorded. This is continued every 10 °C until the water boils. The results are tabulated and a graph of R against θ is plotted. A curve like that in Figure 10.15 is obtained. This graph shows that the rate at which the resistance decreases is initially very rapid and then gradually becomes less.

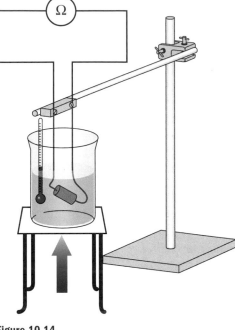

Questions

Some typical results for a range of temperatures from 20 °C to 100 °C are recorded in Table 10.4.

Table 10.4

θ/°C	20	30	40	50	60	70	80	90	100
R/Ω	706	491	350	249	179	135	105	87	74

Figure 10.14

1 Plot a graph of R against θ.
2 What deductions can you make from the shape of the curve?
3 Explain how you can deduce from the graph that the thermistor has a negative temperature coefficient.
4 If you are taking A-level Physics, you need to be familiar with the use of logarithms. Use the data in Table 10.4 to investigate whether the resistance of the thermistor is an exponential function of temperature of the form:

$$R = R_0 e^{-k\theta}$$

where R_0 is the resistance of the thermistor at 0 °C and k is a constant. Taking logarithms to base e, this equation becomes:

$$\ln R = -k\theta + \ln R_0$$

If the equation is valid, a graph of $\ln R$ against θ should be a straight line of gradient $-k$ and intercept $\ln R_0$. (You should find that the equation seems to be valid up to about 65 °C, with k approximately 0.034 K^{-1} and R_0 about 13 kΩ.)

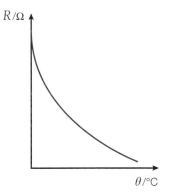

R/Ω

θ/°C

Figure 10.15

5 a) Explain what is meant by the phrase, 'copper has a positive temperature coefficient of resistivity and silicon has a negative temperature coefficient'.

  b) Give another example of each.

  c) Discuss how the equation $I = nAvq$ can be used to explain these terms.

6 A manufacturer states the following: 'Constantan is an alloy of 55% copper and 45% nickel. Its resistivity is high enough to achieve suitable resistance values in even very small lengths and its temperature coefficient of resistivity is not excessive.'

  a) What property will it have if its temperature coefficient of resistivity is not excessive?

  b) The manufacturer sells constantan foil of thickness 0.0125 mm. A 1.0 mm wide strip is cut from the foil. Show that a length of about 25 mm would be needed to make a standard $1.00\,\Omega$ resistor if the resistivity of the constantan is $49.4 \times 10^{-8}\,\Omega$ m.

  c) Discuss why pure copper would a less suitable material for a resistor such as this.

7 Figure 10.16 shows how the resistance of a thermistor varies with temperature.

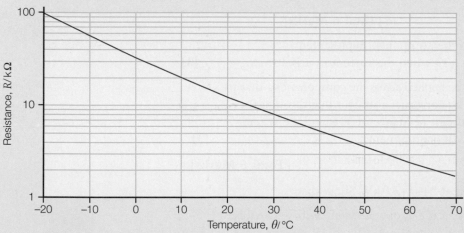

Figure 10.16

  a) To what extent does the graph suggest that the resistance decreases exponentially with temperature?

  b) Copy the table below and use the graph to fill in the missing data.

$\theta/°C$	−20	−16			10	30	
$R/k\Omega$			50	40			2

  c) Use your completed table of data to plot a graph of resistance against temperature using a linear scale for both the resistance and the temperature.

  d) Use this graph to determine the resistance of the thermistor when its temperature is that of

   i) a domestic freezer (−18 °C)

   ii) melting ice (0 °C)

   iii) the human body (37 °C).

Exam practice questions

Use Figure 10.17 to answer Questions 1, 2 and 3.

1 Which graph could be a plot of the number of charge carriers per unit volume in a length of copper wire against temperature? **[Total 1 Mark]**

2 Which graph could be a plot of the resistance of a length of copper wire against temperature? **[Total 1 Mark]**

3 Which graph could be a plot of the resistance of a thermistor against temperature? **[Total 1 Mark]**

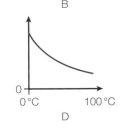
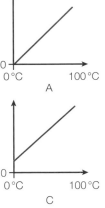

Figure 10.17

4 For the carbon chip shown in Figure 10.6 on page 133, what is the resistance between:

   a) the longer edges **[4]**

   b) the faces of the chip? **[2]**

                         [Total 6 Marks]

5 a) What is the resistance of the component represented in Figure 10.1b on page 129 when the current in it is 40 mA? **[3]**

   b) What p.d. would have to be applied across the component so that its resistance was 50 Ω? (Hint: construct a straight line through the origin to represent a constant resistance of 50 Ω and see where it cuts the *I–V* curve for the component.) **[3]**

                         [Total 6 Marks]

6 a) An electric kettle is rated at 240 V, 3.0 kW. Show that:

     i) the current is such that it can operate safely with a 13 A fuse

     ii) the resistance of the element when it is operating normally is about 19 Ω. **[5]**

   b) If the kettle was operated on a 110 V supply and the resistance of the element was assumed to remain constant, what would be:

     i) the current in the element

     ii) the power dissipated in the element? **[4]**

   c) Discuss whether the assumption that the resistance of the filament remains constant is reasonable. **[2]**

                        [Total 11 Marks]

7 a) Define resistivity. **[2]**

   b) The table below is taken from a data book.

Material	Gauge no.	Diameter/mm	Area/mm²	Ohms per metre	Resistivity/Ω m
copper	24	0.5588	0.2453	(i)	1.72×10^{-8}
constantan	28	(ii)	0.1110	4.41	(iii)
nichrome	32	0.2743	(iv)	18.3	(v)

   Calculate the missing data (i)–(v). **[10]**

                         [Total 12 Marks]

8 The graph in Figure 10.18 shows data from an experiment to investigate how the current, I, varied with the p.d., V, across a 2.00 m length of nichrome wire of diameter 0.25 mm.

a) Explain why the graph shows that the wire obeys Ohm's law. [2]

b) Use the data to calculate a value for the resistivity of nichrome. [6]

c) Nichrome has a negligible temperature coefficient of resistivity. What does this mean? [2]

[Total 10 Marks]

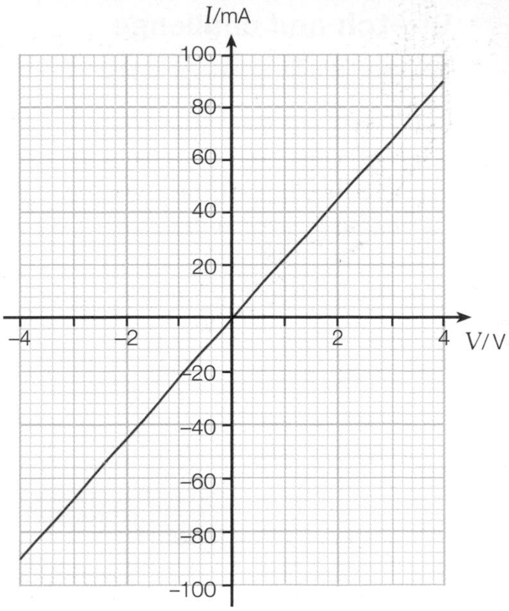

9 a) Sketch a graph to show how the current in a 12 V, 5 W tungsten filament bulb depends on the potential difference across it. [3]

b) Calculate the theoretical resistance of the filament when it is at its operating temperature. [2]

c) Explain, with reference to the equation $I = nAvq$, why the resistance of the filament is much less when the lamp is 'off'. [5]

[Total 10 Marks]

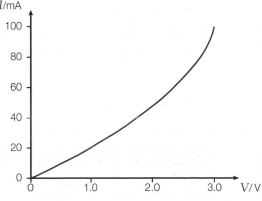

Figure 10.18

10 The label on a reel of constantan resistance wire states that it is 30 swg and has a resistance of $6.29\,\Omega\,m^{-1}$. The resistivity of constantan is $49 \times 10^{-8}\,\Omega\,m$.

a) Show that this is consistent with 30 swg wire having a diameter of 0.315 mm. [4]

b) The diameter is checked with digital vernier calipers reading to a precision of 0.01 mm.

i) Explain what steps should be taken to reduce systematic and random error when determining the diameter.

ii) Estimate the percentage uncertainty in the measurement of the diameter. [5]

c) Show that the length of wire needed to make a 2.2 Ω resistor is about 35 cm. [3]

d) If the resistor is to have a tolerance of 2%, what tolerance is there in cutting the wire to the required length? [2]

[Total 14 Marks]

11 The graph in Figure 10.19 shows the current–voltage characteristic for a thermistor.

a) What is the resistance of the thermistor when the p.d. across it is

i) 1.0 V   ii) 3.0 V? [5]

b) Calculate the rate at which energy is being converted in the thermistor at each of these voltages. [3]

c) i) What does your answer to part (b) tell you about the temperature of the thermistor when the p.d. across it is 3 V compared with when the p.d. is 1 V?

ii) Explain, in terms of the equation $I = nAvq$, why the resistance changes in the way that it does in part (a). [7]

[Total 15 Marks]

Figure 10.19

Stretch and challenge

12 The following passage is taken from an article on carbon nanotubes (CNTs).

Carbon nanotubes are allotropes of carbon with a cylindrical nanostructure. These cylindrical carbon molecules have unusual properties, which are valuable for nanotechnology, electronics and other fields of materials science and technology. In particular, owing to their extraordinary mechanical and electrical properties, carbon nanotubes find applications as additives to various structural materials such as carbon fibre sports equipment and are being used in the latest computer technology.

Their name is derived from their long, hollow structure, with the walls formed by one-atom-thick sheets of carbon, called graphene. The way in which the cylinders are rolled decides the nanotube properties; for example, whether the individual nanotube shell is a metal or semiconductor.

Figure 10.20 Structure of a carbon nanotube

Carbon nanotubes are the strongest and stiffest materials yet discovered in terms of tensile strength and elastic modulus, respectively. In 2000, a multi-walled carbon nanotube was tested to have a tensile strength of 63 GPa. For illustration, this translates into the ability to endure tension of a weight equivalent to a mass of 6422 kg on a cable with cross-section of 1 mm². Under excessive tensile strain, the tubes will undergo plastic deformation but CNTs are not nearly as strong under compression.

Because of their nanoscale cross-section, electrons propagate only along the tube's axis and electron transport involves quantum effects. The minimum electrical resistance of a single-walled carbon nanotube is $\frac{h}{4e^2}$.

a) What is the order of magnitude of structures in nanotechnology (paragraph 1)? **[2]**

b) Discuss the differences in the electrical properties of metals and semiconductors (paragraph 2). **[4]**

c) Explain the terms tensile strength and elastic modulus (paragraph 3). **[2]**

d) Show that a tensile strength of 63 GPa does correspond to the weight of a 6422 kg mass suspended on a cable of cross-section 1 mm² (paragraph 3). **[2]**

e) What is meant by tensile strain and plastic deformation? Suggest why CNTs are not as strong under compression as they are under tension (paragraph 3). **[3]**

f) Estimate the diameter of the CNT shown in Figure 10.20. **[4]**

g) Why is it not possible to use the equation $I = nAvq$ for CNTs (paragraph 4)? **[2]**

h) Show that $\frac{h}{4e^2}$ has the units of resistance (paragraph 4). **[3]**

i) Calculate the minimum resistance of a CNT (paragraph 4). **[2]**

[Total 24 Marks]

11 Internal resistance, series and parallel circuits, and the potential divider

Prior knowledge

In this chapter you will need to:
→ apply and develop your knowledge from the preceding chapters on electricity
→ recall that an ammeter connected in series measures current and that a voltmeter connected across (or in parallel with) a component measures the potential difference across the component
→ remember that charge is conserved at a junction
→ be familiar with the conservation of energy.

The key facts that will be useful are:
→ $R = \dfrac{V}{I}$; $I = \dfrac{V}{R}$; $V = IR$
→ $P = VI$; $P = I^2R$; $P = \dfrac{V^2}{R}$
→ volts = joules per coulomb.

Test yourself on prior knowledge

1 In the circuit shown in Figure 11.1, calculate:
   a) the current in the 2.2 Ω resistor
   b) the current in the 3.3 Ω resistor
   c) the current in the ammeter.

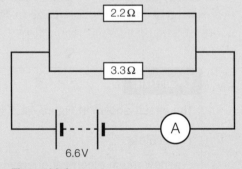

Figure 11.1

2 In the circuit shown in Figure 11.1 calculate:
   a) the power developed in the 2.2 Ω resistor
   b) the power developed in the 3.3 Ω resistor
   c) the rate at which the cell is converting energy
   d) the energy taken from the cell in 5.0 minutes.

11.1 Conservation of energy in circuits

Conservation of energy is a fundamental concept of physics. It is stated in this way: **energy cannot be created or destroyed but merely changed from one form to another.** Consider the circuit in Figure 11.2.

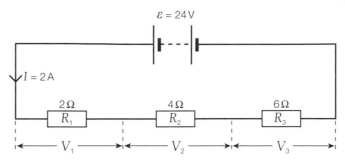

Figure 11.2

By the principle of conservation of energy, when a charge, Q, flows round the circuit:

energy converted by battery = energy dissipated in the three resistors

$$Q\varepsilon = QV_1 + QV_2 + QV_3$$

$$\varepsilon = V_1 + V_2 + V_3$$

As $V = IR$:

$$\varepsilon = IR_1 + IR_2 + IR_3$$

or circuit e.m.f., $\varepsilon = \Sigma IR$

Tip

The symbol Σ ('sigma') means 'sum of'.

If 1 C of charge were to move round the circuit:

energy gained by charge in battery	=	energy lost by charge in R_1	+	energy lost by charge in R_2	+	energy lost by charge in R_3
(24 J)		(4 J)		(8 J)		(12 J)

Example

The circuit shown in Figure 11.2 is left on for 2.0 minutes.

1 Calculate:

   a) how much electrical energy is converted in the cell

   b) how much energy is dissipated in each of the resistors.

2 Explain how your answers show that energy is conserved.

Answers

1 a) Charge flowing $Q = It = 2\,\text{A} \times 120\,\text{s} = 240\,\text{C}$

   Energy converted in cell $= \varepsilon Q = 24\,\text{V} \times 240\,\text{C} = 5.76\,\text{kJ}$

   b) Energy dissipated in $R_1 = I^2R_1t = (2\,\text{A})^2 \times 2\,\Omega \times 120\,\text{s} = 0.96\,\text{kJ}$

   Energy dissipated in $R_2 = I^2R_2t = (2\,\text{A})^2 \times 4\,\Omega \times 120\,\text{s} = 1.92\,\text{kJ}$

   Energy dissipated in $R_3 = I^2R_3t = (2\,\text{A})^2 \times 6\,\Omega \times 120\,\text{s} = 2.88\,\text{kJ}$

2 Add up the energy dissipated in each of the resistors:

$0.96\,kJ + 1.92\,kJ + 2.88\,kJ = 5.76\,kJ = \text{energy converted in cell}$

This shows that energy has been conserved.

11.2 Internal resistance

Unfortunately, not all of the chemical energy converted to electrical energy inside a cell emerges at the terminals of the cell. When a charge flows and produces a current in the cell, some of the energy is used up in 'pushing' the electrons through the cell. In other words, it is used to overcome the **internal resistance** of the cell, which is usually given the symbol r.

Consider a cell of e.m.f., ε, and internal resistance, r, which is connected to a lamp of resistance, R, as in Figure 11.3.

Key term

The **internal resistance** of a cell, or other power supply, opposes the flow of charge through the cell. Some of the energy converted by the cell, or power supply, will be used up inside the cell to overcome this resistance.

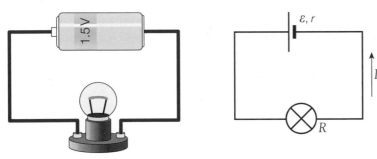

Figure 11.3

If there is a current, I, in the circuit:

rate of energy converted in cell	=	rate of work done against internal resistance	+	rate of work done lighting lamp
εI	=	$I^2 r$	+	$I^2 R$
ε	=	Ir	+	IR

Rearranging: $\qquad IR = \varepsilon - Ir$

and putting $IR = V$: $\qquad V = \varepsilon - Ir$

This is effectively an application of the relationship $\varepsilon = \sum IR$ that we derived in Section 11.1.

The difference between the e.m.f., ε, of a cell and the potential difference, V, at its terminals is $\varepsilon - V = Ir$, and is sometimes called the '**lost volts**'. (Why do you think it is called this?)

An alternative rearrangement of the energy equation that enables us to find the current is:

$$I = \frac{\varepsilon}{R + r}$$

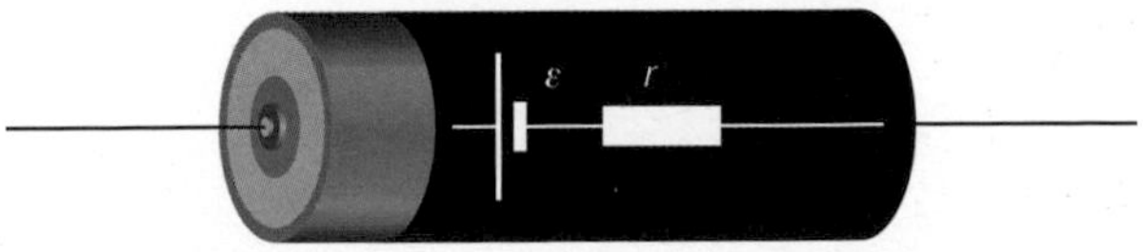

Figure 11.4

Tip

It is helpful to think of a power supply (such as a cell) as being a voltage (the e.m.f.) in series with a resistor (the internal resistance) as shown in Figure 11.4.

Example

A torch battery of e.m.f. 4.5 V and internal resistance 0.4 Ω is connected across a lamp of resistance 6.4 Ω.

1 What is the current in the lamp?

2 How much power is:

  a) dissipated in the lamp

  b) wasted in the cell?

Answers

1 Circuit current, $I = \dfrac{\varepsilon}{R + r} = \dfrac{4.5\,\text{V}}{6.4\,\Omega + 0.4\,\Omega}$

$\qquad\qquad\qquad = 0.66\,\text{A}$

2 a) For power dissipated in lamp:

$\qquad P = I^2R = (0.66\,\text{A})^2 \times 6.4\,\Omega$

$\qquad\quad = 2.8\,\text{W}$

  b) For power wasted in cell:

$\qquad P = I^2r = (0.66\,\text{A})^2 \times 0.4\,\Omega = 0.2\,\text{W}$

> ### Tip
>
> Remember:
>
> $V = \varepsilon - Ir$
>
> and
>
> $I = \dfrac{\varepsilon}{R + r}$

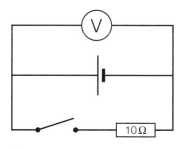

Figure 11.5

Example

In the circuit in Figure 11.5, the high-resistance voltmeter reads 1.55 V when the switch is open and 1.49 V when the switch is closed.

1 Explain why:

  a) the e.m.f. of the cell can be considered to be 1.55 V

  b) the voltmeter reading drops when the switch is closed.

2 Calculate the internal resistance of the cell.

Answers

1 a) As the voltmeter has a very high resistance, it takes virtually no current. Therefore, with the switch open, there is negligible current in the cell, so the reading of 1.55 V can be taken as its e.m.f.

  b) When the switch is closed, the 10 Ω resistor is brought into the circuit. This causes a current, I, in the circuit so that the potential difference, V, across the cell drops to $V = \varepsilon - Ir$. (With the switch open, $I = 0$, so that $V = \varepsilon$.)

2 From $V = \varepsilon - Ir$:

$\qquad Ir = \varepsilon - V$

$\qquad\quad = 1.55\,\text{V} - 1.49\,\text{V}$

$\qquad\quad = 0.06\,\text{V}$

For the 10 Ω resistor:

$\qquad I = \dfrac{V}{R} = \dfrac{1.49\,\text{V}}{10\,\Omega} = 0.149\,\text{A}$

$\qquad r = \dfrac{\varepsilon - V}{I} = \dfrac{0.06\,\text{V}}{0.149\,\text{A}}$

$\qquad\quad = 0.40\,\Omega$

Core practical 3

Investigation of the e.m.f. and internal resistance for a cell

Figure 11.6 shows a standard circuit to find the e.m.f. and internal resistance for a zinc–carbon cell.

Starting with the variable resistor (rheostat) at its highest value (to minimise any heating effects), the current, I, in the cell and the potential difference, V, across its terminals are recorded for different settings of the rheostat.

Rearranging $V = \varepsilon - Ir$:

$$V = -Ir + \varepsilon$$

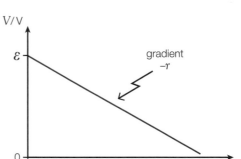

Figure 11.6

If a graph of V against I is plotted, we would expect to get a straight line of gradient $-r$ and intercept c on the y-axis (Figure 11.7). In practice, the line may not be straight because the internal resistance may not be constant – particularly for large currents.

Questions

A typical set of observations is recorded in Table 11.1.

Table 11.1

I/A	0.20	0.40	0.60	0.80	1.00	1.20	1.40	1.60
V/V	1.44	1.32	1.20	1.09	0.95	0.84	0.73	0.59

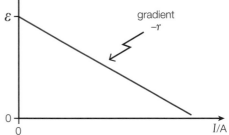

Figure 11.7

1 Plot a graph of these data with V on the y-axis and I on the x-axis.
2 Use your graph to determine values for the e.m.f. and internal resistance of the cell. (You should get values of about 1.56 V and 0.60 Ω, respectively.)

All power supplies – not just cells – have internal resistance, and the value of this internal resistance is often critical for the correct functioning of the power supply. For example, a car battery must have a *very small* internal resistance (typically as low as 0.01 Ω) because the starter motor takes a very large current when the engine is started (perhaps as much as 200 A). This is illustrated in the following example. On the other hand, a laboratory high voltage (E.H.T.) supply, for safety reasons, has a *very large* built-in internal resistance – typically 50 MΩ. This limits the current to a fraction of a milliamp, which is nevertheless still enough to give a nasty shock! E.H.T. supplies must therefore be treated with great care.

Examples

1 A 12 V car battery has an internal resistance of 0.01 Ω.

  a) What is the potential difference across its terminals when the engine is started if the starter motor takes an initial current of 200 A?

  b) Explain why, if the driver has the headlights on, they are likely to go dim when he starts up the engine.

  c) Calculate how much power the battery delivers to the starter motor.

2 A mechanic of resistance 10 kΩ – as measured between his hand and the ground – accidentally touches the 'live' terminal of the battery and experiences a small electric shock. Calculate the power dissipated in his body.

Answers

1 a) Using $V = \varepsilon - Ir$:

$$V = 12\,\text{V} - (200\,\text{A} \times 0.01\,\Omega)$$
$$= 12\,\text{V} - 2\,\text{V}$$
$$= 10\,\text{V}$$

b) The headlights operate at full brightness when the p.d. across them is 12 V. If this is reduced to only 10 V when the engine is started, the headlights will go dim.

c) The power delivered by the battery to the starter motor is given by:

$$P = IV = 200\,\text{A} \times 10\,\text{V}$$
$$= 2000\,\text{W} \text{ (or } 2\,\text{kW)}$$

2 As the internal resistance of the battery is so much less than the resistance of the mechanic, we can ignore any effect of the internal resistance and simply assume that the current in the mechanic is given by

$$I = \frac{\varepsilon}{R} = \frac{12\,\text{V}}{10 \times 10^3\,\Omega} = 1.2 \times 10^{-3}\,\text{A}$$
$$P = IV = 1.2 \times 10^{-3}\,\text{A} \times 12\,\text{V}$$
$$= 0.014\,\text{W} \text{ (14\,mW)}$$

Test yourself

1 A 1.50 V cell having an internal resistance of $0.3\,\Omega$ is connected in series with a $2.2\,\Omega$ resistor and an ammeter. A high-resistance voltmeter is connected across the terminals of the cell.

   a) Draw a circuit diagram of this arrangement.

   b) What will the reading on the ammeter be?

   c) What will the reading on the voltmeter be?

   d) How much power is generated in the resistor?

   e) How much power is lost within the cell?

2 A student sets up the circuit shown in Figure 11.8. The resistor has a tolerance of 5%. Before switching on, she connects a voltmeter across the power supply and adjusts the output to 9.00 V.

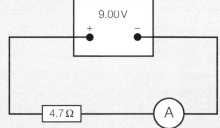

   a) When she switches the supply on, she is surprised **Figure 11.8** that the ammeter only reads 1.50 A. Why do you think she was surprised?

   b) Her teacher suggests that the power supply may have an internal resistance. What do the data suggest is the magnitude of this internal resistance?

3 A car has a battery of e.m.f. 12 V and internal resistance $0.01\,\Omega$. The starter motor has a resistance of $0.04\,\Omega$ and needs an initial current of at least 200 A to start it.

   a) How much current is available to start the motor?

   b) How much power is taken from the battery when the starter motor is first switched on?

   c) What does the terminal p.d. drop to when the motor is switched on?

   d) Discuss what will happen if the terminals corrode and develop a terminal resistance of $0.1\,\Omega$.

11.3 Solar cells

Solar, or photovoltaic, cells are now widely used as a source of electric power. Your calculator is almost certainly powered by a solar cell, and you have probably seen solar cells being used to power road signs. Arrays of solar cells are used to power the electrical systems of satellites and can be used to supply electrical energy to buildings. These must not be confused with solar **heating** panels, in which the energy from the Sun is used to heat water flowing in an array of pipes in the panel. (See example on page 277).

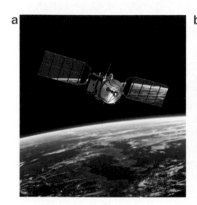

Figure 11.9 a) A satellite using solar cells to power its electrical systems and b) a home with an array of solar cells, and two solar panels on the top left of the roof.

When light strikes the photocell, it gives some of its energy to free electrons in the semiconductor material of the cell (commonly silicon). An electric field within the cell provides a force on the electrons. The electron flow provides the **current** and the cell's electric field causes a **voltage**. With both current and voltage, we have **power**, which is the product of the two.

The characteristics of a solar cell are dependent on the illumination. They can be investigated using the circuit shown in Figure 11.10. The value of the potentiometer required will depend on the type of cell and the amount of illumination.

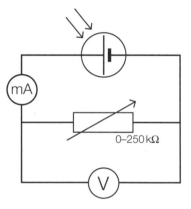

Figure 11.10

Example

The data in the table were obtained for the output of a solar cell using the circuit shown in Figure 11.10 when the cell was illuminated by normal laboratory light.

I/mA	0.010	0.015	0.020	0.025	0.030	0.035	0.040	0.045
V/V	2.07	1.97	1.84	1.72	1.60	1.44	1.28	1.10
R/kΩ								
P/μW								

1 Copy and complete the table of data by adding values for the resistance across the cell, $R = \dfrac{V}{I}$, and the power generated, $P = VI$. You might like to construct a spreadsheet to do this.

2 a) Draw a graph of V against I and comment on the shape of your graph.

b) Use your graph to determine values for:

   i) the e.m.f. of the cell

   ii) the internal resistance of the cell for low current values.

c) Plot a graph of P against R and use it to determine the maximum power generated.

Answers

1

I/mA	0.010	0.015	0.020	0.025	0.030	0.035	0.040	0.045
V/V	2.07	1.97	1.84	1.72	1.60	1.44	1.28	1.10
R/kΩ	207	131	92	69	53	41	32	24
P/μW	21	30	37	43	48	50	51	49

2 a) The graph of V against I should look like Figure 11.11.

The graph is initially a straight line, which shows that the cell has a constant internal resistance for low current values. At larger current values, the internal resistance increases significantly.

b) i) The e.m.f. of the cell is given by the intercept when $I = 0$, that is, e.m.f. = 2.3 V.

ii) The internal resistance of the cell for low current values is equal to the gradient of the straight part of the line. You can determine this for yourself and show that it is about 24 kΩ.

c) If you plot a graph of P against R, you will find that the maximum power generated is about 52 μW.

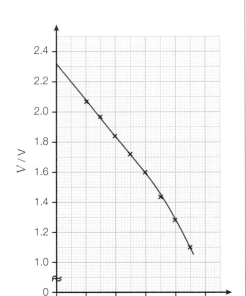

Figure 11.11

If the cell were to be illuminated more strongly, for example by a bench lamp, the e.m.f. of the cell would be much greater and the internal resistance would drop considerably. As a result, the maximum power generated is *much* greater. This is illustrated in Exam practice question 9 on page 169, which shows that when illuminated by a bench lamp, the e.m.f. of the cell increases to more than 6 V, its internal resistance drops to below 4 kΩ and the power increases to over 2 mW.

11.4 Measuring the resistance of a component

We saw in Section 9.4 that resistance was defined as $R = \frac{V}{I}$. This obviously gives a way of measuring resistance, as in the circuit shown in Figure 11.12. The current, I, in the resistor is measured by the ammeter, the potential difference, V, across the resistor is measured with the voltmeter, and then $R = \frac{V}{I}$.

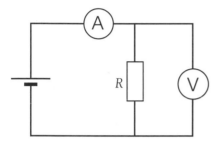

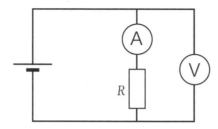

Figure 11.12   **Figure 11.13**

We saw in Section 8.2 that a small amount of current is taken by the voltmeter, so the current recorded by the ammeter is not, strictly speaking, the actual current in the resistor. As long as the voltmeter has a high resistance (or much greater than that of the resistor) we can ignore this effect.

Alternatively, we could set up a circuit as shown in Figure 11.13.

We saw in Section 7.3 that ammeters should have a very small resistance, so they do not affect the current that they are measuring. Nevertheless, the ammeter might have an appreciable effect in the circuit of Figure 11.13, particularly if the resistor also has a low resistance. The ammeter *does*, indeed, measure the current in the resistor, but now the voltmeter measures the p.d. across both the resistor *and* the ammeter. In most cases the circuit shown in Figure 11.12 is preferable.

The voltmeter/ammeter method is rather cumbersome, however, as it requires two meters and a power supply, together with a calculation, to find the resistance. A quicker and easier way is to use a **digital ohmmeter**. As its name suggests, this instrument measures 'ohms' – that is, resistance – directly. Most digital multimeters have an 'ohms' range. Alternatively, digital ohmmeters can be purchased as separate instruments.

It is important to understand that a digital ohmmeter does not actually measure resistance directly in the way that an ammeter measures current. When you turn the dial on a multimeter to the 'ohms' range, you bring a battery into play, which produces a very small current in the component whose resistance is being determined. This current is measured by the meter and then converted into a resistance reading by the 'electronics' inside the meter. The general principle of an ohmmeter can be demonstrated by the Activity 11.3 on page 162, in which a graphical method, rather than electronics, is used to find a resistance value.

Tip

WARNING! Remember, an ohmmeter produces a current. It should therefore never be used to measure the resistance of a component that is connected into a circuit or else damage may be done to the circuit or the ohmmeter.

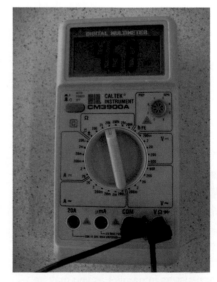

Figure 11.14 Digital multimeter measuring the resistance of a $4.7\,k\Omega$ resistor. The meter is set on the $20\,k\Omega$ range and so the resistance recorded is $4.68\,k\Omega$.

Unfortunately, most school laboratory ohmmeters have a lowest range of $200\,\Omega$ (well $199.9\,\Omega$ actually!), so they can only measure to a precision of $0.1\,\Omega$. This means that for accurate measurements of resistance in the order of an ohm or less, we have to revert back to the voltmeter/ammeter method.

11.5 Resistors in series

Consider the circuit shown in Figure 11.15. If we were to insert an ammeter in the positions A, B, C and D in turn, we would observe that the current, I, was the same at each position. This arises as a result of the **conservation of charge** (Section 7.1) – in simple terms, the rate at which electrons leave each resistor must be the same as the rate at which they enter, as there is nowhere else for them to go!

By connecting a voltmeter across AD and then across each of AB, BC and CD in turn, we would find that:

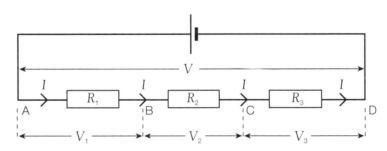

Figure 11.15 Resistors in series.

$$V = V_1 + V_2 + V_3$$

We have seen this result before in Section 11.1, where we discussed how it arises as a result of the application of the conservation of energy.

If the three resistors have a combined resistance R, then by using $V = IR$:

$$IR = IR_1 + IR_2 + IR_3$$

As I is the same throughout, *in series*:

$$R = R_1 + R_2 + R_3$$

11.6 Resistors in parallel

Now consider the parallel arrangement in Figure 11.16.

As the three resistors are connected to the common points A and B, the potential difference, V (which could be measured by connecting a voltmeter between A and B), must be the same across each resistor.

Tip

You must learn how to *derive* the equation $R = R_1 + R_2 + R_3$ for resistors in series as you may be asked to do this in the examination

Tip

When deriving the equations for resistors in series and parallel, remember:

- resistors in **series** have the **same current** in each resistor
- resistors in **parallel** have the **same potential difference** across each resistor

Tip

You must learn how to *derive* the equation

$$\frac{1}{R} = \frac{1}{R_1} + \frac{1}{R_2} + \frac{1}{R_3}$$

for resistors in parallel as you may be asked to do this in the examination

We saw in Section 7.3 that from the conservation of charge at a junction:

$$I = I_1 + I_2 + I_3$$

where I is the total circuit current and I_1, I_2 and I_3 are the currents in R_1, R_2 and R_3, respectively.

If the three resistors have a combined resistance R, using $I = \dfrac{V}{R}$ we have:

$$\frac{V}{R} = \frac{V}{R_1} + \frac{V}{R_2} + \frac{V}{R_3}$$

As V is the same across each resistor, *in parallel*:

$$\frac{1}{R} = \frac{1}{R_1} + \frac{1}{R_2} + \frac{1}{R_3}$$

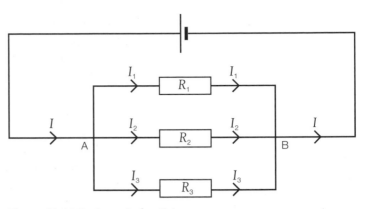

Figure 11.16 Resistors in parallel.

11.7 Series and parallel combinations

Problems involving series and parallel combinations of resistors should be tackled by combining the parallel resistors first, to find the equivalent single resistance, and then adding up the series resistances.

Consider the circuit in Figure 11.17.

For the parallel arrangement of R_1 and R_2:

$$\frac{1}{R} = \frac{1}{R_1} + \frac{1}{R_2}$$

$$\frac{1}{R} = \frac{1}{20\,\Omega} + \frac{1}{30\,\Omega}$$

$$= 0.050\,\Omega^{-1} + 0.033\,\Omega^{-1} = 0.083\,\Omega^{-1}$$

$$R = \frac{1}{0.083\,\Omega^{-1}} = 12\,\Omega$$

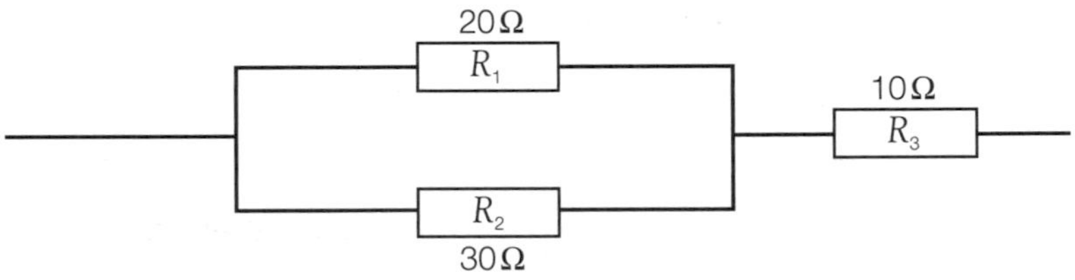

Figure 11.17

We can now add this $12\,\Omega$ to R_3 to give the resistance of the combination of R_1, R_2 and R_3 as $10\,\Omega + 12\,\Omega = 22\,\Omega$. We can represent this calculation diagrammatically as in Figure 11.18.

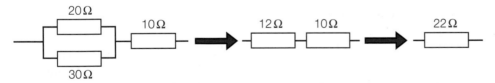

Figure 11.18

Example

Draw each arrangement that can be obtained by using three $10\,\Omega$ resistors either singly or combined in series and parallel arrangements (seven arrangements are possible!). Calculate all possible resistance values.

Answer

Figure 11.19 shows how the seven possible combinations can be obtained.

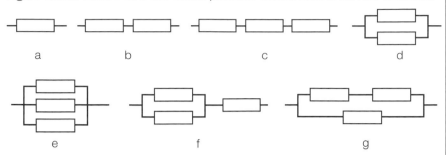

Figure 11.19

The resistances for combinations a–c are very simply $10\,\Omega$, $20\,\Omega$ and $30\,\Omega$, respectively.

For combination d:

$$\frac{1}{R} = \frac{1}{10\,\Omega} + \frac{1}{10\,\Omega}$$
$$= 0.10\,\Omega^{-1} + 0.10\,\Omega^{-1} = 0.20\,\Omega^{-1}$$
$$R = \frac{1}{0.20\,\Omega^{-1}} = 5.0\,\Omega$$

Similarly, for combination e:

$$\frac{1}{R} = \frac{1}{10\,\Omega} + \frac{1}{10\,\Omega} + \frac{1}{10\,\Omega}$$
$$= 0.10\,\Omega^{-1} + 0.10\,\Omega^{-1} + 0.10\,\Omega^{-1} = 0.30\,\Omega^{-1}$$
$$R = \frac{1}{0.30\,\Omega^{-1}} = 3.3\,\Omega$$

To find the resistance of combination f we need to combine the two parallel resistors, which, from combination d, gives $5.0\,\Omega$, and then add this to the third resistor to give:

$$R = 5.0\,\Omega + 10.0\,\Omega = 15.0\,\Omega$$

This is shown in Figure 11.20.

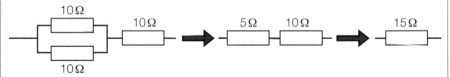

Figure 11.20

To find the resistance of combination g we need to add the two series resistors first, which gives $20\,\Omega$. There is then, effectively, $20\,\Omega$ in parallel with the other $10\,\Omega$ resistor (Figure 11.21), so that:

$$\frac{1}{R} = \frac{1}{20\,\Omega} + \frac{1}{10\,\Omega}$$
$$= 0.05\,\Omega^{-1} + 0.10\,\Omega^{-1} = 0.15\,\Omega^{-1}$$
$$R = \frac{1}{0.15\,\Omega^{-1}} = 6.7\,\Omega$$

This is shown in Figure 11.21.

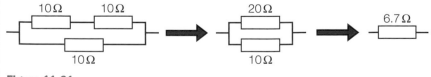

Figure 11.21

Tip

Remember:

- the resistance of a combination of parallel resistors is always less than that of any of the individual resistors
- for two **equal** resistors in parallel, the combined resistance is **half** that of each single resistor
- always work out parallel combinations **separately** and add series values afterwards – **do not** try to do it all in one go!

Activity 11.1

Using an ohmmeter to investigate series and parallel combinations of resistors

A digital multimeter (or ohmmeter) is initially set on its 200 kΩ range. It can then be used to measure the resistance of all possible combinations of three 10 kΩ resistors. The measured values can be compared with the calculated values as shown in Table 11.2. It is advisable to change the range of the ohmmeter to 20 kΩ for the smaller resistance values (unless it is an 'auto-ranging' instrument).

Table 11.2

Calculated resistance/kΩ	10.0	20.0	30.0	5.0	etc.
Measured resistance/kΩ					

Questions

1 Use the example on page 155 to help you work out the calculated values of all the possible combinations, some of which are given already in Table 11.2.
2 Suggest why it is advisable to change the range of the ohmmeter to 20 kΩ for the smaller resistance values.

Activity 11.2 illustrates, in broad terms, the principle of an ohmmeter. It uses a graphical calibration to find an 'unknown' resistance (in this case, the resistance of a series/parallel combination) rather than 'electronics' to convert the measured current into a resistance reading. Such an experiment is typical of the type of activity used for practical assessment. It can be set up very easily using a proprietary circuit board.

Activity 11.2

The principle of an ohmmeter

The circuit shown in Figure 11.22a is set up. The meter should be set on the 2 mA range.

The circuit current, I, recorded by the ammeter is the current in the combination of the three resistors in series – that is, a resistance $R = 1.0\,k\Omega + 2.2\,k\Omega + 4.7\,k\Omega = 7.9\,k\Omega$.

Now a spare lead is connected across the 1.0 kΩ resistor as shown in Figure 11.22b. We say that the 1.0 kΩ resistor has been 'short-circuited'. In effect, a 'bypass' has been built for the current so that there is no current (or at least a negligible amount) in the 1.0 kΩ resistor. The circuit resistance is now just

$R = 2.2\,k\Omega + 4.7\,k\Omega = 6.9\,k\Omega.$

This is then repeated to get values of resistance for all possible series combinations of the resistors, as shown in Table 11.3. Care must be taken not to short circuit all three resistors at once.

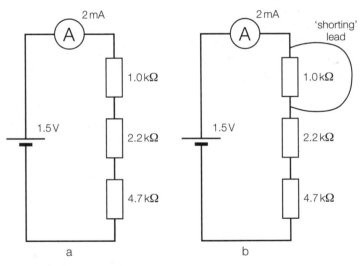

Figure 11.22

Table 11.3

R/kΩ	7.9	6.9	5.7	4.7	3.3	2.2	1.0
I/mA							

If a graph of R against I is plotted, a curve like that in Figure 11.23 should be obtained.

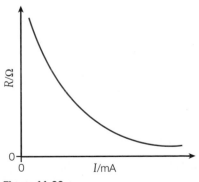

Figure 11.23

The circuit shown in Figure 11.24 is now set up.

The current I is recorded and the graph is used to read off the value R of the resistance corresponding to this current. This value R is the resistance of the series and parallel combination.

Questions
A typical set of observations is shown in Table 11.4.

Table 11.4

R/kΩ	7.9	6.9	5.7	4.7	3.3	2.2	1.0	combination
I/mA	0.19	0.22	0.27	0.33	0.47	0.70	1.54	0.51

1 Plot a graph of R against I.
2 Use your graph to find the experimental value for the combination of resistors.
3 Calculate the theoretical value for the resistance of the combination of resistors shown in Figure 11.24.
4 Determine the percentage difference between the experimental and theoretical values.
5 Suggest possible reasons for this difference.

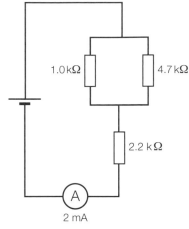

Figure 11.24

Tip

Possible sources of error in such experiments may arise as a result of:

- meter errors
- contact resistance
- resistor values being only nominal (there is commonly a manufacturer's tolerance of perhaps 2%, sometimes as much as 5%)
- the cell running down (especially if the current taken is more than a few milliamps).

11.8 Current and power calculations in series and parallel circuits

Questions involving the calculation of current and power in different series and parallel combinations have to be treated individually. They have to be worked out from first principles using the techniques we have developed in previous examples. There are no short cuts – each problem must be carefully analysed, step-by-step, showing all of your working.

Example

In the circuit shown in Figure 11.25 you may assume that the 6.0V battery has negligible internal resistance.

1 For this circuit, calculate:

   a) the current in each resistor

   b) the power dissipated in each resistor

   c) the power developed by the battery.

   Comment on your answers to parts (b) and (c).

2 Repeat your calculations, but this time assume that the battery has an internal resistance of $r = 0.8\,\Omega$.

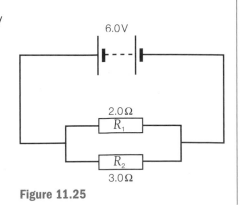

Figure 11.25

Answers

1 a) The 6.0 V of the battery will be the potential difference across each resistor (as its internal resistance is negligible), so:

$$I_1 = \frac{V}{R_1} = \frac{6.0\,\text{V}}{2.0\,\Omega} = 3.0\,\text{A}$$

$$I_2 = \frac{V}{R_2} = \frac{6.0\,\text{V}}{3.0\,\Omega} = 2.0\,\text{A}$$

b) We can now calculate the power dissipated:

$$P_1 = I_1^2 R_1 = (3.0\,\text{A})^2 \times 2.0\,\Omega = 18\,\text{W}$$

$$P_2 = I_2^2 R_2 = (2.0\,\text{A})^2 \times 3.0\,\Omega = 12\,\text{W}$$

c) The power developed by the battery is given by $P = IV$, where the current in the battery is

$$I = I_1 + I_2 = 2.0\,\text{A} + 3.0\,\text{A} = 5.0\,\text{A}$$

$$\text{so, } P = IV = 5.0\,\text{A} \times 6.0\,\text{V} = 30\,\text{W}$$

In other words, the power developed by the battery equals the power dissipated in the two resistors. This should come as no surprise – we would expect this to be the case from the law of conservation of energy.

2 Redraw the circuit to include the internal resistance of the battery as shown in Figure 11.26.

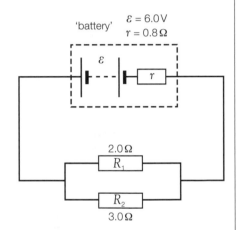

We have to start by calculating the combined resistance of the parallel arrangement:

$$\frac{1}{R} = \frac{1}{2.0\,\Omega} + \frac{1}{3.0\,\Omega} = 0.5\,\Omega^{-1} + 0.33\,\Omega^{-1} = 0.83\,\Omega^{-1}$$

$$R = \frac{1}{0.83\,\Omega^{-1}} = 1.2\,\Omega$$

The total circuit resistance R_T is now calculated as:

$$R_T = 1.2\,\Omega + 0.8\,\Omega = 2.0\,\Omega$$

We can now calculate the circuit current, I, from:

$$I = \frac{\varepsilon}{R} = \frac{6.0\,\text{V}}{2.0\,\Omega} = 3.0\,\text{A}$$

Figure 11.26

The potential difference across the parallel resistors is given by:

$$V = \varepsilon - Ir = 6.0\,\text{V} - (3.0\,\text{A} \times 0.8\,\Omega) = 6.0\,\text{V} - 2.4\,\text{V} = 3.6\,\text{V}$$

Hence:

$$I_1 = \frac{V}{R_1} = \frac{3.6\,\text{V}}{2.0\,\Omega} = 1.8\,\text{A}$$

$$I_2 = \frac{V}{R_2} = \frac{3.6\,\text{V}}{3.0\,\Omega} = 1.2\,\text{A}$$

$$P_1 = I_1^2 R_1 = (1.8\,\text{A})^2 \times 2.0\,\Omega = 6.5\,\text{W}$$

$$P_2 = I_2^2 R_2 = (1.2\,\text{A})^2 \times 3.0\,\Omega = 4.3\,\text{W}$$

The power developed by the battery is $P = \varepsilon I = 6.0\,\text{V} \times 3.0\,\text{A} = 18.0\,\text{W}$, while the power dissipated in the two resistors is only 6.5 W + 4.3 W = 10.8 W. What has happened to the other 7.2 W?

The other 7.2 W is the power 'wasted' in the battery – the work done per second by the charge overcoming the internal resistance as it passes through the battery:

$$\text{'wasted' power} = I^2 r = (3.0\,\text{A})^2 \times 0.8\,\Omega = 7.2\,\text{W}$$

4 Draw each arrangement that can be obtained by using three 15 Ω resistors either singly or combined in series and parallel arrangements (seven arrangements are possible!). Calculate all possible resistance values.

5 The circuit shown in Figure 11.27 is set up. The lamps are each rated 3.0 V, 150 mW and the cell has negligible internal resistance.

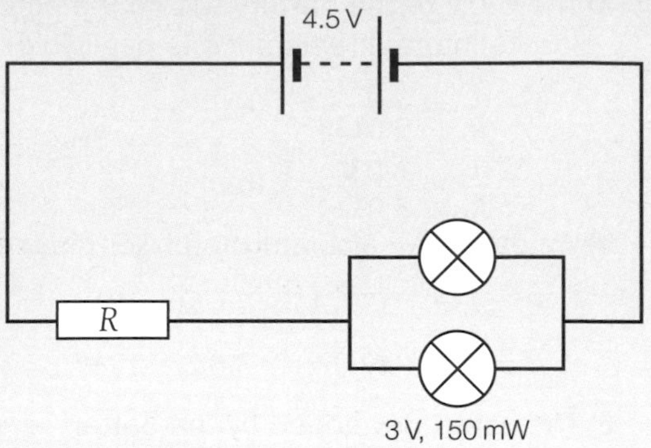

   a) Calculate, when the lamps are operating at normal brightness:

      i)   the current in each lamp

      ii) the resistance of each lamp.

   b) Hence calculate the circuit current.

   c) What must the p.d. across the resistor be for this to be achieved?

   d) Hence calculate the value of the resistor.

   e) If one of the lamps were to blow, determine the initial current in the other lamp. Discuss what may happen subsequently.

6 The maximum power from a solar cell is generated when the load (resistance) connected across the cell has a resistance equal to the internal resistance of the cell. When a particular cell is receiving sunlight of intensity 500 W m^{-2} its maximum power is 60 W at a terminal p.d. of 14 V.

   a) Show that a load resistor of 3.3 Ω with a 2% tolerance would be suitable to achieve this power.

   b) If the cell is 15% efficient, calculate the area of the solar panel that would be needed.

4.5 V

R

3 V, 150 mW

Figure 11.27

11.9 Principle of the potential divider

We can see how a potential divider works by considering the circuit shown in Figure 11.28.

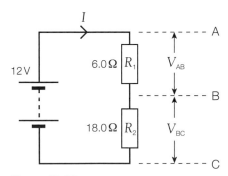

I

12 V

6.0 Ω R₁   V$_{AB}$

A

B

18.0 Ω R₂   V$_{BC}$

C

Figure 11.28

Current, $I = \dfrac{V}{R} = \dfrac{12\,V}{24\,\Omega} = 0.50\,A$

$V_{AB} = IR_1 = 0.50\,A \times 6.0\,\Omega = 3.0\,V$

$V_{BC} = IR_2 = 0.50\,A \times 18.0\,\Omega = 9.0\,V$

The network of resistors R_1 and R_2 has *divided* the potential difference of 12.0 V into 3.0 V and 9.0 V across R_1 and R_2, respectively. Such an arrangement is called a **potential divider**.

If you are mathematically minded, and good at proportions, you will see that the potential difference has been divided in the **ratio** of the resistances (1 : 3). This is a quick and easy way of calculating the potential differences in such an arrangement, but only if the ratio is a simple one!

Example

In the circuits shown in Figure 11.29, you may assume that the battery has negligible internal resistance.

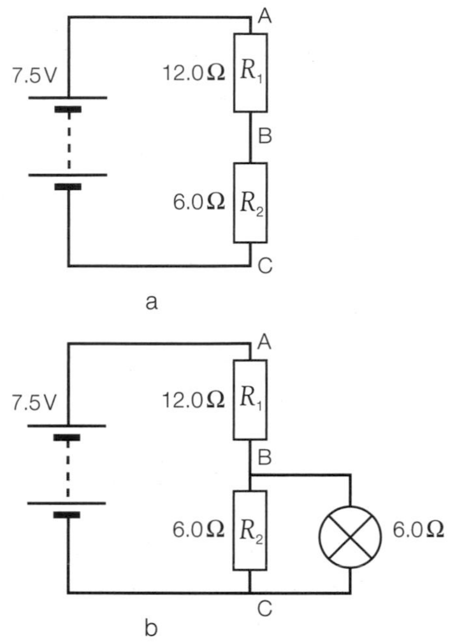

1 Calculate the potential difference across the 6.0 Ω resistor in the circuit shown in Figure 11.29a.

2 A lamp of resistance 6.0 Ω is now connected across the 6.0 Ω resistor as shown in Figure 11.29b. What is:

a) the p.d. across the lamp

b) the current in the lamp?

Answers

1 Circuit current:
$$I = \frac{V}{R} = \frac{7.5\,V}{18.0\,\Omega} = 0.4166 \ldots A$$
Across the 6.0 Ω resistor:
$$V = IR = 0.4166\ldots A \times 6.0\,\Omega = 2.5\,V$$

2 a) It would be tempting (but wrong!) to say that this p.d. of 2.5 V across R_2 would also be the p.d. across the lamp in Figure 11.29b. However, the inclusion of the lamp changes the resistance between B and C:
$$\frac{1}{R_{BC}} = \frac{1}{6.0\,\Omega} + \frac{1}{6.0\,\Omega} = \frac{2}{6.0\,\Omega}$$
$$R_{BC} = 3.0\,\Omega$$

The equivalent circuit is now as shown in Figure 11.30.

The circuit current becomes:
$$I = \frac{V}{R} = \frac{7.5\,V}{15.0\,\Omega} = 0.5\,A$$
$$V_{BC} = IR_{BC} = 0.5\,A \times 3.0\,\Omega = 1.5\,V$$

b) The current in the lamp is:
$$I = \frac{V}{R} = \frac{1.5\,V}{6.0\,\Omega} = 0.25\,A$$

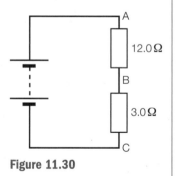

Figure 11.30

Figure 11.29

The above example shows how adding a 'load' to the output can affect the output voltage. In order to minimise this effect, a general rule of thumb is that the resistance of the load should be at least **10 times** greater than the output resistance (R_2 in the example).

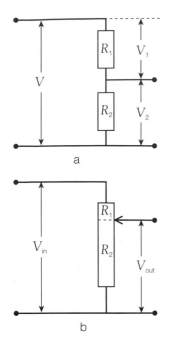

a

b

Figure 11.31

Tip

Although remembering these formulae allows you to work out simple examples very easily and quickly, it is recommended that you also learn to tackle problems from first principles – as shown in the example.

We can now formulate a general expression for a potential divider circuit by considering Figure 11.31a.

If no current is taken by the output:

$$I = \frac{V_2}{R_2} \text{ and also } I = \frac{V}{R_1 + R_2}$$

$$\Rightarrow \frac{V_2}{R_2} = \frac{V}{R_1 + R_2}$$

so:

$$V_2 = V \times \frac{R_2}{R_1 + R_2} \text{ and similarly } V_1 = V \times \frac{R_1}{R_1 + R_2}$$

In Figure 11.31b, the two resistors have been replaced by a three-terminal variable resistor. We can think of the sliding contact as dividing this resistor into two imaginary resistances R_1 and R_2. As the sliding contact is moved up, 'R_2' gets larger and 'R_1' gets smaller (but of course $R_1 + R_2$ remains the same).

As $V_{out} = V_{in} \times \frac{R_2}{R_1 + R_2}$

V_{out} will get progressively larger as the slider is moved up. With the slider right at the top, $V_{out} = V_{in}$; when the slider is right at the bottom, $V_{out} = 0$. Thus a continuously variable range of output voltages from 0 to V_{in} can be obtained. This can be checked by setting up the circuit shown in Figure 11.31b. A 1.5 V cell can be used for V_{in} and a digital voltmeter can be used to measure V_{out}.

11.10 Practical use of a potential divider

Activity 11.3

Investigation of how the potential along a uniform current-carrying wire varies with the distance along it

A 1-metre length of uniform resistance wire, e.g. 30 swg/0.3150 mm diameter nichrome, is taped to a metre rule. A 1.5 V cell is then connected across the wire (not shown). A digital voltmeter, set on the 2 V d.c. range, has its 'common' terminal connected to the zero end of the wire. A long lead terminating in a crocodile clip is connected to the 'volts' terminal as shown in Figure 11.32a.

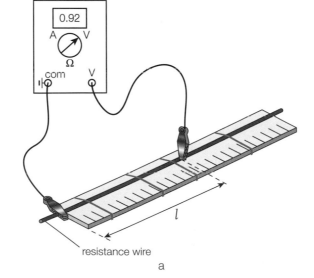

resistance wire

a

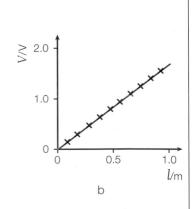

b

Figure 11.32

The zero end of the wire is at a potential of 0 V. The potential at different points along the wire can be read off from the voltmeter at, say, 10 cm intervals by touching the wire firmly with the crocodile clip.

A graph of the potential difference, V, against the length, l, along the wire can then be plotted. This should be a straight line through the origin, as shown in Figure 11.32b. This shows that the potential difference along the wire is proportional to the distance along it, which is the principle of a rheostat when it is being used as apotential divider.

We saw how a **rheostat** (Figure 11.33a) could be used to provide a continuously variable potential difference in Section 9.1. A three-terminal resistor used as a potential divider is called a **potentiometer**. A potentiometer such as that shown in Figure 11.33b is, in effect, a rheostat 'bent' around to form an almost complete circle. It has a fixed contact at each end and a rotating arm that forms the sliding contact.

Although a 'rheostat' is often used in school laboratories as a potential divider, it should really be called a 'potentiometer' when used in this way. Commercial potentiometers can be circular, as shown in Figure 11.33b, or straight – for example radio volume controls can either be rotating knobs or sliders.

We also saw that a rheostat could be used either to vary current (Figure 11.34a) or to vary potential difference (Figure 11.34b).

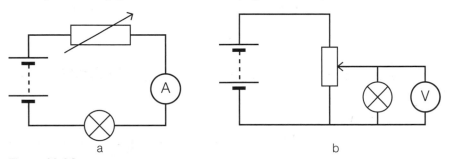

Figure 11.34

a

b

Figure 11.33 a) A rheostat and b) a potentiometer.

We showed by experiment that the circuit in Figure 11.34b was the more useful as we could get a much greater variation in the potential difference – from zero up to the p.d. of the battery. To refresh your memory, look at the following example.

Examples

A student proposes to use the circuit shown in Figure 11.35 to investigate the $I–V$ characteristics of a 220 Ω resistor.

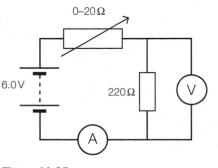

Figure 11.35

1 To decide whether the circuit is appropriate, determine:

a) the maximum voltage that can be obtained across the 220 Ω resistor

b) the minimum voltage that can be obtained across the 220 Ω resistor.

2 The student's teacher suggests that the same apparatus could be rearranged to form a potential divider circuit, which would enable a voltage range of between 0 V and 6.0 V to be achieved. Draw a circuit diagram to show how this could be done.

Answers

1 a) When the rheostat is set to its minimum resistance (0 Ω), the voltage across the 220 Ω resistor will have a maximum value equal to the p.d. of the battery – that is, 6.0 V.

b) When the rheostat is set to its maximum resistance of 20 Ω, the circuit resistance will have a maximum value of:

$$R = 20\,\Omega + 220\,\Omega = 240\,\Omega$$

which will give a minimum circuit current of:

$$I = \frac{V}{R} = \frac{6.0\,\text{V}}{240\,\Omega} = 0.025\,\text{A}$$

and a minimum voltage across the 220 Ω resistor of:

$$V_{min} = IR = 0.025\,\text{A} \times 220\,\Omega = 5.5\,\text{V}$$

The student could therefore only investigate the *I*–*V* characteristics over a voltage range of 5.5–6.0 V – clearly not very satisfactory!

2

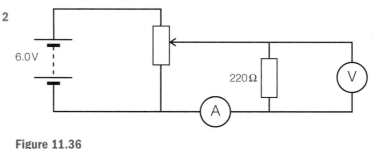

Figure 11.36

11.11 Using a thermistor to control voltage

A thermistor can be used in a potential divider circuit to control the output voltage as shown in Figure 11.37.

From the formula derived in Section 11.9:

$$V_{out} = V_{in} \times \frac{R_2}{R_1 + R_2}$$

if the temperature rises, the resistance R_1 of the thermistor will decrease and so V_{out} will increase.

Looking at the situation from first principles, if R_1 decreases, the circuit resistance will get less and so the circuit current, *I*, will get larger. As $V_{out} = IR_2$, V_{out} will also increase.

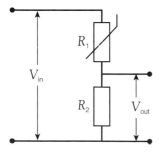

Figure 11.37

> **Tip**
>
> Remember:
>
> - a rheostat controls current
> - a potentiometer controls potential difference.

Example

The thermistor in Figure 11.38 has a resistance of:

1 440 Ω at 20 °C

2 110 Ω at 60 °C.

Calculate the value of the output voltage at each of these resistances.

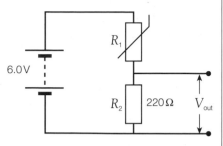

Figure 11.38

Answers

1 At 20 °C, where $R_1 = 440\,\Omega$:

$$V_{out} = V_{in} \times \frac{R_2}{R_1 + R_2}$$

$$= 6.0\,V \times \frac{220\,\Omega}{440\,\Omega + 220\,\Omega} = 6.0\,V \times \frac{1}{3} = 2.0\,V$$

2 At 60 °C, where $R_1 = 110\,\Omega$:

$$V_{out} = V_{in} \times \frac{R_2}{R_1 + R_2}$$

$$= 6.0\,V \times \frac{220\,\Omega}{110\,\Omega + 220\,\Omega} = 6.0\,V \times \frac{2}{3} = 4.0\,V$$

Tip

Core practical 12 actually appears, out of context, in the A-level specification. It is therefore included here as it follows naturally from work on the variation of the resistance of a thermistor with temperature and the application of potential divider circuits.

Core practical 12

Using a thermistor in a potential divider circuit to activate a thermostatic switch

An ohmmeter is used to measure the resistance, R_1, of the thermistor at room temperature.

The arrangement shown in Figure 11.39a is then set up; the circuit diagram for this arrangement is shown in Figure 11.39b.

The resistor is selected such that its resistance, $R_2 \approx \frac{1}{2}R_1$.

The output voltage, V_{out}, for temperatures θ ranging from room temperature to about 60 °C are measured and a graph of V_{out}, against θ is plotted. The graph can then be used to determine the value of, V_{out}, when the temperature is at a particular value, say 40 °C.

A simple electronic circuit can be designed to operate a switch when the output from the circuit in Figure 11.39b reaches a predetermined value. For example, if the switch was set to operate at an output voltage corresponding to 40 °C, the circuit could be used to switch on a warning lamp when the temperature reached 40 °C.

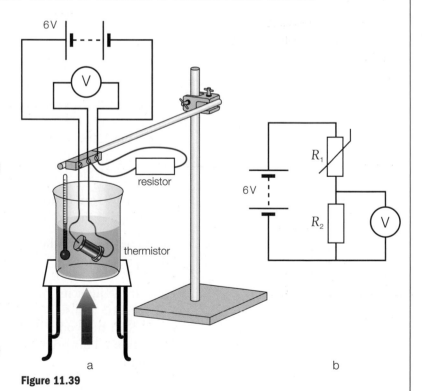

Figure 11.39

Questions

The data in Table 11.5 were obtained from an experiment similar to that described above.

Table 11.5

θ/°C	0.0	10.0	20.0	30.0	40.0	50.0	60.0
V_{out}/V	1.30	1.52	1.78	2.07	2.37	2.70	3.16

1 Plot a graph of output voltage on the *y*-axis against temperature on the *x*-axis.
2 Use your graph to determine the output voltage at a temperature of 37 °C.
3 Describe any experimental technique you would use to improve the accuracy of your data.
4 Describe how you would make measurements at temperatures below room temperature ($\approx$20 °C).
5 Give one safety precaution that you would follow.

11.12 Using a light dependent resistor to control voltage

A light dependent resistor (LDR) is a component that is sensitive to light – when light is incident on an LDR, the value of its resistance falls as the level of light increases. An LDR is made from a semiconductor material with a high resistance, such as cadmium sulfide. It has a high resistance because there are very few electrons that are free and able to conduct – the vast majority of the electrons are locked into the crystal lattice and unable to move.

As light falls on the semiconductor, the light photons (see Chapter 17) are absorbed by the semiconductor lattice and their energy is transferred to the electrons. This gives some of the electrons sufficient energy to break free from the crystal lattice so that they can then conduct electricity. In terms of our equation $I = nAvq$, the number of conducting electrons per cubic metre, n, increases as light shines on the LDR.

This results in a lowering of the resistance of the semiconductor and hence the overall LDR resistance. The process is progressive; as more light shines on the LDR semiconductor, so more electrons are released to conduct electricity and the resistance falls further. The resistance of an LDR can be several MΩ in darkness and then fall to a few hundred ohms in bright light.

An LDR can be used in a potential divider circuit to control the output voltage as shown in Figure 11.40b.

As before, from the formula derived in Section 11.9:

$$V_{out} = V_{in} \times \frac{R_2}{R_1 + R_2}$$

Thus, if the light incident on the LDR falls, the resistance R_2 of the LDR will increase and so V_{out} will increase. This can then be used to activate a switch to turn on a light when it gets dark (and off again when it gets light again). This principle is used, for example, in controlling automatic streetlights.

9.0 V

$R = 10\,k\Omega$

V_{out}

to switching circuit

NORP 12

0 V

b

Figure 11.40 a) Light dependent resistor, b) light control circuit

7 In the circuit shown in Figure 11.41 the battery can be assumed to have no internal resistance.

   a) Calculate the output voltage, V_{out}.

   b) Discuss whether resistors having a power rating of 0.1 W would be suitable.

   c) If the battery actually had an internal resistance of 0.5 Ω, explain what effect, if any, this would have.

   d) If the resistors have a tolerance of 2%, how would you state the value of the output voltage to reflect this?

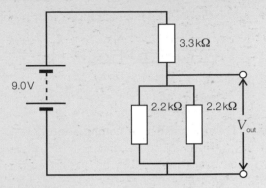

Figure 11.41

8 A student sets up the circuit shown in Figure 11.42 in order to switch on a heater when the temperature falls below 20 °C. The electronic switch needs a potential difference of 6.0 V to activate it. A preliminary test shows that the resistance of the thermistor is about 450 Ω at a temperature of 20 °C.

   a) The student looks in a catalogue and finds a potentiometer rated at 500 Ω, 250 V and 0.5 W. Explain why this would be a suitable choice.

   b) Explain how the student would set the potentiometer to the required value.

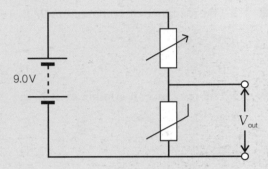

Figure 11.42

9 The resistance of the LDR in the circuit of Figure 11.40b varies with the light falling on it as shown in Figure 11.43.

   a) Estimate the output voltage at

     i) twilight (1.0 lux) and

     ii) in the laboratory (500 lux).

   b) A switch circuit, connected across the output, needs 6.0 V to operate it. Determine the value of the resistor R that would be needed in the circuit to enable a lamp to be switched on at twilight.

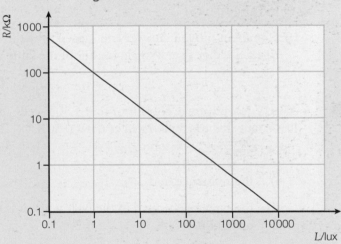

Figure 11.43

Exam practice questions

Use Figure 11.44, which shows four possible ways of connecting three 15 Ω resistors, to answer Questions 1–3.

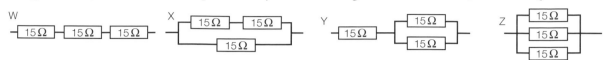

Figure 11.44

1 The resistance of combination Z is:

    A $\dfrac{W}{9}$ **C** 3W

    B $\dfrac{W}{3}$ **D** 9W   **[Total 1 Mark]**

2 The resistance of combination W is:

    A $\dfrac{X}{2}$ **C** $\dfrac{Y}{9}$

    B 2X   **D** 2Y   **[Total 1 Mark]**

3 The resistance of combination X is:

    A $\dfrac{1}{2} \times Y$ **C** $\dfrac{1}{2} \times Z$

    B $2 \times Y$ **D** $2 \times Z$ **[Total 1 Mark]**

4 **a)** Calculate the resistance of the combination of resistors shown in Figure 11.45. **[3]**

    b) Comment on your answer and suggest why this network might be used rather than a single resistor of the same value. **[2]**

                                  [Total 5 Marks]

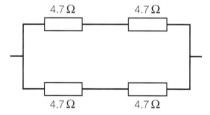

Figure 11.45

5 A 6.0 V torch battery consists of four 1.50 V cells connected in series. It is connected to a bulb rated at 6.0 V, 0.50 A.

    a) Calculate

        i) the resistance, and

        ii) the power generated in the bulb when it is in normal operation. **[2]**

    b) After a time, each cell develops an internal resistance of 0.15 Ω. Show that the p.d. across the bulb falls to about 5.7 V, assuming that its resistance remains the same. **[5]**

    c) How much power is wasted in the battery? **[2]**

    d) To what extent is the assumption in part **b** reasonable? **[2]**

                                    [Total 11 Marks]

6 A student has a 9 V battery in his pocket with some coins. The battery is accidentally 'short-circuited' by a coin. Calculate the current that will flow in the battery if it has an internal resistance of 0.50 Ω and hence explain why the battery gets hot. **[Total 4 Marks]**

7 A laboratory high-voltage supply has a 50 MΩ resistor inside it to effectively give it a large internal resistance.

    a) Explain why it has this resistor. **[2]**

    b) What is the maximum current that the supply can give when the voltage is set to 5 kV? **[2]**

    c) A girl has a resistance of 10 kΩ, as measured between her hand and the ground. Explain why, if she were to accidentally touch the live terminal of the high-voltage supply, the current in her would be virtually the same as in part **b**. **[2]**

    d) Calculate the power that would be dissipated in the girl. Comment on this value compared with the 14 mW from the car battery in the example on page 149. **[3]**

[Total 9 Marks]

8 Copy the table of data from Core practical 3 on page 149 and add columns for the resistance $R = \dfrac{V}{I}$ of the variable resistor and the corresponding power, $P = IV$, dissipated in it. Then plot a graph of P against R. You should find that the power has a maximum value when the resistance of the variable resistor (the 'load') is equal to the internal resistance of the cell (the 'power supply'). This is always the case and is important in electronics for 'matching' the load (such as loudspeakers) to the power supply (such as an amplifier). **[Total 7 Marks]**

9 The data in the table were obtained for the output of a solar cell using the circuit shown in Figure 11.10 on page 151 when the cell was illuminated by a bench lamp.

I/mA	0.10	0.15	0.20	0.30	0.40	0.50	0.60	0.65
V/V	6.07	5.90	5.72	5.35	4.97	4.20	3.00	2.01
R/kΩ								
P/mW								

    a) Copy and complete the table of data by adding values for the resistance across the cell, $R = \dfrac{V}{I}$, and the power generated, $P = IV$. You are recommended to use a spreadsheet for this. **[3]**

    b) i) Draw a graph of V against I and comment on the shape of your graph.

       ii) Use the graph to determine values for the e.m.f. of the cell and the internal resistance of the cell for low current values. **[9]**

    c) i) Plot a graph of P against R.

       ii) Use the graph to determine the maximum power generated. **[4]**

[Total 16 Marks]

10 A student proposes to set up the circuit shown in Figure 11.46. The battery may be assumed to have negligible internal resistance.

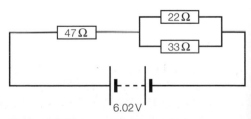

Figure 11.46

a) What would be:

    i) the current in each resistor

    ii) the power generated in each resistor?   **[8]**

b) Discuss whether resistors with a power rating of 500 mW would be suitable for the student to use in this circuit.   **[2]**

[Total 10 Marks]

11 A student sets up the circuit shown in Figure 11.47 with a 10 V analogue voltmeter.

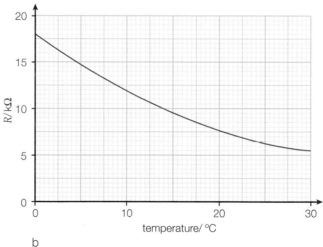

a) Show that the potential difference across the 33 kΩ resistor before the voltmeter is connected is 4.5 V.   **[3]**

b) When the switch is closed, the voltmeter reads 4.0 V.

    i) Account for the difference between this value and your answer to part **a**.

    ii) Calculate the resistance of the voltmeter.   **[7]**

Figure 11.47

c) The voltmeter scale is marked 10 V/100 μA and the resistors have a 5% tolerance. Discuss whether your answer to part **b (ii)** is compatible with these data.   **[4]**

[Total 14 Marks]

12 Figure 11.48a shows a circuit containing a thermistor. Figure 11.48b shows how the resistance of the thermistor varies with temperature.

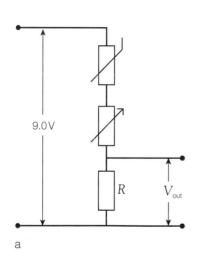

Figure 11.48

a) Explain why V_{out} gets less as the temperature of the thermistor falls. **[3]**

b) The circuit is to be used to switch on a warning lamp when the temperature drops to 0 °C. The switch is activated when V_{out} falls to 5.0 V.

    i) Calculate the value of the resistor R that you would use given that preferred values (that is, standard values supplied by manufacturers) of 10 kΩ, 22 kΩ, 33 kΩ and 47 kΩ are available.

    ii) Explain the purpose of the potentiometer.   **[7]**

[Total 10 Marks]

13 In Figure 11.49, graph X shows how the potential difference across the terminals of a cell depends on the current in the cell. Graph Y is the voltage–current characteristic for a filament lamp.

a) As the current increases, what can be deduced from the graphs about:

   i) the internal resistance of the cell

   ii) the resistance of the filament lamp? **[1]**

b) Use graph X to determine:

   i) the e.m.f. of the cell

   ii) the internal resistance of the cell. **[2]**

c) When the lamp is connected to the cell, what is:

   i) the current in the lamp

   ii) the resistance of the lamp

   iii) the power developed in the lamp? **[3]**

d) Draw a circuit diagram of the circuit you would set up to obtain the data for graph Y using two cells for the power supply. **[3]**

e) Explain

   i) why you would need two cells

   ii) how you would obtain the data. **[3]**

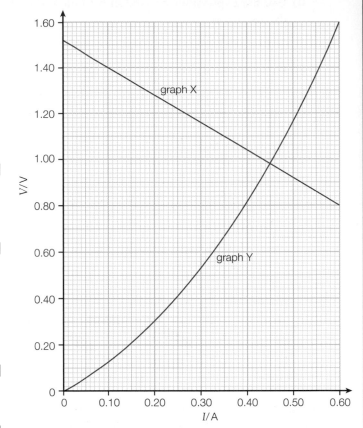

Figure 11.49

[Total 12 Marks]

Stretch and challenge

14 Figure 11.50 shows a resistor value R connected across a cell of e.m.f. ε and internal resistance r. The current in the circuit is I and the potential difference across the resistor is V.

a) The equation governing this circuit is often written $V = \varepsilon - Ir$. Explain what is meant by the terms e.m.f. and potential difference, using this equation to illustrate your answer. **[4]**

b) Derive expressions for

   i) I, and

   ii) V in terms of r and R. **[4]**

c) When $R = 2.2\,\Omega$, the current is $616\,\text{mA}$, and when $R = 4.7\,\Omega$, the current is $308\,\text{mA}$. Use these data to calculate the e.m.f. and internal resistance of the cell. **[4]**

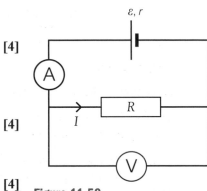

Figure 11.50

d) Use the expressions derived in **b** to show that the power developed in the resistor is

$$P = \frac{\varepsilon^2 R}{(R + r)^2}$$ [2]

e) Hence show that the power is a maximum when $R = r$.   [2]

[Total 16 Marks]

15 Conservation laws are fundamental to physics.

a) Explain what is meant by conservation of charge. Illustrate your answer with reference to the current at a three-way junction.   [3]

b) Three-dimensional integrated circuits are currently being developed to improve existing 2D designs by providing smaller chip areas and higher performance with lower power consumption. Figure 11.51 shows twelve identical resistors, each of resistance R, connected to form a cube.

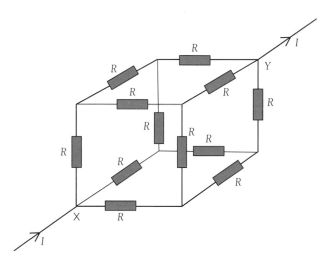

Figure 11.51

Determine the resistance diagonally across the cube, from X to Y.   [7]

[Total 10 Marks]

16 Figure 11.43 on page 167 shows how the resistance R (in kΩ) of an LDR depends on the illumination L (in lux). The resistance is thought to be related to the illumination by an equation of the form

$$R = aL^n$$

where a and n are constants. This equation can be expressed as

$$\log R = n \log L + \log a$$

a) Show that $R = \dfrac{100}{\sqrt[4]{L^3}}$.

b) Hence determine the resistance of the LDR at a distance of 1.00 m from a 60 W lamp, where the illumination is 50 lux.

c) Estimate the resistance of the LDR when it is 0.50 m from the lamp.

d) State any assumptions you have made in arriving at your answer to part **c**.

Fluids

12

Prior knowledge

In this chapter you will need to be able to:
→ be aware of the physical and molecular differences between liquids and gases
→ be familiar with the terms density and pressure.

The key facts that will be useful are:
→ density is the mass per unit volume
→ pressure is the force per unit area
→ fluid pressure acts in all directions
→ fluid molecules do not occupy fixed positions and can move relative to each other.

Test yourself on prior knowledge

1 Calculate the density of a liquid if $230\,cm^3$ has a mass of $250\,g$.

2 The density of air is $1.3\,kg\,m^{-3}$. Calculate the mass of air in a room of length $5.20\,m$, width $4.00\,m$ and height $2.30\,m$.

3 Write down the units of pressure in base units.

4 Calculate the force exerted by the air on a wall of area $11.0\,m^2$ if the atmospheric pressure is $1.0 \times 10^5\,Pa$.

5 Describe the differences in molecular structure of liquids and gases.

12.1 Properties of fluids

A **fluid** is a material that flows. Unlike a solid, in which the atoms occupy fixed positions, the particles of a fluid can move relative to one another. Generally we can consider fluids as liquids or gases, but plasma and some amorphous solids can display fluid behaviour.

This chapter will concentrate on the properties of static fluids such as density, pressure and flotation as well as the motion of objects within fluids and how viscosity affects the flow of liquids and gases.

The study of fluids is important in the food industry where sugar concentrations affect the rate of flow of confectionary, the transportation of oil and gas and the flow of blood through our veins and arteries.

The study of how gases behave when heat energy is transferred in or out is a major topic in physics. Thermodynamics will be studied in detail in the Year 2 Student's book.

12.2 Density, pressure and flotation

Density of fluids

Liquids and gases expand much more than solids when they are heated, so a fixed mass of fluid occupies a bigger volume than the solid form and so its **density** is reduced. Liquids are generally considered to be incompressible, but gases are readily squeezed (try putting your finger over the outlet of a bicycle pump and pushing in the handle). Because of this, the pressure needs to be stated in addition to the temperature when the density of a gas is quoted. Density is given the symbol ρ ('rho').

Table 12.1 gives some examples of the densities of fluids. The values are at 293 K and gas pressure of 1.01×10^5 Pa.

Key term

Density is given by the following expression:

$$\text{density} = \frac{\text{mass}}{\text{volume}}$$

$$\rho = \frac{m}{V}$$

Table 12.1 Densities of some fluids.

Fluid	Density/kg m^{-3}
mercury	13600
water	1000
ethanol	790
carbon dioxide	1.78
air	1.24
helium	0.161
hydrogen	0.081

Activity 12.1

Finding the density of air

A flask and its attachments are placed onto a balance (sensitivity ±0.01 g or less) and the total mass is recorded. A vacuum pump is used to remove as much air as possible from the flask (Figure 12.1). The flask should be encased with a stiff wire mesh as a precaution against implosion and safety goggles must be worn. A protective screen between the flask and observers is also recommended. The flask and attachments are reweighed so that the mass of the gas removed from the flask can be found.

To measure the volume of gas (at its initial pressure), the end of the rubber tube is immersed in a beaker of water and the clip is released. Water is forced into the tube by the external air pressure, and the volume of the water in the flask equals the volume of the evacuated air.

The following results were obtained:

- mass of flask plus attachments plus air = 421.38 g
- mass of flask plus attachments after air removed = 420.80 g
- volume of water used to replace the air removed = 450 ml.

Question

1 Use these readings to determine the density of air.

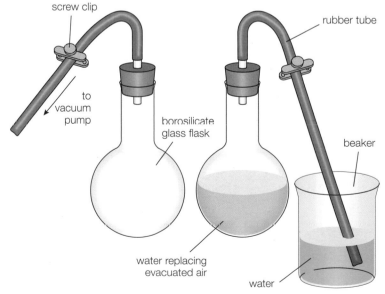
Figure 12.1 Finding the volume of air removed from a flask.

Pressure in fluids

If you dive to the bottom of a swimming pool, you will feel the water pressure pushing into your ears. This pressure is created by the weight of the water above you, and the deeper you go, the greater this pressure. Similarly, the weight of the atmosphere produces an air pressure of about 1.0×10^5 Pa at the Earth's surface.

Figure 12.2 shows a column of fluid of height h, density ρ and area of cross-section A.

$$\text{Pressure at the base} = \frac{\text{weight of column}}{\text{area}} = \frac{mg}{A} = \frac{V\rho g}{A}$$

Volume, V, of the column $= Ah$

hence, $p = h\rho g$

For large values of h, gases compress in the lower regions. This means that the Earth's atmosphere has a lower density at higher altitudes and its pressure is therefore not directly proportional to height above Earth.

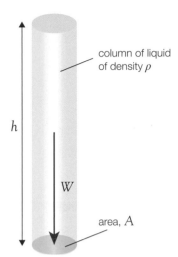

Figure 12.2 Pressure due to a column of liquid.

Example

Estimate the height of the Earth's atmosphere.

Answer

The density of air is about $1.24\,\text{kg}\,\text{m}^{-3}$ at sea level and virtually zero at upper levels of the atmosphere, so the average density can be taken as $0.62\,\text{kg}\,\text{m}^{-3}$.

If air pressure is assumed to be 1.0×10^5 Pa and g to be $10\,\text{N}\,\text{kg}^{-1}$:

$$h = \frac{p}{\rho g} = \frac{1.0 \times 10^5\,\text{Pa}}{0.62\,\text{kg}\,\text{m}^{-3} \times 10\,\text{N}\,\text{kg}^{-1}} \approx 16\,\text{km}$$

Test yourself

1 Estimate the mass of water in an Olympic-size swimming pool.

2 A 500 ml flask containing carbon dioxide has a mass of 180.39 g. A pump is used to remove 80% of the gas and the mass of the flask and gas is now 179.60 g. Calculate the density of carbon dioxide.

3 A mercury barometer shows atmospheric pressure as 772 mm of mercury. Calculate the air pressure in Pa (density of mercury = $13\,600\,\text{kg}\,\text{m}^{-3}$).

4 Calculate the length of a column of water required to give the same pressure at its base as the mercury column in Question 3.

5 Describe an observation that suggests that the pressure of a fluid acts equally in all directions.

Upthrust in fluids

If you are in a swimming pool, you will experience a buoyancy force that enables you to float or swim. This force is called an **upthrust** and it is a consequence of the water pressure being greater below an immersed object than above it.

Consider a cylinder immersed in a liquid, as shown in Figure 12.3. The upthrust is the difference between the force due to water pressure at the bottom of the cylinder, F_2, and that at the top, F_1.

For a fluid of density ρ:

$$F_1 = p_1 A = h_1 \rho g A$$

$$F_2 = p_2 A = h_2 \rho g A$$

$$U = F_2 - F_1 = (h_2 - h_1)\rho g A = (h_2 - h_1) A \rho g = V \rho g = mg$$

The upthrust is equal to the **weight** of the displaced fluid.

Figure 12.3 Upthrust on a cylinder.

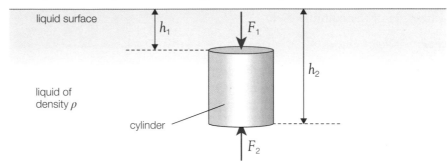

This result is often stated as Archimedes' principle.

Archimedes (who lived around 287–212BC) was seeking a method to verify that the king's crown was made from pure gold and is reported to have leaped from his bath and run through the streets shouting 'Eureka' (I've found it!).

Key term

Archimedes' principle states that when a body is totally or partially immersed in a fluid, it experiences an upthrust equal to the weight of fluid displaced.

Activity 12.2

Estimating upthrust
Estimate the upthrust on your body when it is totally immersed in water and compare it with the upthrust experienced due to the air you displace. You can consider your head as a sphere and your trunk, arms and legs as cylinders.

Flotation

An object will float in a fluid if the upthrust – that is, the weight of the fluid it displaces – is equal to its weight.

Figure 12.4 shows a fully laden ship floating in the cold salt water of the North Atlantic Ocean and the same ship in the warmer fresh water of an upriver tropical port. The ship needs to displace more of the less dense fresh water to balance its weight, and so lies deeper in the fresh water than in the salt water.

Figure 12.4 Flotation of ships.

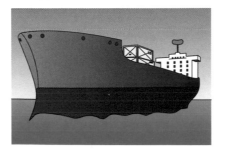

vessel in cold salt water

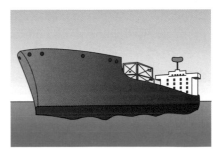

vessel in warm fresh water

Insurance companies require all cargo ships to have maximum load levels on their hulls, and a number of such lines are painted on vessels used for international trade to represent the safety levels in different waters. These lines were introduced as law in the UK in the late nineteenth century by the Member of Parliament Samuel Plimsoll and are still referred to as 'Plimsoll lines'.

Example

Calculate the density of the hot air in a balloon floating at a fixed height close to the ground. The density of the cold air is $1.4\,kg\,m^{-3}$. The total mass of the balloon's fabric, gondola, fuel, burners and occupants is $700\,kg$, and its volume is $2500\,m^3$.

Answer

Upthrust, U = weight of displaced air

$$= 2500\,m^3 \times 1.4\,kg\,m^{-3} \times 9.8\,N\,kg^{-1}$$

$$= 34\,000\,N$$

For the balloon to be in equilibrium:

U = weight of balloon, occupants and accessories + weight of hot air

Weight of hot air = $34\,000\,N - 700\,kg \times 9.8\,N\,kg^{-1} = 27\,000\,N$

Mass of hot air = $2700\,kg$

Density of hot air = $\dfrac{2700\,kg}{2500\,m^3} = 1.1\,kg\,m^{-3}$

Archimedes' ideas are still in common use in industry where the principle of flotation is applied to determine the densities of solids and fluids.

Test yourself

6 State Archimedes' principle.

7 A stone of mass 125 g is attached to a spring balance and lowered into a bowl of water until it is completely immersed. The reading on the balance is seen to be 1.00 N. Calculate:

   a) the upthrust acting on the stone

   b) the volume of water displaced

   c) the density of the stone.

8 A long cylindrical tube is weighted at the bottom so that it floats upright in water at 20 °C. Describe and explain what differences will be observed if the tube is then floated in

   a) ethanol at 4 °C

   b) water at 4 °C.

9 Explain why a hot-air balloon rises when first released.

12.3 Moving fluids – streamlines and laminar flow

Streamlines represent the velocity of a fluid at each point within it. They can be drawn as arrowed lines that show the paths taken by small regions of the fluid.

Figure 12.5a represents the flow of air relative to an aircraft wing and Figure 12.5b the flow of ink from a nozzle into water flowing through a pipe. In Figure 12.5a, the air above and below the aircraft wing exhibits **laminar flow** – that is, adjacent layers of air do not cross into each other. For laminar flow there are no abrupt changes in speed or direction; the velocity at a point is constant. Beyond the wing, the air swirls around and forms **vortices** or **eddy currents**. The streamlines are no longer continuous and the flow is said to be **turbulent**.

Figure 12.5b shows that the flow of water in the pipe is laminar at low rates of flow, but that turbulence occurs when the rate reaches a critical level (see Figure 12.5c). This rate of flow depends on the speed of flow, the radius of the tube, the density and viscosity of the fluid.

Key terms

Laminar flow occurs when adjacent layers of fluid do not cross into each other.

Turbulent flow occurs when layers of fluid cross into each other, resulting in the formation of vortices or eddy currents.

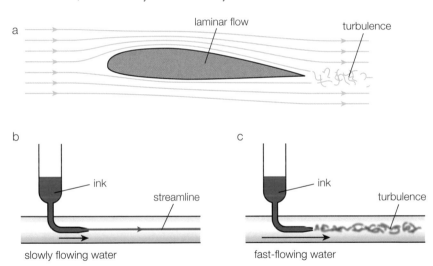

Figure 12.5 Laminar and turbulent flow.

Laminar flow is an important consideration in fluid motion. The uplift on an aeroplane's wings is dependent on laminar flow, and passengers experience a rocky ride when turbulent conditions are encountered. Similarly, the drag forces on a motor car are affected by turbulence, and wind tunnels are used to observe the nature of the air flow over prototype designs.

The efficiency of fluid transfer through tubes is greatly reduced if turbulence occurs, so the rate of flow of oil and gas must be controlled so that the critical speed is not exceeded.

In the food industry, the flow of sweet casings such as toffee and chocolate over nuts or other fillings should be laminar so that air bubbles are not trapped as a result of turbulence.

Viscosity

The **viscosity** of a fluid relates to its stickiness and thus to its resistance to flow. Syrup and engine oil are very viscous, while runny liquids such as water and petrol and all gases have low viscosities.

Viscosity can be described in terms of the resistance between adjacent layers in laminar flow. Imagine two packs of playing cards: one brand new and the other heavily used. The cards in the new pack will slide easily over each other when pushed down at an angle from above (Figure 12.6), but the cards of the old pack will stick together, dragging the lower cards with them.

Comparing viscosities

The viscosities of liquids can be compared by observing their rates of flow through a glass tube. A simple device called a Redwood viscometer can be adapted easily for the laboratory.

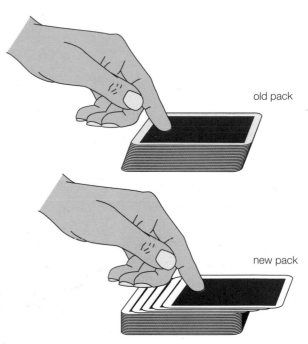

Figure 12.6

Activity 12.3

Comparing viscosities using a viscometer

Using the apparatus in Figure 12.7, fill the funnel with liquid to a level just above the upper mark. Open the clip to allow the liquid to flow through the tube into the beaker. Start timing when the level passes the upper mark and stop as it passes the lower one. Repeat for various samples, and then list your samples in order of increasing viscosity.

For runny liquids, like water or sugar solutions of a low concentration, a capillary tube with a diameter of 1 mm should be used so that very short times (and large percentage uncertainties) are avoided. For viscous liquids, like syrup and honey, tubes with bores of 5 mm or wider can be used.

The viscometer shown in Figure 12.7 can be used to determine the concentration of sugar solutions if a suitable calibration experiment is performed.

The results of such an experiment are given in Table 12.2.

Table 12.2

Sugar solution concentration/%	0	10	20	30	40
Time/s	0.8	1.1	1.5	2.5	4.6

If the conditions are kept the same throughout, the time taken for the same volume of the solutions to pass through the tube will be proportional to the viscosity of the sugar solution.

Questions

1 Draw a graph of time taken against sugar concentration.
2 Describe how the viscosity of the solution depends on the sugar concentration.
3 Use your graph to determine the concentration of sugar solution that takes a time of 2.1 s for the solution level to pass between the two levels.
4 List the physical quantities that need to be kept the same during the experiment.

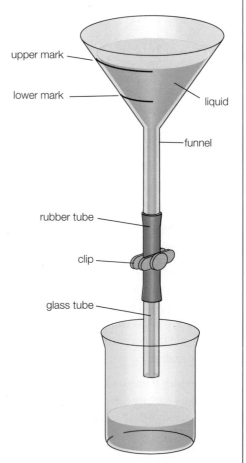

Figure 12.7 Comparing the viscosity of liquids.

The flow of blood in our arteries and veins is important for a healthy cardiovascular system. In the nineteenth century, Jean Louis Marie Poiseuille, a French physician, showed that the rate of flow of a liquid in a uniform tube depended on the pressure per unit length across the tube, the viscosity of the liquid and the fourth power of the radius of the tube.

Although our blood vessels are not uniform pipes and the viscosity of the blood is not constant, it is apparent that the build-up of fatty deposits on arterial walls from excess amounts of some forms of cholesterol will reduce the blood flow and/or increase the blood pressure. If the radius is halved, the rate of flow for a given pressure will reduce 16 times!

In many cases, patients with high blood pressure are prescribed 'blood thinners', such as aspirin, warfarin and pentoxifylline, which reduce the viscosity of the blood.

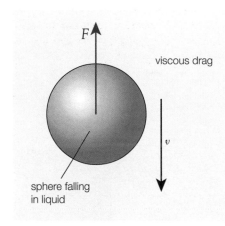

Figure 12.8 Sphere falling through a fluid.

Test yourself

10 Explain what is meant by laminar flow.

11 Explain what is meant by turbulence.

12 Draw the streamlines in the water as it moves relative to a speedboat moving at constant speed on a lake. Label the regions where the flow is laminar and where turbulence occurs.

13 Explain why the rate of flow of gas through pipes of fixed diameter will reach a maximum value even if the pressure across the pipes is further increased.

Stokes' law

When a sphere moves slowly through a fluid, the movement of the fluid relative to the sphere is laminar. The molecules of the fluid adhere to the surface of the sphere and move along with it, creating a viscous drag between the other layers of the fluid (see Figure 12.8).

The Irish physicist George Gabriel Stokes deduced an expression for this force in terms of the radius of the sphere, r, the velocity relative to the fluid, v, and the **coefficient of viscosity**, η:

$$F = 6\pi\eta rv$$

This equation is generally referred to as Stokes' law.

Terminal velocity

When a sphere is released and allowed to fall freely in a fluid, it is subjected to three forces: its weight, W, the upthrust, U, and the viscous drag, F.

Initially, a resultant force, $F_R = W - (U + F)$ will make the sphere accelerate downward. As the velocity of the sphere increases, the viscous drag increases according to Stokes' law until $(U + F) = W$. The resultant force then becomes zero, and the sphere continues to fall at a constant velocity known as the terminal velocity (see Figure 12.9).

Key terms

Stokes' law states that for a sphere moving through a fluid, the viscous drag acting on it is given by the equation $F = 6\pi\eta rv$, provided that the movement of the fluid relative to the sphere is laminar.

A body moving through a fluid reaches its **terminal velocity** when the resultant of all the forces acting upon the body is zero.

Example

A steel ball bearing of mass 3.3×10^{-5} kg and radius 1.0 mm displaces 4.1×10^{-5} N of water when fully immersed. The ball is allowed to fall through water until it reaches its terminal velocity. Calculate the terminal velocity if the viscosity of the water is 1.1×10^{-3} N s m^{-2}.

Answer

$$F = W - U$$
$$6\pi\eta rv = W - U$$
$$v = \frac{W - U}{6\pi\eta r}$$
$$v = \frac{3.3 \times 10^{-5}\,\text{kg} \times 9.8\,\text{N kg}^{-1} - 4.1 \times 10^{-5}\,\text{N}}{6\pi \times 1.1 \times 10^{-3}\,\text{N s m}^{-2} \times 1.0 \times 10^{-3}\,\text{m}} = 14\,\text{m s}^{-1}$$

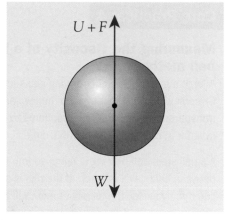

Figure 12.9 Free-body force diagram for a sphere falling in a liquid.

Measuring viscosity using Stokes' law

By measuring the terminal velocity of a sphere falling through a fluid it is possible to determine the **coefficient of viscosity** of the fluid.

For a sphere of radius r and density ρ_s falling through a fluid of density ρ_f and viscosity η with a terminal velocity v, the following equilibrium equation:

$$U + F = W$$

where ,

U = weight of displaced fluid = $m_f g = V\rho_f g = \frac{4}{3}\pi r^3 \rho_f g$

F = viscous drag = $6\pi\eta rv$

W = weight of sphere = $m_s g = V\rho_s g = \frac{4}{3}\pi r^3 \rho_s g$

can be written:

$$\frac{4}{3}\pi r^3 \rho_f g + 6\pi\eta rv = \frac{4}{3}\pi r^3 \rho_s g$$

which gives the viscosity as:

$$\eta = \frac{2(\rho_s - \rho_f)gr^2}{9v}$$

This also tells us that the terminal velocity of a falling sphere in a fluid depends on the square of its radius, so very small drops of rain – and the minute droplets from an aerosol – fall slowly through the air.

The relative viscosities of opaque liquids like molten chocolate can be compared by using a sphere with a thin rod attached so that the time interval between a mark on the lower part of the rod reaching the surface and that of a mark higher up the rod is recorded.

> **Key terms**
>
> **Coefficient of viscosity**, η, is a measure of the resistance to flow for a fluid. It has the units N s m^{-2} which are sometimes given as Pa s.

Core Practical 4

Measuring the viscosity of a liquid using the falling ball method

A large measuring cylinder or long glass tube is filled with a clear, sticky liquid. Glycerol works well, but syrup or honey can be used as long as they are sufficiently transparent to allow the falling spheres to be observed. Lines are drawn around the cylinder at regular intervals of 5.0 cm.

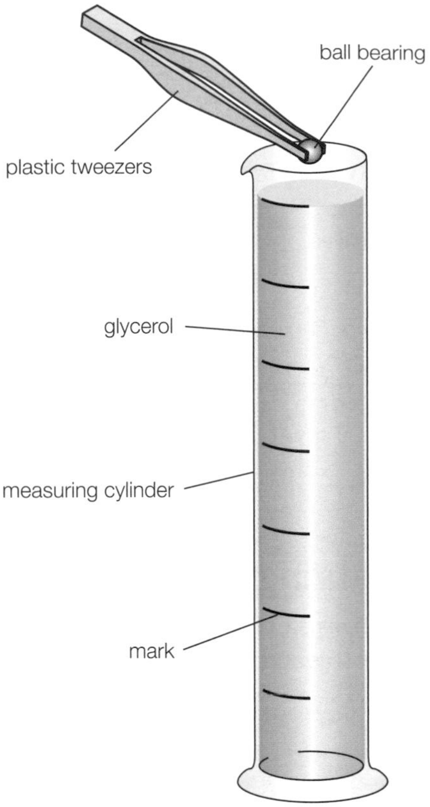

Figure 12.10

A small steel ball bearing is released into the liquid, and the time for the ball to fall between each level is taken. If the ball is small (3 mm in diameter or less) and the liquid fairly sticky, the measurements will all be similar, which indicates that terminal velocity is reached shortly after the sphere enters the liquid.

The timings are repeated and the terminal velocity is calculated by dividing the distance between the markers by the average time taken for the ball to fall between them.

The radius of the ball is found by measuring its diameter using a micrometer. Its density can be determined using the mass of one ball calculated by placing 10–100 balls on a balance. The density of the liquid also needs to be found by taking the mass of a measured volume. The viscosity is found by substituting these values in the expression:

$$\eta = \frac{2(\rho_s - \rho_f)gr^2}{9v}$$

Questions

1 Write a plan for an experiment to determine the viscosity of a liquid using the falling ball method. You will be expected to produce a fully labelled diagram showing all of the apparatus to be used, be able to explain your choice of instrument for all readings to be taken, and describe how you will use your results to achieve your aims.
Your plan should also include details of correct measuring techniques, the independent and dependent variables and steps taken to make a fair test.
Any precautions needed to ensure that the experiment will be safely performed should also be given.

2 The student carried out the experiment and submitted the set of measurements given in Table 12.3.
Distance between marks = 20 cm.

Table 12.3

Diameter, *d*/mm	Average/ mm	Time, *t*/s	Average/s	Terminal velocity
1.98, 1.98	1.98	7.9, 8.1	8.0	2.5
3.02, 3.04	3.03	3.6, 3.6	3.6	5.6
4.01, 3.99	4	2.0, 2.4	2.2	9.1
5.01, 5.01	5.01	1.2, 1.3	1.25	16
6.08, 6.12	6.1	0.9, 0.9	0.9	22

From tables, density of steel = 7800 kg m^{-3} and density of oil = 820 kg m^{-3}.
The student has made several errors in the presentation of these results. Identify these errors and state how they can be remedied.

3 Rearrange the equation for viscosity given above with terminal velocity as the subject.

4 Draw a suitable straight-line graph to show that the relationship between terminal velocity and the radius of the ball bearings agrees with your equation.

5 One point on the graph appears to be anomalous. Identify this point and state which of the readings should have been repeated.

6 Find the gradient of the graph and use this to calculate the viscosity of the oil.

7 State how you would determine the percentage uncertainty of the diameter and time readings. Calculate the percentage uncertainties for the smallest diameter and the shortest time readings. Which of these will have the greatest effect on the accuracy of the value of viscosity?

8 Suggest how the experiment may be modified to reduce errors.

12.4 Variations in viscosity

The viscosity of a fluid is very dependent on temperature. When the temperature of a fluid rises, its internal energy increases. The adhesive forces between molecules and the cohesive forces at solid surfaces decrease causing a reduction in the viscosity. Hot fluids become more 'runny'. This is the case for lubricating oil in car engines; as the oil gets hotter, its viscosity is reduced and the frictional forces between moving parts get smaller, reducing wear.

For most of our calculations we have assumed that the flow is laminar and that Stokes' law equation applies. In practice, the behaviour of most fluids is much more complex. The viscosity of molten chocolate does not depend only on its temperature – it reduces as the rate of flow increases. This is an example of **thixotropy**. Many gels and colloids have this property: for example, toothpaste flows out of the tube when pressure is applied but keeps its shape on the toothbrush. Butter, margarine and other spreads are normally solid but will flow under the pressure of a knife onto the bread.

Stirring or vibration can also affect thixotropic substances. Drip-dry paints become runny when stirred or applied to a surface but not when held on the brush. Some landslides are caused when normally stable mixtures of clay and water are disturbed.

Test yourself

14 Draw a diagram showing the forces acting on a sphere falling through a fluid.

15 Explain the relationship between the forces drawn in Question 14 when the sphere has reached its terminal velocity.

16 What conditions must apply for the equation viscous drag, $F = 6\pi\eta rv$ to apply?

17 Use the equation $F = 6\pi\eta rv$ to show that the units of viscosity can be given as $N\,s\,m^{-2}$.

18 Give three reasons why the rate of flow of oil pumped ashore from a North Sea oil field will be different from that pumped through a desert region to a port in the Middle East.

Exam practice questions

1 Figure 12.11 shows a free-body force diagram of a sphere falling through a fluid and a velocity–time graph starting from the moment the sphere is released, where U is the upthrust on the sphere, W is its weight and F is the viscous drag.

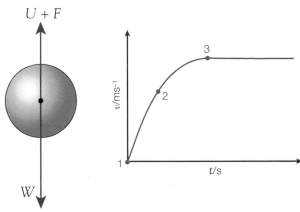

Figure 12.11

a) The acceleration of the sphere at point 1 on the graph is

    A $0\,\mathrm{m\,s^{-2}}$ **C** $4.9\,\mathrm{m\,s^{-2}}$

    B $2.5\,\mathrm{m\,s^{-2}}$ **D** $9.8\,\mathrm{m\,s^{-2}}$ **[Total 1 Mark]**

b) The acceleration at point 3 is

    A $0\,\mathrm{m\,s^{-2}}$ **C** $4.9\,\mathrm{m\,s^{-2}}$

    B $2.5\,\mathrm{m\,s^{-2}}$ **D** $9.8\,\mathrm{m\,s^{-2}}$ **[Total 1 Mark]**

c) Between points 1 and 2 on the graph

    A $W = F$ **C** $W = U + F$

    B $W = U$ **D** $W > U + F$ **[Total 1 Mark]**

d) At point 3

    A $W = F$ **C** $W = U + F$

    B $W = U$ **D** $W > U + F$ **[Total 1 Mark]**

2 The resistive force on a ball bearing falling through oil does not depend on:

    A   its diameter   **C**   its velocity

    B   its mass   **D**   the viscosity of the oil. **[Total 1 Mark]**

3 The principle of a liquid barometer is to 'balance' the air pressure with the pressure due to a column of the liquid. Calculate the column length of a mercury barometer when the atmospheric pressure is $1.01 \times 10^5\,\mathrm{Pa}$.   **[2]**

4 A spherical balloon of radius 20 cm contains 50 g of air.

    a) Calculate the density of air in the balloon.

    b) Comment on why it differs from that given in Table 12.1 (page 174)

                                               [Total 3 Marks]

5 The density of the water in the Dead Sea is about $1200\,\text{kg}\,\text{m}^{-3}$.

    a) Calculate the volume of water that needs to be displaced for a
person of weight $600\,\text{N}$ to float in this sea.   **[2]**

    b) How does this compare with the volume that needs to be displaced
for the person to float in a swimming pool?   **[1]**

                                     [Total 3 Marks]

6 Oil is often pumped underground or undersea through pipes that are
hundreds of kilometres long.

    a) State how the diameter of the pipe, the place of origin and the
temperature of the oil affect the rate of flow through the pipe.   **[4]**

    b) The oil companies need to transport the oil as quickly as possible.
Why does the flow rate need to be restricted?   **[1]**

                                     [Total 5 Marks]

7 Figure 12.12 shows the movement of air relative to an aircraft wing.
The lines represent the motion of layers of air particles close to the wing.

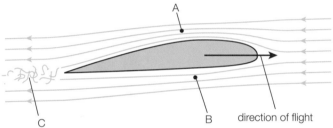

Figure 12.12

At A and B, the layers of air do not cross over each other, but the air at
C swirls around and the particles mix together.

    a) Write down a word that describes the flow of air at A and B and
another that represents the movement at C.   **[2]**

    b) The uplift on an aeroplane is a consequence of the faster-moving air
at A having a lower pressure than that at B. For a Boeing 737 cruising
at $270\,\text{m}\,\text{s}^{-1}$ (Mach 0.8), this pressure difference is about $4 \times 10^3\,\text{Pa}$.
Calculate the uplift if the underwing area is $120\,\text{m}^2$.   **[2]**

    c) Estimate the mass of the plane and its load.   **[1]**

                                     [Total 5 Marks]

8 Droplets in a deodorant spray have a mass of about $4 \times 10^{-12}\,\text{kg}$ and a
radius of around $0.1\,\text{mm}$.

    a) Estimate the terminal velocity of the droplets in air of viscosity
$2 \times 10^{-5}\,\text{N}\,\text{s}\,\text{m}^{-2}$.   **[3]**

    b) How does your estimate compare with the speed of
falling raindrops?   **[1]**

                                     [Total 4 Marks]

9 Figure 12.13 shows a sphere of weight W, and radius r, falling through a liquid.

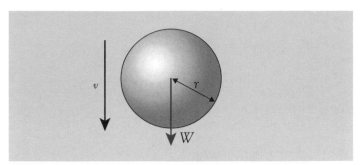

Figure 12.13

a) Copy and add labelled arrows to the diagram to represent the upthrust, U, and the viscous drag, F. **[2]**

b) The radius of the sphere is 1.5 mm and its density is 7800 kg m⁻³. Show that its weight is about 1.1×10^{-3} N. **[2]**

c) What condition is necessary for the sphere to reach its terminal velocity? **[1]**

d) If the upthrust on the sphere is 0.20×10^{-3} N, calculate the value of the terminal velocity of the sphere in a liquid of viscosity 0.35 Pa s. **[3]**

e) In an investigation into the effect of the temperature on the viscosity of oil, the terminal velocity of a ball bearing is measured as it falls through the oil over a range of temperatures. Figure 12.14 shows the apparatus used and a graph of the terminal velocity against the temperature of the oil.

  i) Use the graph to find the value of the terminal velocity of the ball at 30 °C.

  ii) Use the graph to describe how the viscosity of the oil changes between 10 °C and 80 °C.

  iii) Some motor oils are termed 'viscostatic'. What property of the oil is described by this statement? **[5]**

[Total 13 Marks]

10 A sample of builders' aggregate contains pebbles of various sizes, coarse sand and fine sand. Some aggregate is added to a large jar of water. The mixture is shaken and the jar is then placed on a flat surface and the particles are allowed to settle.

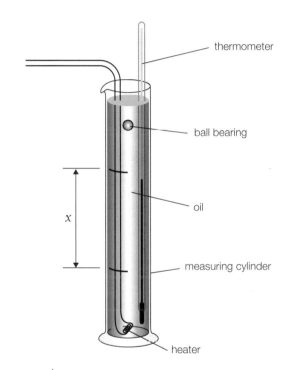

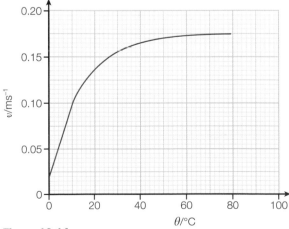

Figure 12.14

After some time it is observed that the largest pebbles have settled at the bottom of the jar, with layers of small pebbles, coarse sand and fine sand, in that order, on top of the pebbles. A short time later it is observed that some of the fine sand is still in suspension in the lower part of the water.

a) Explain, in terms of the size of the particles, why they settle in this order. **[6]**

b) If you assume that the smallest sand particles have a diameter of about 0.2 mm and the depth of the water is 22 cm, estimate the time taken for the water to become completely clear. (Assume that the density of the sand particles is 1200 kg m^{-3} and the viscosity of water is 0.8 N s m^{-2}.) **[4]**

[Total 10 Marks]

Stretch and challenge

11 Figure 12.15 is a schematic diagram of a vortex-shedding flowmeter.

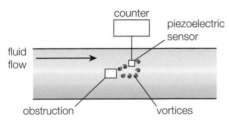

Figure 12.15

A non-streamlined object is situated in the moving fluid and a series of vortices is created behind it. The pressure changes in the vortices are detected using piezoelectric sensors. The frequency of the pulses generated by the sensors is proportional to the rate of flow of the fluid.

a) Explain why the non-streamlined object creates the vortices, and how the frequency of the pulses can be used to determine the rate of flow. **[3]**

Water flows to a domestic water heater at a rate of 12 litres per minute through a tube of diameter 22 mm.

b) Calculate the speed of the water in the tube. **[4]**

Poiseuille's formula states that the rate of flow of fluid in a pipe is
$$\frac{V}{t} = \frac{\pi \Delta p r^4}{8 \eta l}$$
where Δp = pressure difference, η = viscosity, r = radius of tube, l = length of tube.

A vortex-shedding flowmeter in the pipe carrying water of viscosity 0.80 N s m^{-2} emits pulses of frequency 20 Hz.

c) Show that the frequency of the emitted pulses for the same flowmeter as used for oil of viscosity 2.5 N s m^{-2} flowing through a pipe of diameter 60 cm at a rate of 39 cubic metres per minute is about 90 Hz. **[4]**

d) State any assumptions made about the flow of fluid in the tube. **[2]**

e) Explain why it is unlikely for such a flow rate to be achieved using a pipe of the same length with a diameter of 30 cm. **[3]**

[Total 16 Marks]

Solid materials

13

Prior knowledge

In this chapter you will need to:
→ be aware that solids have a fixed shape and, unlike liquids and gases, the atoms in a solid occupy fixed positions
→ understand that solids have a wide range of properties and react differently when subjected to deforming forces.

The key facts that will be useful are:
→ atoms in a solid are held in position by forces called bonds
→ the extension of a spring is proportional to the applied force
→ work done = average force × distance moved in the direction of the force
→ work is needed to stretch or compress a solid
→ elastic potential energy in a stretched wire = average force × extension.

Test yourself on prior knowledge

1 A spring is extended by 2.0 cm when a stretching force of 5.0 N is applied. Calculate the extension
   a) if a force of 12.5 N is applied
   b) if an identical spring is attached to one end of the spring (end to end) and a force of 10 N is applied
   c) if an identical spring is attached in parallel (side by side) and a force of 5.0 N is applied.
2 Calculate the tension in a spring fixed at one end with a mass of 120 g suspended from the other end.
3 Calculate the work done when the point of application of a force of 4.00 N is moved through a distance of 2.5 m.
4 Determine the elastic potential energy stored in a wire held in a clamp if a 5.0 kg mass attached to the other end makes the wire extend by 2.0 mm when it is released.

13.1 Elastic and plastic deformation

A material undergoing **elastic deformation** will return to its original dimensions when the deforming force is removed. A **plastic** material will remain deformed.

A rubber band is often called an 'elastic' band. It can be stretched to several times its normal length and will return to its starting length when the tension is removed. Less noticeably, steel wires or cables can also behave elastically.

Modelling clay is a simple example of a material that deforms plastically. It can be pulled, squeezed and twisted into the desired shape.

Some materials can behave in an elastic or plastic manner depending on the nature of the deforming forces. A thin steel sheet will deform elastically when small forces are applied to it, but the huge forces of a hydraulic press will mould the sheet into car panels. Once shaped, the panels regain their elastic properties for everyday stresses.

You will investigate the elastic and plastic properties of materials using force–extension graphs in Section 13.3.

13.2 Properties of solid materials

Solid materials play a vital role in our lives. Engineering components, sports equipment and the bones in our bodies have been designed or have evolved to be ideally suited for their purpose. In this section, you will study the properties of a range of materials and consider their significance in a variety of applications.

A simple example of how confectionery products can be compared is by grading some of their properties on a scale of 1–10. A group of testers can judge how hard, smooth, chewy, sticky, crunchy, creamy, sweet or brittle each product is, and the results may be analysed to give a taste profile.

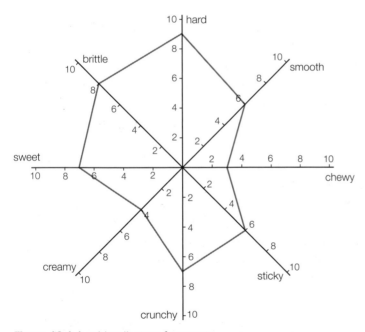

Figure 13.1 A spider diagram for sweets.

Activity 13.1

Creating a spider diagram
Use the properties given in Figure 13.1 – or any others that you think could be used – to describe a number of common confectionery products. Find the average grade for each property and construct a spider diagram to display the taste profile.

Tip

Please note that these properties (stiffness, toughness, brittleness etc.) are not essential material for the exam, but have been provided here as extension material.

Most of these properties are subjective and loosely described. In physics and engineering, the properties of solid materials need to be defined clearly and universally accepted.

Hardness

Hardness is a surface phenomenon. The harder the material, the more difficult it is to indent or scratch the surface. A simple method to compare the hardness of two materials involves finding out which scratches the surface of the other.

Activity 13.2

Comparing the hardness of a range of solids

Collect a number of blocks of different woods, metals and plastics (for example, from a materials kit) and use a corner of each block to try to score the other blocks. Grade the materials in terms of hardness.

Table 13.1 The Mohs scale of hardness.

Mohs	Mineral	R
1	talc	1
2	gypsum	3
3	calcite	9
4	fluorite	21
5	apatite	48
6	feldspar	72
7	quartz	100
8	topaz	200
9	corundum	400
10	diamond	1600

The **Mohs scale** of hardness shown in Table 13.1 grades ten minerals from the softest – talc, which is rated as 1 on the scale – to the hardest – diamond, which has a rating of 10. R is the relative hardness of the minerals (e.g. diamond is 1600 times harder than talc and 4 times as hard as corundum). In engineering, the hardness of a metal is measured using the Brinell Hardness Number (BHN), which is the ratio of the load applied to a small steel sphere to the area of the indentation it makes in the surface of the metal being tested.

Wall and floor tiles are glazed so that the surface has a hardness of about 7 on the Mohs scale. In order to snap the tiles, the surface is scored with a tungsten-hardened steel knife edge (Mohs number 7.5–8) or a diamond-crusted cutting wheel. A sharp iron nail (Mohs number 5) would not scratch the surface.

Stiffness

A **stiff** material exhibits very small deformations even when subjected to large forces. It would require a great force to bend the upright of a laboratory stand even by a few millimetres, whereas a polythene metre rule can be bowed easily. Similarly, a steel piano wire is much stiffer than a rubber band, as it needs a very large force to produce a small extension.

The stiffness of a material is measured in terms of its **modulus of elasticity**. You will learn more about stiffness in Section 13.4.

Toughness

A **tough** material is able to absorb the energy from impacts and shocks without breaking. Tough metals usually undergo considerable plastic deformation in order to absorb the energy.

Car tyres are made from tough rubber/steel compositions. They are designed to withstand impacts from irregular surfaces. The absorbed energy is transferred to the tyres as internal energy – that is, the tyres get hot after a journey.

The energy per unit volume absorbed when materials are deformed will be studied in Section 13.4.

Brittleness

A **brittle** object will shatter or crack when subjected to dynamic shocks or impacts. Brittle materials undergo little or no plastic deformation before breaking. The glass used for car windscreens is brittle, as are cast iron gratings, concrete and biscuits.

Strength

An object is **strong** if it can withstand a large force before it breaks. The strength of a material will depend on its size – for example, a thick cotton thread requires a bigger breaking force than a thin copper wire.

The strength of a material is therefore defined in terms of its breaking **stress**, where stress is the force per unit cross-sectional area.

Malleability

A **malleable** material can be hammered out into thin sheets. Gold is very malleable and can be hammered into 'gold leaf', which is used to decorate pottery and picture frames, for example.

Ductility

Ductile materials can be drawn into wires. Copper wires are used extensively for electrical connections and are produced by drawing out cylinders to the desired thickness.

Although most ductile materials are also malleable, the reverse is not always true. Many malleable materials will shred or break when extended.

Test yourself

Select the most appropriate property from those given in Section 13.2 to copy and complete the following sentences.

1 Mild steel can withstand a heavy blow from a large hammer because it is _____, whereas cast iron is likely to shatter as it is _____.

2 Lead is easier to hammer out into thin sheets than aluminium, because it is more _____ than the aluminium.

3 A glass rod will undergo very little extension when loaded because it is _____ and, being _____, is likely to snap, with little further extension, when the load is increased.

4 Copper is a _____ material and so it can easily be drawn into thin wires.

5 A steel cable is able to withstand a higher load before breaking than an aluminium cable of the same dimensions because steel is _____ than aluminium.

13.3 Hooke's law

You may be familiar with a simple experiment investigating the extension of a spring. The spring is loaded by adding masses to a weight hanger and the extensions are measured as the load increases until the spring is 'uncoiled'. The apparatus for the experiment and a sketch of the resulting load–extension graph are shown in Figure 13.2.

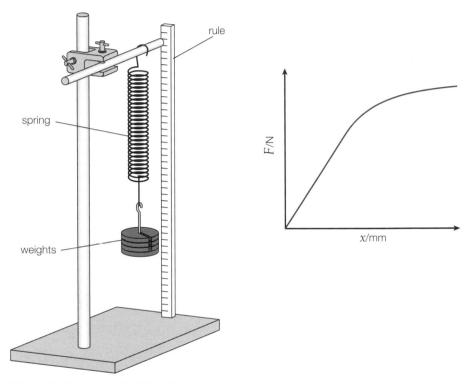

Figure 13.2 Investigating Hooke's law.

Hooke's law states that up to a given load, the extension of a spring is directly proportional to the force applied to the spring:

$$F = k\Delta x$$

where k is the stiffness of the object.

The graph initially has a linear region when the spring is still coiled. As the spring loses its 'springiness', the extension increases disproportionately.

The constant k represents the 'stiffness' of the spring and is called its **spring constant**.

> ## Key term
>
> **Hooke's law** states that up to a given load, the extension of a spring is directly proportional to the force applied to the spring and is given by $F = k\Delta x$.

> ## Example
>
> Estimate the spring constant of:
>
> **a)** a 0–10 N newton meter
>
> **b)** a suspension spring of a family car.

Answers

a) Assume the scale is 10 cm long:
$$k = \frac{F}{\Delta x} = \frac{10\,N}{0.10\,m} = 100\,N\,m^{-1}$$

b) If the mass of the car is 1600 kg, the force on each of the four springs will be $(400\,kg \times 9.8\,m\,s^{-2}) \approx 4000\,N$.

To take the weight off the spring, the car body needs to be raised about 10 cm:
$$k = \frac{F}{\Delta x} = \frac{4000\,N}{0.10\,m} = 4 \times 10^4\,N\,m^{-1}$$

The spring experiment is a useful introduction to this section. Most materials behave like springs to some degree, as the bonds between atoms and molecules are stretched when they are loaded. By studying the force–extension graphs of materials, we can see whether the material obeys Hooke's law and examine many other properties.

Activity 13.3

Investigating the properties of copper wire

A thin copper wire is clamped between wooden blocks at one end of a bench and passed over a pulley wheel at the opposite end of the bench (Figure 13.3). The wire should be at least two metres long. (The longer the wire, the bigger the extension for a given load, which means that more precise readings can be taken.)

A weight hanger is attached to the wire close to the pulley and a metre rule is fixed to the bench as shown in Figure 13.3. A strip of sticky tape is attached to the wire so that its edge is on the lower end of the metre scale. (The length of the wire from the fixed end to the tape and the diameter of the wire can be taken now, as they may be used at a later stage.)

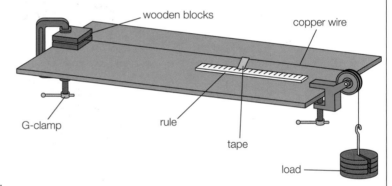

Figure 13.3 Stretching copper wire.

Weights are placed on the hanger in 200 g increments (2.0 N), and the corresponding extensions are read off the scale. When the copper starts to 'give', a few weights should be removed (tricky) and the wire will continue stretching until it breaks. **Safety note:** It is important that safety glasses are worn to reduce the risk of eye damage when the wire breaks. Place a tray continuing sand or foam rubber beneath the weights to prevent them from landing on your feet should they fall.

A graph of force against extension is plotted. A typical graph for a three-metre length of 26 swg (standard wire gauge) wire is shown in Figure 13.4.

Questions

1. Explain why it is advisable to use such a long piece of wire for this experiment.
2. The edge of the tape marker is kept very close to the scale on the rule throughout. Explain why this is necessary for taking precise readings of the extension.
3. The dotted lines on the force–extension graph show how the extension changes when the weights are removed (a) just after point B and (b) after a large extension. Describe the behaviour of the wire in each situation.

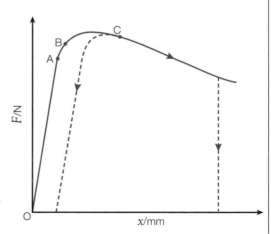

Figure 13.4 Force–extension curve for copper wire.

Force–extension graphs

It is not always possible to use the spring set-up to investigate the tensile properties of materials. Metals, for example, will often require very large forces to produce measurable extensions, and so different arrangements or specialised equipment are needed.

Elastic and plastic behaviour during stretching

Figure 13.4 shows a steep linear region followed by a region of large extension with reducing force. In the initial section (O–A), the extension is proportional to the applied force, so Hooke's law is obeyed.

If the load is removed from the wire up to the **limit of proportionality**, or even a little beyond this, the wire will return to its original length. This is known as the *elastic* region of the extension, in which loading and unloading are **reversible**. Arrows are drawn on the graph to illustrate the load–unload cycles.

The atoms in a solid are held together by **bonds**. These behave like springs between the particles and as the copper wire is stretched, the atomic separation increases. In the elastic region, the atoms return to their original positions when the deforming force is removed.

Point B on the graph is known as the **elastic limit**. Beyond this point the wire ceases to be elastic. Although the wire may shorten when the load is removed, it will not return to its original length – it has passed the point of reversibility and has undergone **permanent deformation**.

As the load is increased, the wire **yields** and will not contract at all when the load is removed. Beyond this **yield point** – point C on the graph – the wire is *plastic* and can be pulled like modelling clay until it breaks. If the broken end of the wire is wound around a pencil, the plasticity can be felt when the wire is extended. In the plastic region, the bonds between the atoms are no longer being stretched and layers of atoms slide over each other with no restorative forces.

A very strange effect is noticed if the load is removed during the plastic phase and the wire is reloaded: the wire regains its springiness and has the same stiffness as before (Figure 13.5).

The ability of some metals to be deformed plastically and then regain their elasticity is extremely important in engineering. A mild steel sheet can be pressed into a mould to the shape of a car panel. After the plastic deformation, the stiffness and elasticity of the steel is regained and further pressings of the panel are also possible.

Key term

A material is said to be **elastic** if it regains its original dimensions when the deforming force is removed.

Key term

A material is **plastic** if it retains its shape after a deforming force is removed.

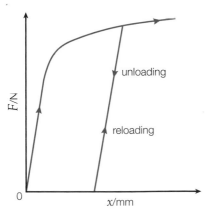

Figure 13.5 Repeated loading and unloading of copper wire.

Test yourself

6 State Hooke's law.

7 a) Calculate the value of the spring constant of a spring if a mass of 10 kg suspended from it gives an extension of 5.0 cm.

   b) Determine the spring constant for two springs, both identical to the one in a, joined together

   i) in series (end to end)

   ii) in parallel (side by side).

8 Explain the meaning of the following terms relating to a copper wire that has been loaded to breaking point:

a) limit of proportionality

b) elastic limit

c) yield point

d) plastic flow.

Steel wires

It is possible to use a similar set-up to the copper wire experiment for steel wires, but as these are much stiffer and yield after relatively small fractional increases in length, it is best not to attempt the investigation without specialist equipment.

Steel is produced by mixing iron with small quantities of carbon. The properties of the steel depend on the percentage of carbon and the heat treatment of the steel. Mild steel contains less than 0.25% carbon and exhibits similar plastic behaviour to copper, whereas high-carbon steel is usually quench-hardened (rapidly cooled by dipping into oil or water) and is quite brittle.

Figure 13.6 shows force–extension curves for mild steel and high-carbon steel wires of similar length and diameter as the copper wire.

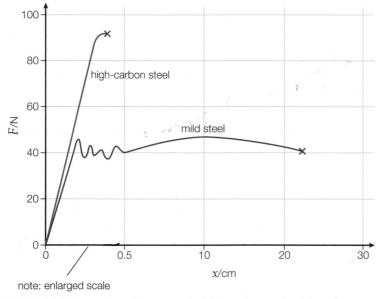

note: enlarged scale

Figure 13.6 Force-extension curves for high-carbon and mild steel.

The graph for mild steel shows a similar trend to that for copper, but the yield point is at a lower fractional extension than for copper and the curve is initially quite 'jerky'. Considerable extension is seen after the mild steel yields, but the final break occurs at a much lower percentage increase than for copper. As more carbon is added to the steel, the plastic region is reduced, as shown in the second graph.

Natural rubber

A force–extension graph for a rubber band can be obtained in a similar manner to the experiment with the spring, but an alternative method is shown in Figure 13.7. **Safety note:** It is important that safety glasses are worn to reduce the risk of eye damage when the band breaks. It is best to use a short, thin rubber band, as it will stretch to several times its original length and thick bands are difficult to break. To alter the length of the band, the boss is loosened and moved up the stand. The force at each extension is read off the newton-meter.

The rubber band stretches very easily at first, reaching a length of three or four times its original value (Figure 13.8). It then becomes very stiff and difficult to stretch as it approaches its breaking point.

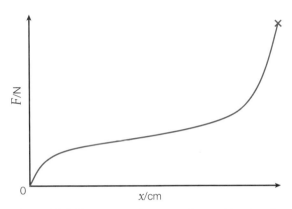

Figure 13.8 Force-extension curve for a rubber band.

Natural rubber is a polymer. It contains long chains of atoms that are normally tangled in a disordered fashion (like strands of spaghetti). Relatively small forces are needed to 'untangle' these molecules so a large extension is produced for small loads. When the chains are fully extended, additional forces need to stretch the bonds between the atoms, so much smaller extensions are produced for a given load; the rubber becomes stiffer.

It should be noted that the band returns to its original length if the force is removed at any stage prior to breaking – that is, the rubber band is elastic.

Force–compression graphs

Up to now we have considered only the behaviour of materials that have been subjected to stretching – or **tensile** forces. If weights were placed on top of a large rectangular sponge, the sponge would be noticeably squashed and force–compression readings easily measured. This is much more difficult for metals. Whereas long thin samples of copper wire can be extended by several centimetres with tensile forces of 50 N or less, the shorter, thicker samples needed for compression tests require much larger forces to produce measurable compressions.

In engineering laboratories, large hydraulic presses are used, but a school's compression testing kit, shown in Figure 13.9, can demonstrate the effects of compression on a range of sample materials.

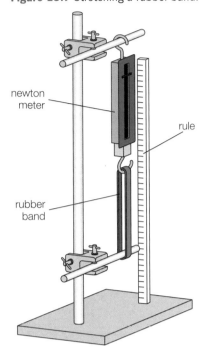

Figure 13.7 Stretching a rubber band.

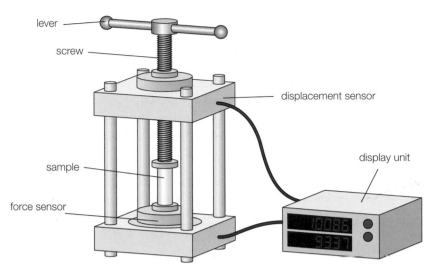

Figure 13.9 Compression testing kit.

The sample is placed into the press and the screw is tightened to hold it firmly in position. Force and position sensors are connected to a display module or to a computer via a data-processing interface and the readings are zeroed.

The sample is compressed by rotating the lever clockwise. A series of values of force and the corresponding compression is taken and a force–compression graph drawn.

The elastic region is similar to that seen in the tensile tests. The bonds are 'squashed' as the atoms are pushed together, and the particles move back to their original position when the force is removed.

Plastic behaviour is much more difficult to examine as the samples twist and buckle at failure.

Some kits allow the sample to be stretched as well as compressed, so that a comparison may be made.

Elastic strain energy

The concept of potential energy was introduced in Chapter 6. This relates to the ability of an object to do work by virtue of its position or state. Elastic potential energy – or **elastic strain energy** – is therefore the ability of a deformed material to do work as it regains its original dimensions. The work done stretching the rubber of a catapult (slingshot) is transferred to elastic strain energy in the rubber and then to kinetic energy of the missile on its release.

The work done during the stretching process is equal to the average force times the distance moved in the direction of the force:

$$\Delta W = F_{av}\Delta x$$

The work done on a wire, and hence its elastic strain energy, can be obtained from a force–extension graph.

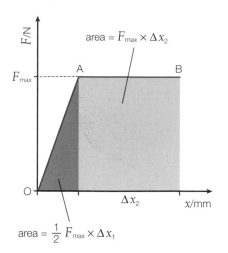

Figure 13.10 Elastic strain energy.

For the Hooke's law region of the graph (O–A) in Figure 13.10, the average force is $\frac{F_{max}}{2}$, so the work done is:

$$\Delta W = \frac{1}{2}F_{max}\Delta x$$

This represents the area between the line and the extension axis – that is, the area of the triangle made by the line and the axis. Similarly, the work done when the force is constant (A–B on the graph) will be the area of the rectangle below the line.

For any force–extension graph, the elastic strain energy is equal to the area under the graph. To calculate the energy for non-linear graphs, the work equivalent of each square is calculated and the number of squares beneath the line is counted. For estimated values, the shape can be divided into approximate triangular or rectangular regions.

Example

Estimate the elastic strain energy in the material that produces the force–extension graph in Figure 13.11.

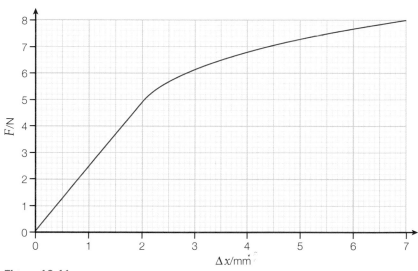

Figure 13.11

Answer

Area of 'Hooke's law region' $= \frac{1}{2}$ (5 N)(2 × 10⁻³ m) = 5 × 10⁻³ J

Area of trapezium $= \frac{1}{2}$ (5 N + 8 N)(5 × 10⁻³ m) = 4 × 10⁻² J

Elastic strain energy = 5 × 10⁻² J

For a material that is extended within the Hooke's law limit, the strain energy can be calculated in terms of the spring constant of the material:

$$\Delta W = \frac{1}{2}F\Delta x \text{ and } F = k\Delta x$$

$$\Delta W = \frac{1}{2}(k\Delta x)\Delta x = \frac{1}{2}k\Delta x^2$$

9 Force–extension graphs are drawn for two different wires.

a) How would you determine

   i) which wire was stiffer

   ii) which was tougher?

b) If the wires were loaded until they snapped, how would you know if either wire was brittle?

10 A load of 12 kg causes a wire to extend by 2.5 mm. If the wire obeys Hooke's law

a) calculate the stiffness of the wire

b) calculate the elastic potential energy transferred to the wire.

11 Explain the statement, 'natural rubber is a polymer'.

12 How would you estimate the value of the work done stretching a rubber band from the resulting force–extension graph for the band?

13.4 Stress and strain: the Young modulus

In everyday terms, one might say that students put their teacher under stress and the teacher takes the strain. Likewise, when a **stress** is applied to a material, the **strain** is the effect of that stress.

Stress is usually written as $\sigma = \dfrac{F}{A}$, and has the unit pascal, Pa. Strain is written as $\varepsilon = \dfrac{\Delta l}{l}$ and, because it is a ratio of two lengths, it has no unit.

A property of materials that undergo tensile or compressive stress is the **Young modulus**, E.

The Young modulus can be represented by the equation

$$E = \frac{\sigma}{\varepsilon} = \frac{Fl}{A\Delta l}$$

and is measured in pascals, Pa.

Key terms
Stress = force/cross-sectional area; $\sigma = \dfrac{F}{A}$
Strain = extension/original length; $\varepsilon = \dfrac{\Delta l}{l}$
Young modulus = stress/strain

Example

A steel wire of length 2.00 m and diameter 0.40 mm is extended by 4.0 mm when a stretching force of 50 N is applied. Calculate:

a) the applied stress

b) the strain on the wire

c) the Young modulus of steel.

Answers

a) Stress, $\sigma = \dfrac{F}{A} = \dfrac{50\,\text{N}}{\pi(0.20 \times 10^{-3}\text{m})^2} = 4.0 \times 10^8\,\text{Pa}$

b) Strain, $\varepsilon = \dfrac{\Delta l}{l} = \dfrac{4.0 \times 10^{-3}\,\text{m}}{2.00\,\text{m}} = 2.0 \times 10^{-3}\ (0.20\%)$

c) Young modulus, $E = \dfrac{\sigma}{\varepsilon} = \dfrac{4.0 \times 10^8\,\text{Pa}}{2.0 \times 10^{-3}} = 2.0 \times 10^{11}\,\text{Pa}$

Tip

Stress-strain graphs are useful because they will be the same for any sample of a given material and the gradient of the proportional region equals the Young modulus of the material.

The importance of stress and strain as opposed to force and extension is that they are properties of the material: a stress–strain graph is always the same for a given material, whereas a force–extension graph depends on the dimensions of the sample used.

For most investigations, the cross-sectional area of the sample is a few square millimetres at most $(1\,\text{mm}^2 = 1 \times 10^{-6}\,\text{m}^2)$, so the stress is very large (up to several hundred MPa). As strains are often very small, the Young modulus can be hundreds of GPa for metals.

Stress–strain graphs

The shape of stress–strain graphs is much the same as that of force–extension graphs, although for large extensions a reduction in cross-sectional area will result in an increase in the stress for a particular load. The main advantage is that the information gained from the graph relates to the properties of the material used and not just those of the particular sample tested.

Figure 13.12 shows a stress–strain graph for the copper wire used earlier in the stretching experiment. It should be noted that the strain scale is expanded over the initial 1% so that the full extent of the plastic region can be included.

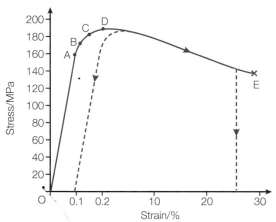

Figure 13.12 Stress-strain graph for copper wire.

- O–A represents the **Hooke's law region.** Strain is proportional to stress up to this point. The **Young modulus** of copper can be found directly by taking the gradient of the graph in this section.
- B is the **elastic limit.** If the stress is removed below this value, the wire returns to its original state.
- The stress at C is termed the **yield stress.** For stresses greater than this, copper will become **ductile** and deform plastically.
- D is the maximum stress that the copper can endure. It is called the **ultimate tensile strength (UTS)** or simply the strength of copper.
- E is the breaking point. There may be an increase in stress at this point due to a narrowing of the wire at the position on the wire where it breaks, which reduces the area at that point.

Core practical 5

Experiment to determine the Young modulus of copper
The Young modulus of copper can be determined using the results taken during Activity 13.3 on page 193 in which the properties of copper wire were investigated. In addition to the force and extension readings, the length of the wire from the clamp to the marker and the diameter of the wire are needed.

Questions
1 a) Use the values of length and diameter from the earlier activity to calculate the stress and strain for each reading of force and extension, OR
   b) Use the following typical set of results from a similar experiment: length of wire = 3.00 m; mean diameter of wire = 0.46 mm. Copy and complete Table 13.2.

Table 13.2

Force/N	Extension/mm	Stress/Pa	Strain/%
0	0		
5.0	0.5		
10.0	1.0		
15.0	1.5		
20.0	2.0		
25.0	2.5		
30.0	3.0		
35.0	4.0		
40.0	6.0		
35.0	300		
30.0	750		

2 Identify the independent and dependent variables.

3 Explain the choice of instrument used to measure the length, extension and diameter of the wire. (Reference should be made to the percentage uncertainty in one reading for each of the length, extension and diameter.)

4 State any precautions or special techniques used to ensure the readings were as precise as possible.

5 Comment on any precautions taken to ensure that the experiment was conducted in a safe manner.

6 Plot a graph of stress against strain and use the graph to determine a value of the Young modulus of copper.

7 Suggest any modifications that could be made to improve the accuracy of the value of the Young modulus.

Figure 13.13 provides stress–strain graphs to show the comparative properties of high-carbon steel, mild steel and copper. The early part of the strain axis is extended to show the Hooke's law region more clearly.

The graphs illustrate the different behaviour of the three materials. The gradient of the Hooke's law region is the same for the steels and is greater than that for copper. The steels have a Young modulus of about 200 GPa and are stiffer than copper ($E \approx 130$ GPa).

High-carbon steel is the strongest as it has the greatest **breaking stress** (UTS), but it fractures with very little **plastic deformation** – it is brittle. Quench-hardened high-carbon steel is commonly used for cutting tools and drill bits.

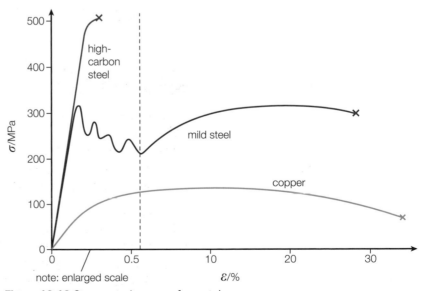

Figure 13.13 Stress-strain curves for metals.

Mild steel has an upper and lower yield point and a fairly long plastic region. Because it regains its high stiffness when the deforming force is removed, it is ideal for pressing into car body parts.

The long plastic region of copper means that it is very ductile and so is easily drawn into wires. This, together with its low resistivity (see Chapter 10), makes copper an invaluable material for the electrical industry.

Compressive stress–strain graphs

If the initial length and cross-section of the sample used for the force–compression graph are measured, compressive stresses and strains can be calculated. The Young modulus of the material under compression can then be found by taking the gradient of the stress–strain graph. It is generally the same as the modulus found in tensile tests.

For artificial joints used in hip replacement surgery, it is important that the Young modulus of the artificial joint, as well as its hardness and durability, is similar to that of the bone it replaced; otherwise differences in compression for similar stresses could lead to deterioration of the bone around the artificial joint.

Example

Estimate the stress on the lower leg bones of a stationary erect human being. Explain why your value is likely to be much less than the ultimate compressive stress of bone.

Answer

Assume that the person has a mass of 60 kg and that the bones have a supporting area of cross-section of about 5 cm².

Force on each bone, $F = 30\,\text{kg} \times 10\,\text{m s}^{-2} = 300\,\text{N}$

Cross-sectional area, $A = 5 \times 10^{-4}\,\text{m}^2$

$$\text{Stress} = \frac{F}{A} \approx 600\,\text{kPa}$$

When walking, running and jumping, the stresses on the bones will be much bigger.

Energy density

The **energy density** is the work done in stretching a specimen (the strain energy stored) per unit volume of the sample.

For a wire that obeys Hooke's law:
$$\text{energy density} = \frac{\text{work done}}{\text{volume}} = \frac{\Delta W}{Al} = \frac{F_{\text{av}}\Delta l}{Al} = \frac{1}{2}\frac{F\Delta l}{Al}$$
$$= \frac{1}{2}\frac{F}{A} \times \frac{\Delta l}{l}$$
$$= \frac{1}{2}\,\text{stress} \times \text{strain}$$

This is represented on the stress–strain graph by the area between the line and the strain axis.

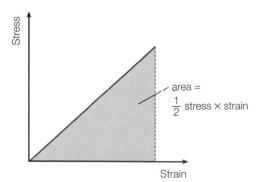

Figure 13.14 Energy density.

area = $\frac{1}{2}$ stress × strain

As with the force–extension graphs and energy stored, the area under any stress–strain graph represents the energy density.

The ability of a specimen to absorb a large amount of energy per unit volume before fracture is a measure of the toughness of the material. Mild steel and copper with 20–30% plastic strains absorb large amounts of energy before breaking and so are tough materials. High-carbon steel fractures with little plastic deformation and the area under the stress–strain graph is small, so the material has a low energy density and is brittle.

Car tyres need to absorb energy as they roll over uneven surfaces and so are very tough, whereas glass and ceramic materials show very little plastic extension and are brittle.

Hysteresis in rubber

The elastic property of rubber is complex. When a rubber band is stretched and relaxed it does return to its original length, but the manner by which it does so is very different to that of a metal.

The experiment carried out to obtain a force–extension graph for a rubber band can be changed so that the band is loaded as before, but this time the force is removed a little at a time before the rubber band breaks, until the force is zero and the original length of the rubber band is regained.

Figure 13.15 shows such a variation of stress for a rubber band that is extended to four times its length before it is relaxed.

The arrows indicate the loading and unloading of the rubber band. The area under the loading curve represents the work done per unit volume *on* the band as it stretches, and the area beneath the unloading line is the work done per unit volume of the band *by* the band as it relaxes. This difference is known as **hysteresis** and the shaded area enclosed by the loading and unloading curves is called a hysteresis loop.

The hysteresis loop for a stress–strain graph represents the energy per unit volume transferred to internal energy during the load–unload cycle.

The force–extension graph for rubber is identical in shape to the stress–strain graph, but the area of the hysteresis loop now represents the total energy transferred to internal energy for each cycle.

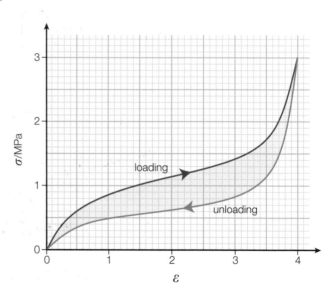

Figure 13.15 Hysteresis loop for rubber.

If a rubber band is repeatedly stretched and relaxed in a short time, it will become warm. A simple experiment to demonstrate this effect involves rapidly stretching and relaxing a rubber band several times and then placing it against your lips.

Much of the kinetic energy transferred by a moving vehicle occurs during the hysteresis loops in the tyres, which become quite warm after a journey. This is called 'rolling' resistance and is an important factor in determining the fuel consumption and hence carbon dioxide and other exhaust emissions from the vehicle.

Example

Estimate the energy per unit volume transferred to internal energy for each load–unload cycle of the rubber band used for the stress–strain graph in Figure 13.15.

Answer

About six squares are enclosed by the loop.

Each square is equivalent to $0.5\,MPa \times 0.5 = 0.25\,MJ\,m^{-3}$

shaded area $\equiv 1.5\,MJ\,m^{-3}$

A more accurate value can be obtained by counting small squares. As examination questions usually require an estimated value, such a time-consuming method is inappropriate during a timed test.

Test yourself

13 a) Define the following terms:

    i) stress

    ii) strain.

  b) State the unit of

    i) stress

    ii) strain.

14 Calculate the Young modulus for steel if a wire of length 2.20 m and diameter 0.56 mm is extended by 2.2 mm when a load of 5.0 kg is applied to it.

15 When a polymer is loaded and unloaded it undergoes hysteresis.

  a) Explain what is meant by hysteresis.

  b) What is represented by the area between the loading and unloading curves on a stress–strain graph for a polymer?

Summary of the properties of solid materials

In this section, several properties that describe the behaviour of solid materials have been introduced. Table 13.3 lists these, together with definitions and examples where appropriate.

Table 13.3

Property	Definition	Example	Opposite	Definition	Example
strong	high breaking stress	steel	weak	low breaking stress	expanded polystyrene
stiff	gradient of a force–extension graph; high Young modulus	steel	flexible	low Young modulus	natural rubber
tough	high energy density up to fracture: metal that has a large plastic region	mild steel, copper, rubber tyres	brittle	little or no plastic deformation before fracture	glass, ceramics
elastic	regains original dimensions when the deforming force is removed	steel in Hooke's law region, rubber	plastic	extends extensively and irreversibly for a small increase in stress beyond the yield point	copper, modelling clay
hard	difficult to indent the surface	diamond	soft	surface easily indented/ scratched	foam rubber, balsa wood
ductile	can be readily drawn into wires	copper	hard, brittle		
malleable	can be hammered into thin sheets	gold	hard, brittle		

Example

Graphs A, B and C in Figure 13.16 represent the variations of strain with stress for three different materials.

a) Grade the samples in terms of:

  i) strength

  ii) stiffness

  iii) toughness.

b) Suggest a possible material for each of A, B and C.

Answers

a) i) B, C, A

  ii) A, B, C

  iii) C, B, A

b) A: cast iron, glass or ceramic; B: mild steel, brass; C: rubber or other similar polymer.

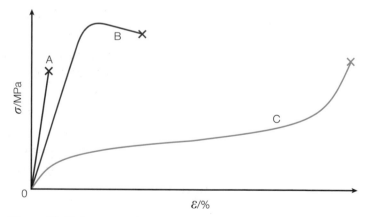

Figure 13.16 Stress–strain curves.

Exam practice questions

1 Figure 13.17 shows the stress–strain graphs for four different materials.

   a) Which material is toughest?

      A   B   C   D   **[Total 1 Mark]**

   b) Which material is strongest?

      A   B   C   D   **[Total 1 Mark]**

   c) Which material is a polymer?

      A   B   C   D   **[Total 1 Mark]**

   d) Which material is brittle?

      A   B   C   D   **[Total 1 Mark]**

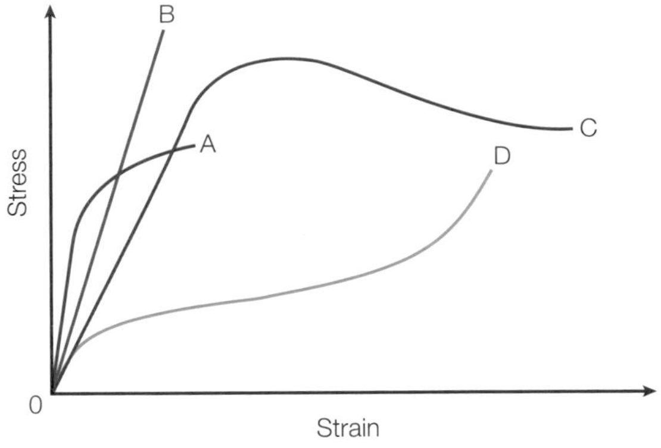

Figure 13.17

2 a) State Hooke's law. **[1]**

   b) A spring of length 10.0 cm extends to 12.5 cm when an 8.0 N weight is suspended from it. Calculate the spring constant of the spring. **[2]**

                                                           [Total 3 Marks]

3 Metals may be ductile and malleable. Explain these terms, and state two examples of metals that usually possess such properties. **[Total 2 Marks]**

4 The following is a passage from a report on the advantages and disadvantages of a stainless steel kettle compared with a plastic one. Copy and complete the passage by selecting appropriate words from the list below.

brittle dense elastic harder ions plastic polymers stronger tougher

If the plastic kettle is dropped, or receives a sharp blow, it is more likely to crack or shatter than the steel one. This is because the plastic is a _____ material. The steel is much _____ and is capable of absorbing much more energy, and may only suffer from minor indentations due to _____ deformation. The surface of the steel is _____ than the plastic and so is less likely to be scratched. The plastic kettle will retain its heat better than the steel one. It consists of long-chain molecules called _____ which give it a low thermal conductivity. The plastic is also less _____ than steel and so the kettle is lighter. **[Total 3 Marks]**

5 a) Identify on the force–extension curve for copper Figure 13.18:

     i) the yield point

     ii) the Hooke's law region

     iii) the plastic region. **[3]**

   b) How could you estimate the work done on the wire? **[1]**

   c) Describe what would happen if the load were removed:

     i) at A

     ii) at D **[2]**

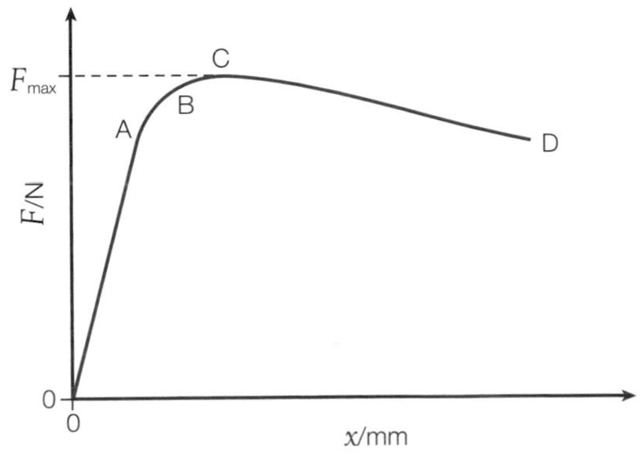

Figure 13.18

d) After removing the load at D, describe the behaviour of the copper wire when weights are added until the force again reaches F_{max}. **[3]**

[Total 9 Marks]

6 a) Define:

   i) stress

   ii) strain

   iii) Young modulus. **[3]**

b) Calculate the Young modulus of a wire of diameter 0.60 mm and length 2.00 m if a load of 50 N extends the wire by 2.5 mm. **[2]**

[Total 5 Marks]

7 The stress–strain graph in Figure 13.19 shows the hysteresis curve for a rubber band.

a) What is meant by the term hysteresis? **[1]**

b) Which line represents the loading curve? **[1]**

c) What does the area within the loop represent? **[1]**

d) Describe the changes in the molecular structure of the rubber during one loading–unloading cycle. **[6]**

[Total 9 Marks]

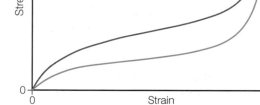

Figure 13.19

8 A copper wire and a steel wire, both of length 2.20 m, are connected side by side and securely clamped to a ceiling beam. A load of 4.00 kg is attached to the lower ends so that both wires extend by the same amount. The copper wire has a diameter of 0.56 mm and a Young modulus of 1.2×10^{11} Pa; the diameter of the steel wire is 0.46 mm and its Young modulus is 2.0×10^{11} Pa.

a) Calculate the value of the tension in both the copper and the steel wires. **[6]**

b) Determine the value of the extension of the wires. **[2]**

[Total 8 Marks]

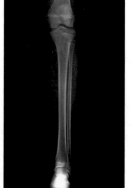

Figure 13.20a

9 Figures 13.20a and 13.20b show two of the bones in a human leg and a graph of tensile or compressive stress against strain for the bones of a 30-year-old person. The bone breaks at point X.

a) Use the straight-line part of the graph to calculate the Young modulus for the bone. **[2]**

b) What is the maximum compressive or tensile stress for the bone? **[1]**

A man of mass 70 kg jumps from a platform 2.00 m above the ground. He lands, stiff legged, onto both feet and, unfortunately, breaks the tibia bone in one leg. The impact of his feet on the ground lasts 0.30 s.

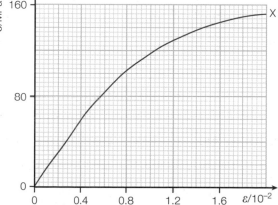

Figure 13.20b

c) Calculate the average force of impact of the ground on the man.　　**[3]**

d) If we assume that the tibia is approximately circular in cross-section, has a diameter of 25 mm and takes the bulk of the compressive force of impact, show that the compressive stress on each tibia will be about 1.5 MPa. Explain why it is unlikely that this stress has caused the fracture.　　**[3]**

e) In fact bones break more often because they are bent than because of a simple compression or tension. Draw a diagram of a bent bone and mark the regions where tension and compression would occur. **[2]**

[Total 11 Marks]

10 Figure 13.21 shows a woman in a Pilates class, exercising with a resistive stretch band.

The resistive band used is 1.50 m long, 15 cm wide and 0.4 mm thick and is made of latex.

For the exercise shown, the woman extends the band from its initial length by pushing her feet forwards and pulling back with her arms. She applies a force of 200 N to the band by increasing the distance between her hands and feet by 40 cm.

a) Show that the average value of the Young modulus of the latex is about 3 MPa.　　**[3]**

b) The Young modulus of the latex actually varies during the stretching. Explain why it changes, with reference to the molecular structure of latex.　　**[3]**

Figure 13.21

The same exercise is repeated using two identical bands, both held at each end. The woman applies the same force and the bands now extend by 30 cm.

c) Explain how this shows that Hooke's law is not obeyed when the bands are stretched.　　**[4]**

d) Does the woman do more, the same or less work during the second exercise? Give a reason for your answer.　　**[2]**

At the end of the Pilates session, the woman notices that the bands used have become slightly warmer than at the start.

e) Explain why the temperature of the bands will increase after being stretched and released many times.　　**[3]**

[Total 15 Marks]

Stretch and challenge

11 The spreadsheet shown approximately models the behaviour of a rubber band when it is loaded with increasing weights. The aim of the model is to determine the total work done on the band by adding the work done for each additional force.

	A	B	C	D	E
1	Force/N	Length of band/m	Increase in length/m	Work done/J	Total work done/J
2	0.00E+00	1.20E−01	0.00E+00	0.00E+00	0.00E+00
3	4.00E+00	1.32E−01	1.20E−02	4.80E−02	4.80E−02
4	8.00E+00	1.50E−01	1.80E−02	1.44E−01	1.92E−01
5	1.20E+01	1.78E−01	2.80E−02	3.36E−01	5.28E−01
6	1.60E+01	2.88E−01	1.10E−01	1.76E+00	2.29E+00
7	2.00E+01	3.00E−01	1.20E−02	2.40E−01	2.53E+00
8	2.40E+01	3.08E−01	8.00E−03	1.92E−01	2.72E+00
9	2.80E+01	3.14E−01	6.00E−03	1.68E−01	2.89E+00
10	3.20E+01	3.19E−01	5.00E−03	1.60E−01	3.05E+00
11	3.60E+01	3.24E−01	5.00E−03	1.80E−01	3.23E+00
12	4.00E+01	3.28E−01	4.00E−03	1.60E−01	3.39E+00

a) i) Show how cell C4 is calculated.

  ii) Show how cell D6 is calculated.

  iii) Show how cell E8 is calculated. [3]

b) The method used to determine the work done on the band does not give an accurate value. Explain the reason for this, and suggest how the cells in column D could be calculated to improve the accuracy of the values of the work done for each additional load. [3]

c) The work done on the band can also be determined from a force–extension graph. Use the values given in columns A and C to plot such a graph and use your graph to estimate the work done. [4]

d) Comment on the value of your estimate and the value given in cell E12. [2]

[Total 12 Marks]

Nature of waves

14

Prior knowledge

In this chapter you will need to:
→ be familiar with the terms frequency, period and amplitude of oscillating (vibrating) objects
→ be aware that energy from such vibrations can be transmitted in the form of waves
→ know that waves can be transverse or longitudinal
→ have an understanding of the properties of waves that make up the electromagnetic spectrum.

The key facts that will be useful are:
→ frequency is the number of cycles every second
→ period is the time for one oscillation
→ amplitude is the maximum displacement of an oscillation
→ wavelength is the distance between successive crests (or troughs) of a wave
→ wave speed = frequency × wavelength
→ electromagnetic waves can travel through a vacuum.

Test yourself on prior knowledge

1 A mass on a spring is made to oscillate up and down. The distance between the top and bottom of the oscillations is 10 cm and the mass completes 20 oscillations in 4.0 s. Calculate
  a) the amplitude   b) the frequency
  c) the period of the oscillations.
2 Give one example of
  a) a longitudinal wave   b) a transverse wave
  c) an electromagnetic wave.
3 Calculate the frequency of a wave travelling at 20 m s^{-1} if the wavelength is 80 cm.
4 State one similarity and one difference between microwaves and X-rays.

14.1 Mechanical oscillations and waves

Mechanical waves

Mechanical waves require a medium for transmission and are generated by vibrating sources. The energy from the vibrations is transferred to the medium and is transmitted by the particles within it.

The passage of a wave through a medium can be demonstrated using a 'slinky' spring. The vibrations can be applied to the spring either parallel to the direction of transmission of the wave or perpendicular to it, as shown in Figure 14.1. The movement of the 'particles' can be observed by marking one loop of the slinky with a coloured pen. The marked loop moves back and forth or side to side about its initial position.

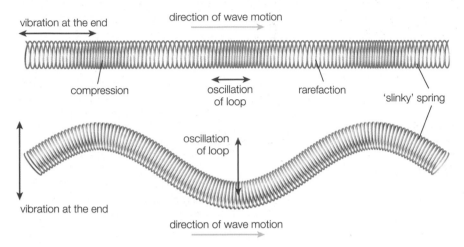

Figure 14.1 Longitudinal and transverse waves.

Key terms

A **longitudinal** wave consists of particles that oscillate parallel to the direction of propagation of the wave.

A **transverse** wave consists of particles that oscillate perpendicular to the direction of propagation of the wave.

Waves with the particles oscillating parallel to the wave motion are called **longitudinal** waves. Those with oscillations perpendicular to the direction of progression are called **transverse** waves.

Sound is transmitted using longitudinal waves. This can be demonstrated using a lighted candle placed close to a large loudspeaker connected to a signal generator, as shown in Figure 14.2. If the frequency of the signal generator is set to 50 Hz or less, the candle flame will be seen to flicker back and forth by the oscillating air molecules. It should be noted that the bulk of the air does not move with the wave, as the particles vibrate about a fixed mean position.

Close inspection of the longitudinal wave in the slinky in Figure 14.1 shows the movement of tightly packed coils followed by widely spaced sections. These are called **compressions** and **rarefactions**. For sound waves in a gas, they create high and low pressure regions (Figure 14.3).

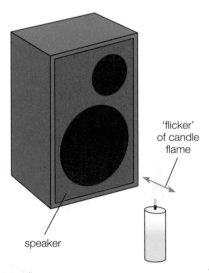

Figure 14.2 Demonstrating that sound is transmitted using longitudinal waves.

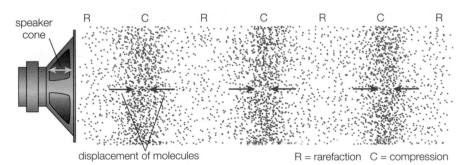

Figure 14.3 Sound waves in air.

If you throw a stone into a still pond, you will see waves on the surface spreading outwards in a circular fashion. An object floating on the water will bob up and down as the wave passes, but does not move along the wave (Figure 14.4). The water wave is an example of a transverse wave.

Activity 14.1

Wave animations

To envisage the motion of the particles in longitudinal and transverse waves is quite difficult using only static diagrams. Many animations that give a much clearer picture of the waves are available. Use a search engine to find 'wave animations' and compare several simulations. (The website phet. colorado.edu is a good starting point for a wide range of free animations.). Make a note of the site that you believe gives the best animation.

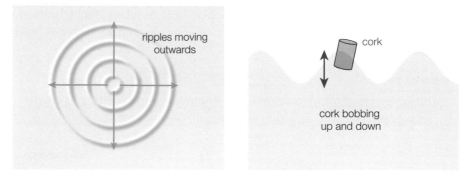

Figure 14.4 Water waves are transverse waves.

Oscillations

You have seen that mechanical waves are produced by vibrations. A detailed study of oscillations is given in the Year 2 Student's book, but a brief outline is useful to explain the particle behaviour in waves.

A simple pendulum, a mass on a spring and a rule clamped on a bench are familiar examples of oscillating systems (Figure 14.5). In all cases, the motion is repetitive about a fixed point, with the object at rest at the extremes of the motion and moving at maximum speed in either direction at the midpoint.

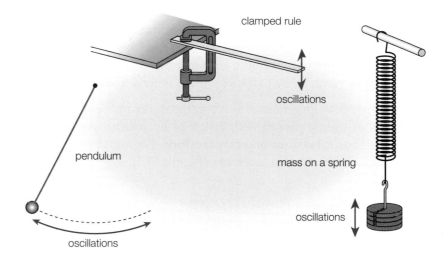

Figure 14.5 Oscillating systems.

Key terms

Amplitude is the maximum displacement from the mean position (metres, m).

Period is the time taken for one complete oscillation (seconds, s).

Frequency is the number of complete oscillations per second (hertz, Hz).

Three properties can be used to describe an oscillation. They are the **amplitude**, A, the **period**, T, and the **frequency**, f.

If an oscillator has a frequency of 10 Hz, it will complete ten oscillations every second, so it follows that each oscillation will take 0.1 s. The relationship between the frequency and the period can therefore be written as:

$$f = \frac{1}{T}$$

Displacement–time graphs for oscillating particles

If a soft marker pen is attached to a mass on a spring so that it moves up and down with the mass and a piece of white card is moved at constant speed across the tip of the pen, a trace of the position of the mass over a period of time will be drawn onto the card (Figure 14.6). A more accurate method of plotting the displacement against time can be achieved using a motion sensor with its associated software.

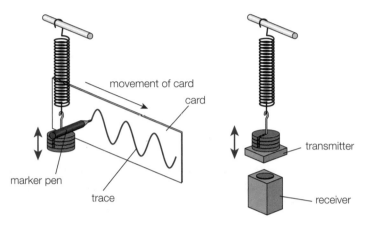

Figure 14.6 Oscillations.

The graph in Figure 14.7 shows that the amplitude of the oscillation is 2.0 cm and the period is 0.8 s. The shape of the repetitive graph is the same as that of a graph of sin θ against θ, so the motion is termed sinusoidal. This variation of displacement with time of the particles within a wave is referred to as a **waveform**.

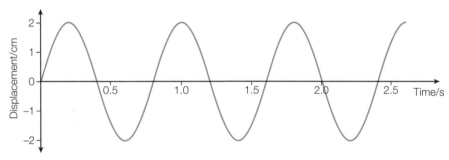

Figure 14.7

Test yourself

1 Explain the difference between a longitudinal wave and a transverse wave.

2 Figure 14.8 shows a sinusoidal wave travelling from left to right along a rope. The dashed line represents the position of the wave at $t = 0$ and the solid line is its position after a time of 50 ms.

   a) For this wave, what is
      i) the amplitude   ii) the wavelength?
   b) For this wave, what is
      i) the speed   ii) the frequency   iii) the period?

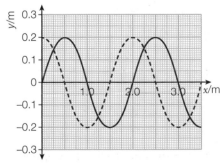

Figure 14.8

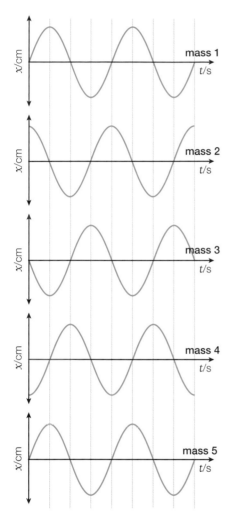

Figure 14.10 Phase differences.

Key term

The **phase** of an oscillation refers to the position within a cycle that the particle occupies relative to the onset of cycle.

Phase

Figure 14.9 shows a mass on a spring undergoing one complete oscillation and the positions of the mass after successive intervals of one quarter of a period. Halfway through the cycle, the mass is moving through the midpoint in the opposite direction to the starting position. The oscillations at these points are out of step with each other or 'out of **phase**'. Any two oscillations that are half a cycle out of step are said to be in **antiphase**. At the end of the cycle, the mass is moving through the midpoint in the same direction as at the start and the vibrations are now back in phase.

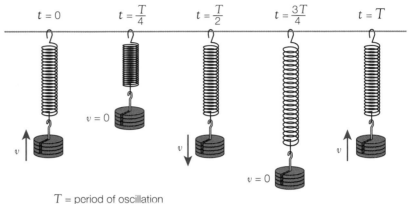

T = period of oscillation

Figure 14.9 Phases of an oscillation.

If you consider the oscillations in Figure 14.9 to be five separate masses, you will note that masses 1 and 5 are always in phase, mass 3 is in antiphase with both 1 and 5 and masses 2 and 4 are one quarter of a cycle and three quarters of a cycle out of phase with mass 1.

For a sinusoidal waveform, one cycle represents 2π radians (360°), so oscillations that are in antiphase are said to have a **phase difference** of π radians (180°), while those that differ by one quarter and three quarters of a cycle are $\frac{\pi}{2}$ radians (90°) and $\frac{3\pi}{2}$ radians (270°) out of phase, respectively. The phase relationships of masses 1–5 are represented on the displacement–time graphs in Figure 14.10.

Displacement–distance graphs

Imagine a water wave travelling along the surface of a glass-sided aquarium. A snapshot of the edge of the wave viewed through the glass would reveal a **wave profile** showing the shape of the surface at that instant. This profile shows the positions of the particles along the surface at that time and can be represented on a displacement–distance graph (see Figure 14.11).

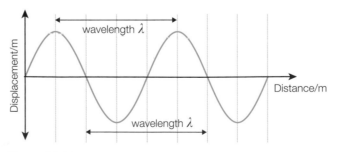

Figure 14.11 Displacement–distance graph.

Although this looks similar to the displacement–time graph, it differs in that it represents the positions of all the particles along that section of the wave as opposed to the motion of a single particle within the wave.

The vibrations of the particles along the wave are all a little out of phase with their neighbours. A Mexican wave at a sports ground illustrates this quite well. When the person to your right starts standing up, you follow a moment later. The person to your left repeats the motion after you, and so the wave moves around the stadium. After a certain length of the wave, the particles come back into phase – for example, the particles at the 'crest' of the wave are all at the top of their oscillation and those in the 'trough' are all at the lower amplitude. The distance between two adjacent positions that are in phase is the **wavelength**, λ, of the wave.

For longitudinal waves such as sound, the displacement of the particles is parallel to the motion, so a wave profile is more difficult to draw, but a displacement–position graph identical to that for transverse waves can be drawn with positive displacements to the right of the mean positions and negative displacements to the left (Figure 14.12).

Tip

It is worth remembering that

- angle in radians

  $= \dfrac{\pi}{180°} \times$ angle in degrees
- angle in degrees

  $= \dfrac{180°}{\pi} \times$ angle in radians.

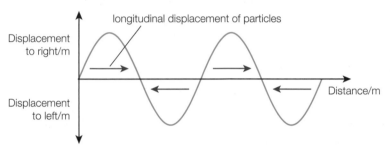

Figure 14.12 A distance–displacement graph for a longitudinal wave.

Test yourself

3 Convert the following angles to radians
  a) 180°   b) 270°   c) 57.3°   d) 120°.

4 Convert the following angles to degrees
  a) $\frac{\pi}{6}$ rad   b) $\frac{\pi}{2}$ rad   c) $\frac{5\pi}{2}$ rad   d) 3.5 rad.

5 Figure 14.13 shows two waves at an instant in time. State the phase difference in the waves at
  a) position A   b) position B   c) position C   d) position D.

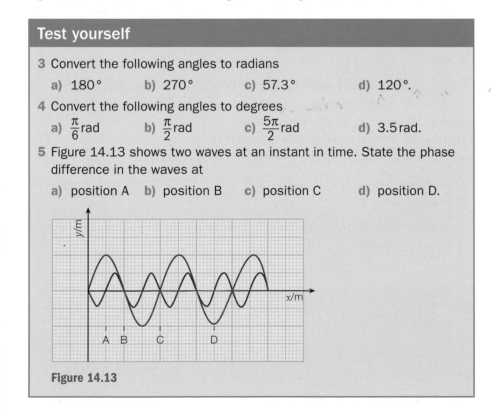

Figure 14.13

14.2 Electromagnetic waves

Mechanical waves are initiated by vibrating objects passing on some of their energy to the atoms or molecules of a material medium. Electromagnetic waves are created when charged particles are accelerated.

Radio waves are an example of electromagnetic radiation. Electrons in an aerial are made to oscillate using electronic circuitry, and this produces a wave of continuously varying electric and magnetic fields with the same frequency as the oscillator. The electric field variations are in the plane of the antenna, with the magnetic field in a plane at right angles to it, as shown in Figure 14.14.

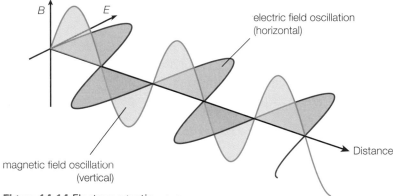

Figure 14.14 Electromagnetic wave.

Electric and magnetic fields will be studied in some detail in the Year 2 Student's book, but you should be aware that the fields do not require a material medium and that varying electric fields cannot exist without associated magnetic fields and vice versa. Electromagnetic waves can be detected when they pass over charged particles. Some of their energy is transferred to the particles, which makes them vibrate at the frequency of the wave.

Electromagnetic waves can also occur as a result of 'quantum jumps' of electrons in atoms or from 'excited' nuclei. The waves produced in this manner are emitted as **photons**, which are 'bundles' of waves of a few nanoseconds duration that have wavelengths dependent on their energy. The photon nature of light will be investigated in Chapter 17.

The complete range of electromagnetic waves and some of their properties are given in Table 14.1.

You will have come across this material during your GCSE studies, so this is a useful reminder!

Table 14.1 Waves in the electromagnetic spectrum.

Type of wave	Wavelength range/m	Method of production	Properties and applications
γ rays	10^{-16}–10^{-11}	excited nuclei fall to lower energy states	highly penetrating rays used in medicine to destroy tumours, for diagnostic imaging and to sterilise instruments
X-rays	10^{-14}–10^{-10}	fast electrons decelerate after striking a target	similar to γ rays but the method of production means that their energy is more controllable used in medicine for diagnosis and therapy and in industry to detect faults in metals and to study crystal structures
ultraviolet	10^{-10}–10^{-8}		stimulates the production of vitamin D in the skin, which results in a tan makes some materials fluoresce used in fluorescent lamps and to detect forged banknotes
visible light	4–7×10^{-7}	electrons in atoms raised to high-energy states by heat or electric fields fall to lower permitted energy levels	light focused onto the retina of the eye creates a visual image in the brain can be detected by chemical changes to photographic film and electrical charges on the charge-coupled devices (CCDs) in digital cameras essential energy source for plants undergoing photosynthesis
infrared	10^{-7}–10^{-3}		radiated by warm bodies used for heating and cooking and in thermal imaging devices
microwaves	10^{-4}–10^{-1}	high-frequency oscillators such as a magnetron background radiation in space	energy is transferred to water molecules in food by resonance at microwave frequencies used in mobile phone and satellite communications
radio	10^{-3}–10^{5}	tuned oscillators linked to an aerial	wide range of frequencies allows many signals to be transmitted groups of very large radio-telescopes can detect extremely faint sources in space

14.3 The wave equation

The speed of a travelling wave depends on the nature of the wave and the medium through which it is passing. However, for all waves there is a relationship between the speed, wavelength and frequency:

$$\text{speed, } v = \frac{\text{distance}}{\text{time}}$$

A wave will travel a distance of one wavelength, λ, in the time it takes to complete one cycle, T:

$$v = \frac{\lambda}{T}$$

In Section 14.1, you saw that the frequency and period of an oscillation were related by the expression:

$$f = \frac{1}{T}$$

It follows that:

$$v = f\lambda$$

that is, wave speed = frequency × wavelength

This is known as the **wave equation**.

Examples

1 Calculate the wavelengths of:

a) the sound from a trumpet playing a note of frequency 256 Hz

b) ultrasound of frequency 2.2 MHz passing through body tissue.

2 Calculate the frequencies of:

a) radio waves of wavelength 246 m

b) microwaves of wavelength 2.8 cm.

Answers

1 a) Speed of sound in air (at 20 °C) is 340 m s^{-1}

$$\lambda = \frac{v}{f} = \frac{340 \,\text{m s}^{-1}}{256 \,\text{Hz}} = 1.33 \text{ m}$$

b) Speed of sound in soft tissue is 1540 m s^{-1}

$$\lambda = \frac{v}{f} = \frac{1540 \,\text{m s}^{-1}}{2.2 \times 10^6 \,\text{Hz}} = 0.70 \text{ mm}$$

2 a) Speed of electromagnetic waves (in a vacuum) is 3.0×10^8 m s^{-1}

$$\lambda = \frac{v}{f} = \frac{3.0 \times 10^8 \,\text{m s}^{-1}}{246 \,\text{m}} = 1.2 \text{ MHz}$$

b) $$\lambda = \frac{v}{f} = \frac{3.0 \times 10^8 \,\text{m s}^{-1}}{0.028 \,\text{m}} = 11 \text{ GHz}$$

Test yourself

6 From which parts of the electromagnetic spectrum do waves of the following wavelengths come?

a) 5×10^{-11} m

b) 600 nm

c) 70 mm

d) 300 m

7 a) Which has the longer wavelength, blue or red light?

b) Which has the higher frequency, green light or yellow light?

8 A radio station is transmitted on two channels: 909 kHz/330 m and 693 kHz/433 m. Show that both signals are transmitted at the same speed.

9 Calculate the frequency of

a) a 3.0 cm microwave transmitter

b) 598 nm yellow light

c) X-rays of wavelength 4.2×10^{-12} m.

10 Calculate the wavelength of ultrasound of frequency 1.4 MHz when it passes through

a) soft tissue at 1540 m s^{-1}

b) bone at a speed of 4080 m s^{-1}.

Exam practice questions

1 Two oscillations are said to be in antiphase if their phases differ by:

   A $\dfrac{\pi}{2}$ radians   **C** $\dfrac{3\pi}{2}$ radians

   B π radians   **D** 2π radians.   **[Total 1 Mark]**

2 Which of the following does not apply to sound waves?

   A They always travel at $340\,\mathrm{m\,s^{-1}}$ in air.   **C** They result from vibrations.

   B They are longitudinal waves.   **D** They transmit energy.

                                                       [Total 1 Mark]

3 Which of the following does **not** apply to X-rays?

   A They are produced by fast-moving electrons striking a target.

   B They can be detected using a photographic film.

   C They have a longer wavelength than visible light.

   D They travel through a vacuum with a speed of $3 \times 10^8\,\mathrm{m\,s^{-1}}$.   **[Total 1 Mark]**

4 Microwaves of wavelength 12 cm have a frequency of:

   A 25 MHz   **C** 2.5 GHz

   B 36 MHz   **D** 3.6 GHz   **[Total 1 Mark]**

5 Define the terms amplitude, frequency and period of an oscillation. Give the SI unit for each.   **[Total 3 Marks]**

6 Describe the differences between longitudinal and transverse waves. Give one example of each.   **[Total 4 Marks]**

7 In what ways do electromagnetic waves differ from mechanical waves?   **[Total 2 Marks]**

8 The oscilloscope traces in Figure 14.15 are for two oscillations of the same frequency. On the y-axis, one division represents a displacement of 1.0 cm; the time base (x-axis) is set at 10 ms per division.

Use the traces to find:

   a) the frequency of the oscillations   **[1]**

   b) the amplitude of each oscillation   **[2]**

   c) the phase difference between the oscillations.   **[1]**

                                              [Total 4 Marks]

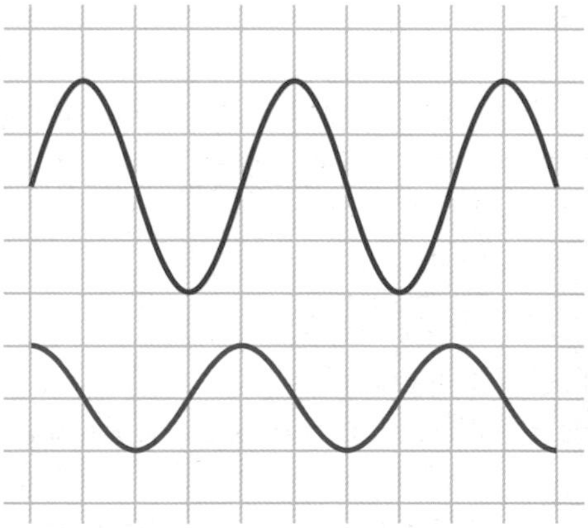

9 Explain the term wavelength of a wave. Show that speed, v, frequency, f, and wavelength, λ, are related by the expression $v = f\lambda$.   **[Total 3 Marks]**

Figure 14.15

10 **a)** Calculate:

    i) the speed of water surface waves of frequency 50 Hz and wavelength 1.5 cm   **[1]**

    ii) the wavelength of electromagnetic radiation of frequency $5.0 \times 10^{15}\,\mathrm{Hz}$ **[1]**

    iii) the frequency of microwaves of wavelength 2.8 cm.   **[1]**

b) In what region of the electromagnetic spectrum are the waves in part **ii**? [1]

[Total 4 Marks]

11 Two speakers are placed side by side. One emits a sound wave of frequency 189 Hz and the other a wave of frequency of 283 Hz. The speed of sound in air is 340 m s^{-1}.

a) Calculate the wavelengths of the sound waves emitted by each speaker. [2]

b) At one instant both waves are transmitted in phase. Determine the distance from the speakers at which these waves will next be

    i) in phase

    ii) π radians out of phase [4]

c) Determine the phase difference between waves transmitted in phase at distances of

    i) 4.5 m from the speakers

    ii) 9.9 m from the speakers [4]

[Total 10 Marks]

Stretch and challenge

12 Energy generated by earthquakes travels through the Earth's crust in the form of *primary* (P or 'Push') waves and *secondary* (S or 'Shake') waves.

a) State the usual terms for these types of wave and describe how the method of propagation differs in each case. [3]

A seismometer in a recording station detects the initial tremors from an earthquake. The time difference between the arrival of the first P wave and the first S wave is 88 s. The speed of P waves through the crust is 7800 m s^{-1} and that of the S waves is 4200 m s^{-1}.

b) Show that the distance from the epicentre of the earthquake to the recording station is about 800 km. [4]

c) Explain why, in order to determine the position of the epicentre of the earthquake, at least three recording stations are needed. [2]

The energy from earthquakes beneath the oceans can be transmitted along the surface by waves known as tsunamis. For seas of depth about 2 km these waves travel at speeds around 140 m s^{-1} with a period of 20 minutes. The speed of waves on the surface of the sea is related to the depth by the equation $c = \sqrt{(gd)}$.

d) Show that the wavelength of the tsunami changes from about 170 km in the deep water to about 20 km as it approaches land on water of depth 20 m. [6]

e) If the energy carried in each wave of the tsunami is assumed to be constant, describe the effect on the wave of this reduction in wavelength. [3]

[Total 18 Marks]

14 Nature of waves

15 Transmission and reflection of waves

> **Prior knowledge**
>
> In this chapter you will need to:
> → be aware that mechanical waves are transmitted by oscillating particles within a medium and that the medium itself does not move as the wave passes through it
> → know the difference in the mode of progression of transverse and longitudinal waves and be able to explain it in terms of the oscillating particles
> → understand that visible light is part of the electromagnetic spectrum.
>
> The key facts that will be useful are:
> → wave speed = frequency × wavelength
> → electromagnetic waves travel through a vacuum with a speed of $3.0 \times 10^8 \, \mathrm{m\,s^{-1}}$
> → electromagnetic waves consist of transverse variations of electric and magnetic fields.

Test yourself on prior knowledge

1 Explain the difference between longitudinal and transverse waves.
2 Why can't sound waves pass through a vacuum?
3 Why is the lightning flash from a distant storm seen several seconds before the clap of thunder is heard?
4 What is ultrasound?
5 Calculate the wavelength of a sound of frequency 256 Hz travelling through air at a speed of $340 \, \mathrm{m\,s^{-1}}$.
6 Calculate the frequency of ultraviolet light of wavelength 360 nm.
7 The speed of microwaves is reduced when they pass through a block of wax. Explain what effect this has on the wavelength of the waves.

15.1 Transmission

In Chapter 14, you saw that waves can transfer energy from a source to an observer. In this chapter you will look into the ways different types of wave are affected by the medium through which they travel and what can occur when a wave is incident on an interface between different media.

You will study the properties reflection, refraction and polarisation of waves and how these are used in the food industry and for ultrasound imaging in medicine.

Longitudinal waves

Longitudinal waves progress by the interaction of particles oscillating along the direction of travel of the waves. For sound in air, the successive compressions and rarefactions were shown in Section 14.1 by the candle flame flickering in front of the loudspeaker.

The method of energy transfer can be demonstrated using mechanics trolleys connected together by springs, as shown in Figure 15.1. A pulse applied to the first trolley is transferred from trolley to trolley by the springs and is seen to move along the line of the trolleys.

Activity 15.1

Transmission of pulses through trolleys

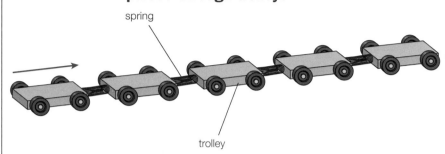

Figure 15.1

The factors affecting the speed of transmission of a longitudinal wave can be investigated using the arrangement shown in Figure 15.1. The speed of the pulse is calculated by dividing the distance between the first and last trolley by the time it takes to travel from start to finish.

The experiment is repeated using two springs in parallel between each trolley and then doubling the mass of each trolley by putting a second (identical) trolley on top of all of them.

Questions
1 How will the spring constant of the coupling be affected by using two springs in parallel?
2 The results of such an experiment are shown in Table 15.1.

Table 15.1

Arrangement	Time/s
one trolley, one spring	1.1; 1.3; 1.2; 1.1
one trolley, two springs	0.9; 0.8; 0.8; 0.9
two trolleys, one spring	1.8; 1.7; 1.7; 1.7

The distance travelled by the pulse = 1.20 m.
  a) Calculate the speed of the pulse for each arrangement.
  b) Suggest a possible relationship between the speed of the pulse, the mass of the trolley and the spring constant of the coupling.
  c) Determine which set of readings will give the greatest uncertainty in the value of the speed.

You will probably already have measured the speed of sound in air by timing the delay between the flash of a gun and the bang over a measured distance. The value is about $340\,\mathrm{m\,s^{-1}}$ at $20\,°C$. For solids, such as steel, the value is much bigger as strong bonds between the particles enable the energy to be transmitted more effectively.

Core practical 6

Finding the speed of sound in air using a 2-beam oscilloscope

A signal generator is connected to a speaker and to one of the inputs of a double-beam cathode ray oscilloscope (CRO). A microphone, attached to a rectangular block, is placed in front of the speaker on a metre rule. The output from the microphone is connected to the other input of the CRO. When the signal generator is switched on to give an audible sound from the speaker, the sensitivity and time base settings of the oscilloscope are adjusted so that two waveforms are displayed on the screen. One wave represents the sound given off by the speaker and the other that received by the microphone.

If the microphone is moved towards or away from the speaker, the waveforms will move relative to each other showing the difference in phase of the sound wave emitted by the speaker and that received by the microphone.

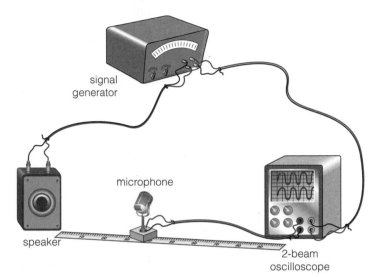

Figure 15.2 Speed of sound in air.

The microphone is placed close to the speaker and then moved away until the waveforms are in antiphase. This can be seen more easily if the two traces are superimposed on each other. The position of the microphone on the rule is noted. The microphone is moved slowly away from the speaker until the waveforms are again in antiphase. The scale reading is taken so that the distance moved by the microphone between the two positions is found. This distance represents one complete wavelength.

This is repeated for a range of frequencies so that a table of frequency and the associated wavelength can be drawn.

The wave equation, $c = f\lambda$, can be rearranged as $\lambda = \dfrac{c}{f}$. A graph of λ against $\dfrac{1}{f}$ is drawn.

The gradient of the graph will give the magnitude of the speed of sound in air.

Questions

A set of readings taken by a student is shown in Table 15.2.

Table 15.2 Experimental results.

Frequency/Hz	400	500	600	700	800
Wavelength/m	0.850	0.680	0.567	0.486	0.425

1 When the frequency was 800 Hz, the student observed that there were four complete waves shown across the 10 divisions drawn on the screen. Determine the time base setting. Give your answer in μs per division.

2 Copy and complete the table to show the values of $\dfrac{1}{f}$.

3 Plot a graph of λ/m against $\dfrac{1}{f}$/s

4 Use the graph to calculate the value of the speed of sound in air.

5 It is suggested that it would be beneficial to measure the distance between two or more positions of antiphase, particularly at higher frequencies. Give a reason why this may be so, and what adjustments would need to be made to the oscilloscope controls.

Transverse waves

How does a particle moving sideways enable a wave to progress perpendicularly to the oscillation? This question can be answered using the model of masses connected by springs and the resolution of forces discussed in Chapter 14.

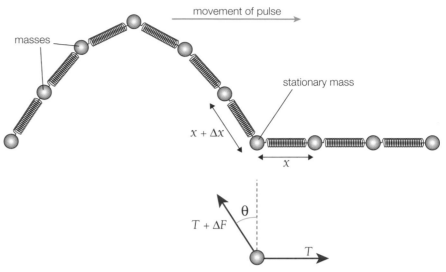

Figure 15.3

Vertical component of tension $= (T + \Delta F) \cos \theta$

Vertical component of 'extra' tension $= \Delta F \cos \theta$

There is an increase in tension between the displaced mass and the stationary mass just ahead of it. Both experience a force that can be represented by two components at right angles. The stationary particle is pulled upwards by the component of the extra tension ($\Delta F \cos \theta$). An analysis using Newton's laws of motion (too complex to include here) shows that the pulse moves along the row of masses with a speed dependent on the magnitude of the masses and the tension in the springs.

The speed of a transverse wave moving along a stretched wire is given by

$$c = \sqrt{\frac{T}{\mu}}$$

where T is the tension in the wire and μ is its mass per unit length.

This relationship is important when studying the factors that affect the frequency of stringed instruments, which will be investigated in Chapter 16.

Water waves are more complex, because the particles near the surface rotate so that up and down motion of the surface will progress. The speed of surface waves also depends on the elastic properties and mass of the particles, as well as the depth of the liquid.

Activity 15.2

Studying the relationship between the speed of surface waves and the depth of the water

Pour water into a tray to a depth of one centimetre (Figure 15.4). A pulse can be given to the water by rapidly raising and lowering one end of the tray. Measure the time taken for the pulse to travel to the end of the tray and back again. Find the length of the tray and calculate the speed of the wave. Repeat several times to find an average speed and take further measurements for a range of depths.

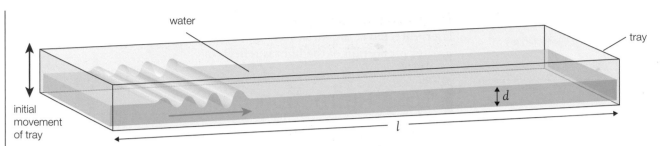

Figure 15.4 Speed of water waves.

Questions

It is suggested that the speed of water waves can be found using the equation $c = \sqrt{(gd)}$, where d is the depth of water and g is the gravitational field strength.

1 Use the results of the experiment, or those given in Table 15.3, to plot a suitable graph to check the validity of the equation.

2 Use the graph to obtain a value for g.
Length of tray = 82 cm

Table 15.3

d/cm	2.0	4.0	6.0	8.0	10.0
t/s	3.7	2.6	2.1	1.8	1.7

Electromagnetic waves

Unlike mechanical waves, electromagnetic waves do not require a medium for their transmission. All the regions of the electromagnetic spectrum travel through a vacuum with a speed of $3 \times 10^8 \, \text{m s}^{-1}$. The radiation interacts with charged particles and it may be reflected, absorbed or transmitted through different media. Light, for example, can pass through glass and some plastics. The speed of the light is reduced by its interaction with electrons and is dependent on the atomic structure of the medium. The frequency of the wave affects its speed, with blue light travelling more slowly than red light in media other than a vacuum.

Some materials reflect or absorb a range of wavelengths but transmit others. For example, a blue stained-glass window transmits blue light and absorbs light of all other colours.

The greenhouse effect is due to the upper atmosphere being transparent to short-wave infrared radiation from the Sun but reflecting back the longer wavelength rays that are re-radiated from the warm surface of the Earth.

The power per unit area of the radiation falling on to a surface is known as the **intensity** of the radiation. This is also referred to as the radiant flux, especially when considering radiation received on the Earth from stars. Radiant flux will be explained more fully in the space topic in the Year 2 Student's book.

> **Key term**
>
> $\text{intensity} = \dfrac{\text{power}}{\text{area}}$ or $I = \dfrac{P}{A}$ W m^{-2}

15.2 Reflection

When a wave is incident on an interface between two different media, the energy may be absorbed, transmitted or reflected. The direction of the reflected wave is given by the **law of reflection**.

The fraction reflected depends on the nature of the two media and the angle at which the wave strikes the interface.

> **Key term**
>
> The **law of reflection** states that the angle between the incident ray and a normal drawn at the point of reflection is equal to the angle between the reflected ray and the normal, in the plane of the reflection.

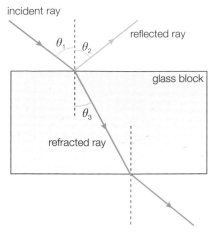

incident ray
reflected ray
θ_1 θ_2
glass block
θ_3
refracted ray

Figure 15.5

Key term

A **wavefront** is a line, or surface, in a wave, along which all the points are in phase.

Figure 15.5 shows the paths taken by a ray of light (for example, from a laser) directed at the surface of a glass block. As the angle is increased, a greater proportion of the light is reflected.

15.3 Refraction

Figure 15.5 illustrates how a ray of light at an angle to the normal changes direction when it passes from one medium to another. This effect is due to the change in wave speed and is known as **refraction**. The cause of the deviation can be explained using a ripple tank as shown in Figure 15.6. Waves are generated on the surface of water in a flat-bottomed glass dish using an oscillating horizontal bar just touching the surface of the liquid.

In Figure 15.6, a series of straight ripples move away from the oscillator. A shadow image of the ripples can be observed on a screen under a tank. Each bright line represents the crest of a wave, so all points along the line are in phase. Such a line is called a wavefront. The distance between successive bright lines is therefore one wavelength. If a strobe lamp of frequency close to or equal to that of the oscillator is used, the movement of the wavefronts can be slowed down or frozen.

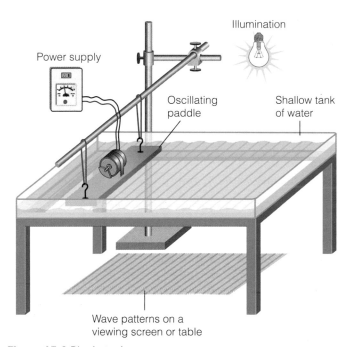

Illumination

Power supply

Oscillating paddle

Shallow tank of water

Wave patterns on a viewing screen or table

Figure 15.6 Ripple tank.

A piece of clear Perspex® placed in the dish creates a region of shallow water above it. The waves travel more slowly over the shallow water and as the waves have a fixed frequency, the wavelength is reduced $\left(\lambda = \frac{v}{f}\right)$.

The reduction in wavelength causes the wavefronts to change direction (Figure 15.7). As the wavefronts are perpendicular to the motion, the path of the waves is deviated towards the normal when the speed is reduced and away from the normal when it is increased.

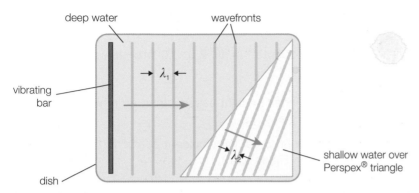

Figure 15.7 Refraction of wavefronts.

Test yourself

1 Demonstrate that the equations $c = \sqrt{\dfrac{T}{\mu}}$ for waves in a string and $c = \sqrt{(gd)}$ for surface water waves both show that the unit of c is m s^{-1}.

2 Suggest a reason that the speed of sound in a steel rod is greater than that in a lead rod.

3 Figure 15.8 shows the progression of circular waves on the water in a ripple tank up to the boundary where they meet a region of shallow water. Some of the wave is reflected from the interface and the rest continues to travel over the shallow water.

Draw diagrams showing

a) the wavefronts of the reflected wave

b) the wavefronts of the refracted wave.

Figure 15.8

Refraction of light

For all refracted waves the path is deviated *towards the normal* when the wave is *slowed down* and *away from the normal* when the speed *increases*. In Figure 15.5, the light entering the glass block bends towards the normal on entering the block and away from the normal on leaving. The light travels more slowly in the glass than in the air. The size of the deviation of the wavepath depends on the relative speeds in the two media. The ratio of the speeds is called the **refractive index** between the media:

refractive index from medium 1 to medium 2 $= \dfrac{\text{speed in medium 1}}{\text{speed in medium 2}}$

$$_1n_2 = \frac{v_1}{v_2}$$

Analysis of the wavefront progression shows that the ratio of the speeds is equal to the ratio of the incident angle and the refracted angle:

$$_1n_2 = \frac{\sin \theta_1}{\sin \theta_2}$$

This is known as **Snell's law**.

> **Key term**
>
> **Snell's law** states that the refractive index for a wave travelling from one medium to another is given by the expression:
>
> $$_1n_2 = \frac{\sin \theta_1}{\sin \theta_2} = \frac{v_1}{v_2}$$

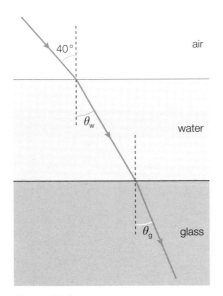

Figure 15.9

Examples

The speed of light is $3.0 \times 10^8 \, \text{m s}^{-1}$ in air, $2.3 \times 10^8 \, \text{m s}^{-1}$ in water and $2.0 \times 10^8 \, \text{m s}^{-1}$ in glass.

1 Calculate the refractive index for light passing from air to water, air to glass and from water to glass.

2 Calculate the angles θ_w and θ_g for light incident at $40°$ to the normal at the air–water interface in Figure 15.9.

Answers

1 $\quad _a n_w = \dfrac{3.0 \times 10^8 \, \text{m s}^{-1}}{2.3 \times 10^8 \, \text{m s}^{-1}} = 1.3$

$\quad _a n_g = \dfrac{3.0 \times 10^8 \, \text{m s}^{-1}}{2.0 \times 10^8 \, \text{m s}^{-1}} = 1.5$

$\quad _w n_g = \dfrac{2.3 \times 10^8 \, \text{m s}^{-1}}{2.0 \times 10^8 \, \text{m s}^{-1}} = 1.2$

2 $\quad 1.3 = \dfrac{\sin 40°}{\sin \theta_w}$

$\quad \theta_w = 30°$

$\quad 1.2 = \dfrac{\sin 30°}{\sin \theta_g}$

$\quad \theta_g = 25°$

Refraction occurs for all waves. Sound can be deviated as it passes from warm air to cooler air and microwaves can be refracted by wax. It is simpler to demonstrate the refraction of light, however, so the bulk of the applications and effects discussed in this section will relate to the visible region of the electromagnetic spectrum.

In most cases, we observe the effects of light passing across an interface between air and the refracting medium. It is convenient to ignore any reference to the air and state the value as the refractive index of the material (sometimes called the **absolute refractive index**).

The angle of incidence is usually represented by i and the angle of refraction by r, so Snell's law gives the refractive index of a medium by the expression:

$$n = \frac{\sin i}{\sin r} = \frac{c}{v}$$

where c is the speed of light in a vacuum (or air) and v is the speed in the medium.

For light travelling from a medium of refractive index n_1 to one of refractive index n_2 at angles θ_1 and θ_2 (Figure 15.10), a more general expression for Snell's law can be derived using the speeds v_1 and v_2:

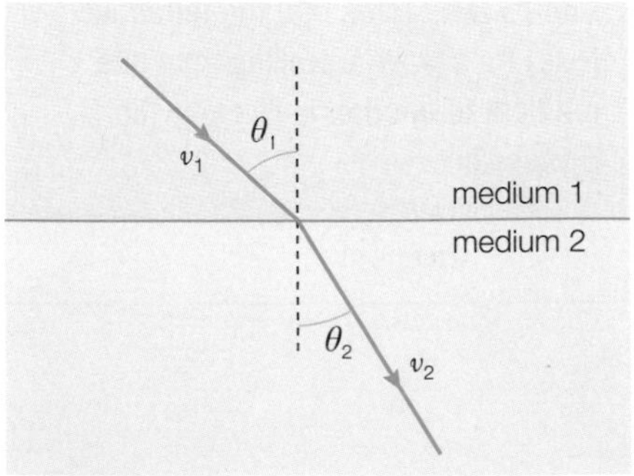

$$n_1 = \frac{c}{v_1} \qquad n_2 = \frac{c}{v_2}$$

where c = speed of light in air.

$$\frac{n_2}{n_1} = \frac{v_1}{v_2} = {}_1 n_2 = \frac{\sin \theta_1}{\sin \theta_2}$$

$$n_1 \sin \theta_1 = n_2 \sin \theta_2$$

Figure 15.10

This is a very useful alternative equation for snell's law,

Activity 15.3

Measuring the refractive index of glass

The glass block in Figure 15.11 is placed on a sheet of plain paper. By tracing the rays through the block, a range of values of i and the corresponding values of r are measured. The gradient of a graph of $\sin i$ against $\sin r$ is the average value of the refractive index of the glass.

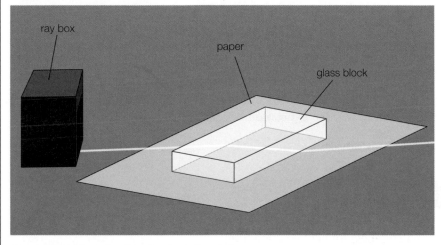

Figure 15.11 Measuring the refractive index of glass.

Total internal reflection

When light travels from glass to air or glass to water, it speeds up and bends away from the normal. The passage of light from a medium with a high refractive index to one with a lower refractive index is often referred to as a 'dense-to-less-dense' or 'dense-to-rare' transition. This relates to optical density, which is not always equivalent to physical density.

If the angle of incidence at the glass–air interface is increased, the angle of refraction will approach $90°$. If the angle is increased further, no light can leave the glass and so it is all reflected internally according to the laws of reflection (Figure 15.12).

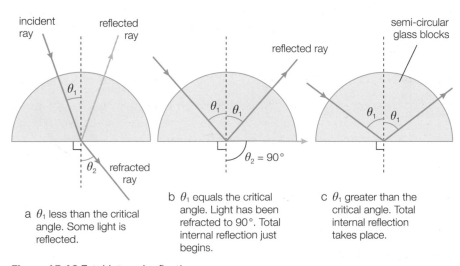

a θ_1 less than the critical angle. Some light is reflected.

b θ_1 equals the critical angle. Light has been refracted to $90°$. Total internal reflection just begins.

c θ_1 greater than the critical angle. Total internal reflection takes place.

Figure 15.12 Total internal reflection.

Key term

The **critical angle** is the angle of incidence above which total internal reflection occurs.

The angle at which **total internal reflection** just occurs is termed the **critical angle**.

Applying the general form of Snell's law:

$$n_1 \sin \theta_1 = n_2 \sin \theta_2$$

where n_1 = refractive index of glass, n_2 = refractive index of air = 1, θ_1 = critical angle, C, and $\theta_2 = 90°$, leads to:

$$\frac{n_1}{n_2} = \frac{\sin \theta_2}{\sin \theta_1}$$

Hence, for a glass–air interface:

$$n_1 = \frac{1}{\sin C}$$

For light travelling from a medium of refractive index n_1 to one of lower refractive index n_2, the expression becomes

$$\frac{n_2}{n_1} = \frac{\sin C}{\sin 90°} \rightarrow \sin C = \frac{n_2}{n_1}$$

Example

The refractive index is 1.50 for glass and 1.33 for water. Calculate the critical angle for light passing from:

a) glass to air

b) water to air

c) glass to water.

Answers

a) $\sin C = \dfrac{1}{1.50}$ $C = 42°$

b) $\sin C = \dfrac{1}{1.33}$ $C = 49°$

c) $\sin C = \dfrac{1.33}{1.50}$ $C = 62°$

Total internal reflection plays a big part in our lives. Figure 15.13 shows light being reflected back using a prism and the passage of a ray of light through a glass fibre. The prism principle is used for the reflection of light from car headlamps (cat's eyes) and glass fibre is used extensively in communications – a topic that will be studied in detail in Year 2 Student's book.

Tip

The passage of light from one medium to another is reversible. So, when light strikes a glass surface at grazing incidence, i.e. at an angle of almost 90 °, the angle of refraction will be equal to the critical angle.

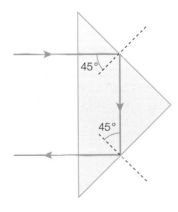

45° is greater than the critical angle so total internal reflection occurs at both surfaces.

Figure 15.13

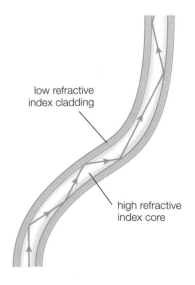

low refractive index cladding

high refractive index core

Activity 15.4

Experiment to measure the refractive index of a liquid

It is difficult to trace a ray of light through a liquid without the light having to pass through a container. A method involving the measurement of the critical angle at a glass–liquid interface may be used (Figure 15.14).

A mark is drawn on one edge of a glass block. The block is placed on a table and the mark is covered with a layer of liquid as shown in Figure 15.14. In practice the liquid layer will be much smaller than shown in the diagram.

When the mark is viewed through the bottom end edge of the block it is clearly seen because light passing through the liquid (and the mark) enters the block and exits at the viewing edge as shown in diagram (a).

As the eye is moved upwards, it is seen that the mark 'vanishes' at a point P shown in diagram (b). If the position of the eye is raised further, the mark will not be seen.

This is because, as the angle of incidence of light from the liquid into the glass increases, the refracted angle increases. When the angle of incidence is 90 °, the refracted ray enters at the critical angle as seen in diagram (b).

When the eye looks into the block at positions higher than P, no light will have entered through the liquid, and the mark cannot be seen. The only light that will emerge from the edge is that from the opposite side that has hit the upper surface at an angle greater than the critical angle and has been totally internally reflected (diagram (c)).

The position of P is marked on the edge of the block. By tracing round the block and marking the positions of the marks on to a piece of paper, the critical angle C and the angles i and r can be measured.

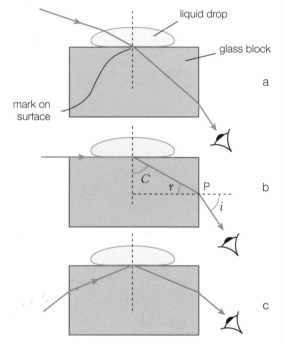

Figure 15.14

Using the values of i and r, the refractive index of the glass is found using $n_g = \dfrac{\sin i}{\sin r}$. The refractive index of the liquid is then calculated using C:

$$n_l = n_g \sin C$$

This is the principle of a **refractometer** – a device that is used in the food industry to determine the concentrations of sugar solutions, for example, by measuring the refractive index.

Questions

1 When pure water is used in the above experiment, values of i = 48 °, r = 29 ° and C = 61 ° are obtained. Use these values to calculate the refractive index of water.
2 When a sugar solution is used, the light emerges from the block at P at an angle of 31 ° to the normal. Determine the refractive index of the sugar solution.

Test yourself

4 State Snell's law of refraction.
5 a) Calculate the speed of light in:
   i) ethanol (n = 1.37)
   ii) cellulose (n = 1.46)
   iii) diamond (n = 2.39).

b) Calculate the angles of refraction for light incident at 30° at the following interfaces:

    i) air to cellulose

    ii) ethanol to cellulose

    iii) diamond to ethanol.

6 What is meant by the critical angle?

7 Calculate the critical angle for light travelling from

    a) diamond to air

    b) cellulose to ethanol

    c) diamond to ethanol.

15.4 Lenses

Lenses focus light by refraction. The surfaces of lenses are shaped so that parallel rays of light passing through the lens will either converge to a single point or diverge from a single point. Figure 15.15 shows the passage of parallel rays through a **converging** and **diverging** lens.

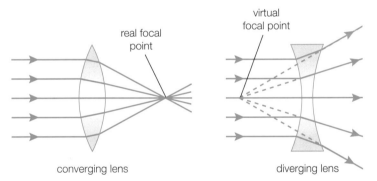

Figure 15.15 Converging and diverging lenses.

The points where parallel rays meet or appear to diverge from are called the focal points of the lenses.

Figure 15.15 shows how a converging lens forms a **real focus** where the rays actually meet. The focal point for the diverging lens is on the same side as the source of the rays and forms a **virtual focus** where no rays intersect.

The power of a lens

The power of a lens relates to the ability of the lens to deviate rays of light through large angles. A powerful lens will have a focal point that is close to the centre; it will have a short focal length.

The power of a lens is not related to the rate of conversion of energy and is given the unit **dioptre** (m^{-1}).

A lens of focal length 0.25 m has a power of 4.0 dioptres.

The power of a converging lens is given a positive value while that of a diverging lens is always negative. If a combination of two or more lenses is used, the total power, P_c is the sum of the powers of the lenses:

$$P_c = P_1 + P_2 + P_3$$

> ### Key term
>
> The **focal point** of a lens is the point where parallel rays of light will meet, or appear to diverge from, after passing through the lens.

> ### Key terms
>
> The **focal length** of a lens is the distance between the optical centre of the lens and the focal point.
>
> The **power of a lens** is $\dfrac{1}{\text{focal length}}$;
>
> $$P = \dfrac{1}{f}$$
>
> when f is in metres.

Ray diagrams

We can see objects because they give off light, either by direct emission or by scattering or reflecting light that falls on them. When some of this light passes through a lens, an **image** of the object is formed. The image can be **real** or **virtual**.

A real image is formed by the actual intersection of rays of light and so can be projected onto a screen. A virtual image can only be seen when looking through the lens and appears to be at the point where the rays originate.

The positions, size and nature of images produced by lenses can be found by drawing ray diagrams. There are three predictable rays from a point on an object that pass through a lens. These are shown in Figure 15.16 and are used to determine the position of the image of that point.

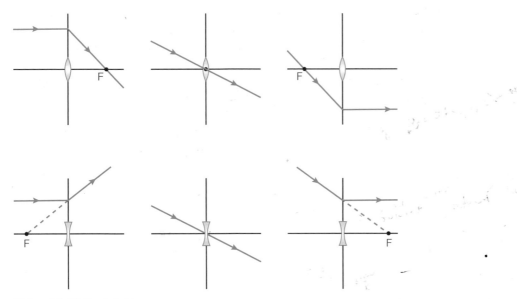

Figure 15.16 Predictable rays.

Ray diagrams are drawn using the predictable rays from a single point on the object (usually the top) so that the position of the image will be where the rays meet, or appear to originate from, after passing through the lens. It is usual to represent the object as an arrow with its base on the **principle axis** with the head vertically above it. A ray of light from the foot of the arrow along the principle axis will pass through the **optical centre** of the lens without deviation, so it follows that the image of this point will be somewhere on the principle axis.

Key terms

The **principle axis** is a line that passes normally through the optical centre of the lens.

The **optical centre** of a lens is the point through which rays of light will pass without deviation.

Examples

1 Draw ray diagrams to determine the position, size and nature of the image formed by a converging lens of focal length 10 cm when an object of height 2.0 cm is placed the following distances from the lens:

   a) 5.0 cm   b) 15 cm   c) 50 cm.

2 Draw a ray diagram to determine the position, size and nature of the image formed by a diverging lens of focal length 10 cm when an object of height 2.0 cm is placed 20 cm from the lens.

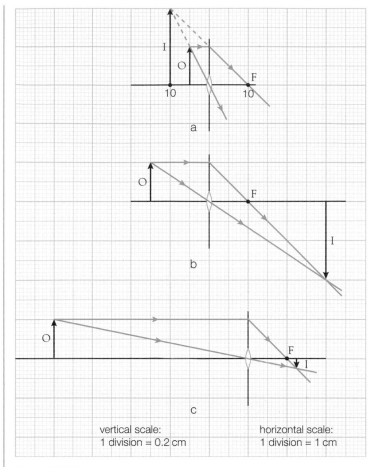

vertical scale:
1 division = 0.2 cm

horizontal scale:
1 division = 1 cm

Figure 15.17

Answers

(see Figure 15.17.)

1 a) The image is 10 cm from the lens, 4.0 cm high, virtual and erect.

b) The image is 30 cm from the lens, 4.0 cm high, real and inverted.

c) The image is 13 cm from the lens, 0.5 cm high, real and inverted.

2 The image is 7 cm from the lens, 0.7 cm tall, virtual and erect (Figure 15.18).

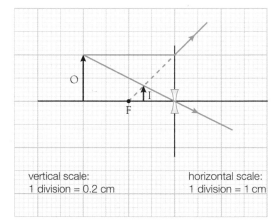

vertical scale:
1 division = 0.2 cm

horizontal scale:
1 division = 1 cm

Figure 15.18

The lens equation

The distance of the object from the optical centre of the lens (the object distance) is represented as u, the image distance as v and the focal length as f, as shown in Figure 15.19.

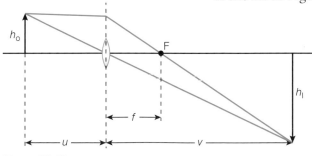

Figure 15.19

Using the similar triangles in Figure 15.19, the following ratios can be deduced:

$$\frac{h_O}{u} = \frac{h_I}{v} \quad \text{and} \quad \frac{h_O}{f} = \frac{h_I}{v-f}$$

The linear magnification of the image, m, is the ratio of the size of the image to the size of the object.

$$m = \frac{h_I}{h_O} = \frac{v}{u} = \frac{v-f}{f}$$

Rearranging the above gives us **the lens equation**:

$$\frac{1}{u} + \frac{1}{v} = \frac{1}{f}$$

The lens equation can be used to determine the image position for any type of lens if the **real is positive** sign convention is used. All distances associated with real images and focal points are given positive values and those associated with virtual images and focal points are negative.

It follows that converging lenses will have a positive value for focal length and for power, and the values for the focal length and power of a diverging lens will be negative.

Example

Find the position, nature and linear magnification of the image produced when an object is placed 30 cm from

a) a converging lens of focal length 20 cm

b) a diverging lens of focal length –20 cm.

Answers

a) $\dfrac{1}{v} = \dfrac{1}{f} - \dfrac{1}{u} = \dfrac{1}{20\,\text{cm}} - \dfrac{1}{30\,\text{cm}} = \dfrac{1}{60\,\text{cm}}$

giving $v = +60$ cm, and so it is a real image.

$m = \dfrac{v}{u} = \dfrac{60\,\text{cm}}{30\,\text{cm}} = 2.0$

The image is 2.0 times the size of the object.

b) $\dfrac{1}{v} = \dfrac{1}{f} - \dfrac{1}{u} = \dfrac{1}{-20\,\text{cm}} - \dfrac{1}{30\,\text{cm}} = \dfrac{-5}{60\,\text{cm}}$

giving $v = -12$ cm, and so it is a virtual image.

$m = \dfrac{v}{u} = \dfrac{-12\,\text{cm}}{30\,\text{cm}} = -0.4$

The image is 0.4 times the size of the object.

Activity 15.5

Experiment to find the focal length of a converging lens

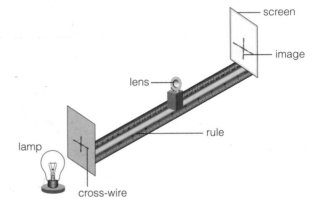

Figure 15.20

> **Tip**
>
> It is helpful to have some idea as to the value of the focal length before you start the experiment proper so that you don't waste time with values of u that would give virtual images. This can be done by focusing a distant object (e.g. a ceiling light) onto the screen and measuring the distance from the lens to the screen. As the rays from the light are approximately parallel, this distance will give a rough idea of the focal length.

A converging lens is held in a holder and placed between an illuminated cross-wire and a screen as shown in Figure 15.20.

With the background lighting subdued, the cross-wire, lens and screen are adjusted so that a clear image of the cross-wire is seen on the screen. A metre rule is placed alongside the arrangement so that the distance from the cross-wire to the lens (the object distance, u) and the distance of the lens to the screen (the image distance, v) can be measured.

> **Tip**
>
> A good technique of determining v is to move the screen *towards* the lens until a sharp image is obtained and noting this point on the metre rule. Then move the screen nearer the lens so that the image goes out of focus. Now move the screen *back* until a sharp image is obtained again and note this point. Both readings should be recorded and the average taken. Hence determine the distance to the *centre* of the lens to find the value of v.

The object distance is set over a range of values and the corresponding values of image distance are measured.

A graph of $\dfrac{1}{v}$ against $\dfrac{1}{u}$ is plotted.

The intercepts on both axes are equal to $\dfrac{1}{f}$ since when $\dfrac{1}{u} = 0$, $\dfrac{1}{v} = \dfrac{1}{f}$, from the lens equation.

Questions

The results of an experiment using a particular lens are shown in Table 15.4.

Table 15.4

u/cm	30.0	35.0	40.0	45.0	50.0	60.0
v/cm	59.8	47.1	40.0	35.9	33.4	29.7

1 Use the values of u and v to plot a graph of $\frac{1}{v}$ against $\frac{1}{u}$.

2 Use the graph to find a value of the focal length of the lens.

3 Suggest any special methods or techniques that could be used to reduce the uncertainty in the measurements of the object and image distances.

Test yourself

8 Describe how you would estimate the value of the focal length of a converging lens if the only equipment you have is a rule.

9 Explain the difference between a real and a virtual image.

10 a) Calculate the power of

   i) a converging lens of focal length 20 cm

   ii) a diverging lens of focal length 50 cm.

  b) If the two lenses in part **a** are combined to form a single lens, calculate

   i) the total power of the combination

   ii) the focal length of the combined lenses.

11 Sketch ray diagrams to show how the image is formed by

  a) the lens of a camera

  b) a magnifying glass.

  In each case state if the image is real or virtual.

12 Calculate the position, linear magnification and nature of the images formed by a converging lens of focal length 15 cm for an object placed

  a) 10 cm from the lens

  b) 20 cm from the lens.

15.5 Polarisation

Transverse waves with the particles oscillating in one plane only are said to be plane polarised. If a rope is fixed at one end, a wave can be passed along it by an up-and-down or side-to-side motion of the hand (Figure 15.21). The waves will lie in a vertical or horizontal plane and are said to be vertically or horizontally **plane polarised**.

The plane of polarisation can be varied by moving the hand in different directions.

If a rope is passed through a vertical slit in a wooden board (Figure 15.22), horizontal vibrations would be stopped and only vertical ones transmitted. Such an arrangement acts as a **polarising filter**.

Key term

Plane polarised waves are transverse waves in which the oscillations occur in a single plane.

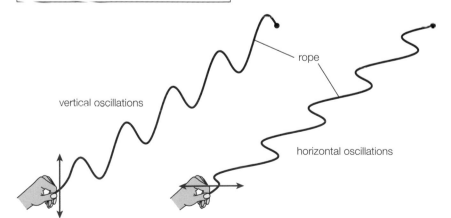

vertical oscillations

rope

horizontal oscillations

Figure 15.21 Polarised waves.

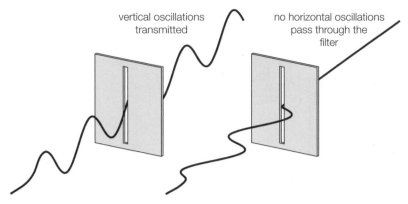

vertical oscillations transmitted

no horizontal oscillations pass through the filter

Figure 15.22 Polarising filter.

Longitudinal waves have particles that have no components of oscillations in the planes perpendicular to the direction and so cannot be polarised. A sound wave, for example, would pass through the slit whatever its orientation.

Radio waves and microwaves transmitted from an aerial have variations in the electric field in the plane of the aerial and in the magnetic field at right angles to it. To pick up the strongest signal, the receiving aerial must be aligned in the same plane as the transmitter (Figure 15.23). When the receiver is rotated, the strength of the received microwaves falls until it reaches zero when it is at right angles to the transmitting aerial. The receiver will only pick up the component of the signal in the plane of its antenna.

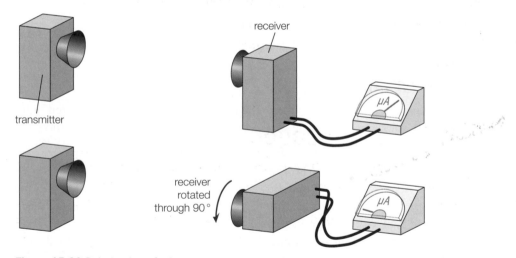

receiver

transmitter

receiver rotated through 90°

Figure 15.23 Polarisation of microwaves.

Polarised light

Most sources of light are unpolarised. The light emitted from the Sun or a lamp consists of variations of electric and magnetic fields in all planes.

Light may be polarised using filters that behave like the slit for the rope and by reflection (Figure 15.24). The filter is usually made of transparent polymers with the molecular chains aligned in one direction. If two filters are held together and rotated until the transmitting planes are at right angles, a source of light viewed through them will reduce in intensity and finally disappear.

Light falling at an angle onto a transparent material will undergo reflection and refraction at the surface. At a particular angle of incidence that depends

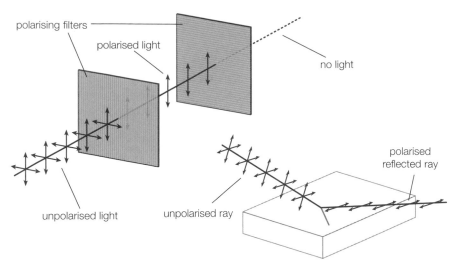

Figure 15.24 Polarisation of light.

on the refractive index of the medium, the reflected light will be completely plane polarised in the plane of the reflecting surface. For glass, this angle is about 56°. A polarising filter that allows through vertically polarised light will, therefore, cut out the horizontally polarised reflected light. This is the principle of antiglare sunglasses.

Light can also be polarised by scattering. The sky seems to be blue because the short-wave blue region of the visible spectrum is scattered much more than the red. If you look at the sky through a polarising filter, a difference in intensity can be seen when the filter is rotated. Bee's eyes are able to detect polarised light and this aids their navigation.

Optical activity

Some complex molecules rotate the plane of polarisation of transmitted light. In sugar solutions, the angle of rotation is dependent on the concentration of the solution.

In liquid crystal displays (LCDs), optically active crystals are aligned by an electric field between a polarising filter and a reflecting surface. The plane of polarisation is rotated by the aligned crystals so that, when it has been reflected back to the filter, its plane of polarisation is at right angles to that of the filter and no light passes out. The element of the seven-segment display unit therefore appears black.

Activity 15.6

Effect of the concentration of a sugar solution on the plane of polarisation of light

Two polarising filters and a 360° protractor can be used to measure the effect of the concentration of a sugar solution on the plane of polarisation of light, but simple polarimeters are widely available (Figure 15.25).

Distilled water is put into the cell to check that the scale reading is zero when the filters are crossed and the light-emitting diode (LED) blacked out. A number of sugar solutions of different concentrations (20–100 g in 100 ml of distilled water) are prepared. The cell is filled to a fixed level with the solution. The upper filter is rotated until the light source disappears. The angle of rotation is measured. The angle is found for all the known concentrations and some unknown solutions, including those containing clear honey and syrup.

A graph of the angle of rotation against the concentration is plotted and used to find the values for the unknown concentrations.

Question
The results of such an experiment are given in Table 15.5.

Table 15.5

Concentration/$g\,ml^{-1}$	0.20	0.40	0.60	0.80
Rotation angle/°	15	30	46	60

Plot a graph of the angle of rotation against concentration and hence deduce the concentration of a similar length of sugar solution that rotates the plane of polarisation through 33 °.

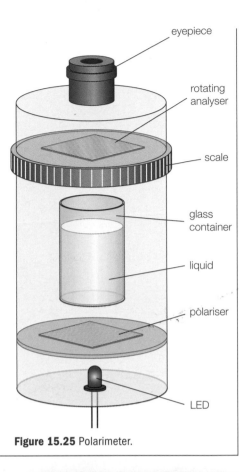

Figure 15.25 Polarimeter.

Stress analysis

Long-chain polymers like Perspex® can also rotate the plane of polarisation. The degree of rotation depends on the strain on the molecules and the wavelength of the light. This is particularly useful in industrial design. Models of load-bearing components are made from Perspex® and are put under stress and viewed through crossed sheets of Polaroid®. The multicoloured interference stress patterns (Figure 15.26) are analysed to detect potential regions of weakness.

Test yourself

13 What is plane polarised light?

14 Why is it not possible to polarise sound waves?

15 Describe the effect of an optically active medium on polarised light passing through it.

16 State three applications of polarised waves.

Figure 15.26 Interference stress patterns.

15.5 Pulse-echo techniques

One method of finding the speed of sound in air is to bang a drum while standing a measured distance (100 m or more) from the wall of a large building and timing the period between striking the drum and hearing the echo (Figure 15.27).

The speed is calculated by dividing the distance travelled by the sound (to the wall and back) by the time taken. If the speed of the sound in air is known, the same method can be used to find the distance of the observer from the wall.

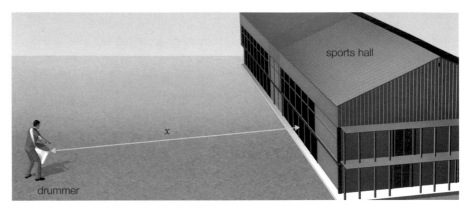

Figure 15.27

$$v = \frac{2x}{t}$$

hence:

$$x = \frac{vt}{2}$$

Sonar and radar are methods developed during the Second World War that are still widely used to gauge the positions of ships and aircraft. They achieve this by sending out pulses of radio and sound waves and noting the times and direction of the reflected pulses. Bats and dolphins are examples of animals that emit and receive high-frequency sounds to navigate or detect food sources.

A more recent development of the pulse-echo technique is **ultrasound** imaging in medicine. Ultrasound describes sound waves of frequency greater than the upper threshold of human hearing (about 20 kHz). In practice, frequencies in the range 1–3 MHz are generally used for medical images.

Amplitude scans (A-scans) are used to determine the depth of boundaries between tissues or bone and tissue (Figure 15.28). Pulses of ultrasound are emitted by a transducer and directed into the body at the region to be investigated. A coupling gel is smeared onto the body at the point of entry so that very little ultrasound is reflected from the skin. Some of the pulse's energy is reflected at the boundaries and received by the transducer.

In Figure 15.28, the reflections from the inner abdomen wall, the front and back of the organ and the spinal column are shown on the screen of a CRO. The time between the reflections and the entry of the pulse can be measured using the time base of the CRO, and the depth of the boundaries calculated using the pulse–echo formula.

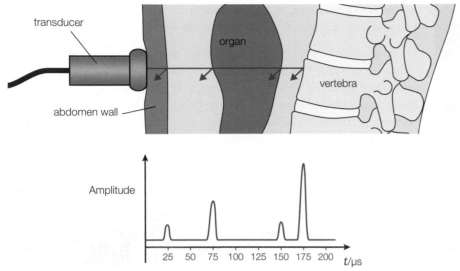

Figure 15.28 Ultrasound scan.

The fraction of sound that is reflected depends on the difference in a property known as the **acoustic impedance** of the tissue on each side of the interface. The acoustic impedance depends on the density of the medium so a much bigger reflection occurs at a tissue–bone boundary than at a tissue–muscle interface. The amplitude of the reflections received by the transducer will be reduced by attenuation – that is, energy absorbed or scattered within the body. Amplification of the reflected pulses by a factor depending on the distance travelled compensates for the attenuation, and the size of the peaks on the monitor indicates the relative fractions of the pulse reflected at each boundary.

Example

Use the timescale of the A-scan in Figure 15.28 to determine the distance of the organ from the inner abdominal wall and the width of the organ. The speed of sound is $1500\,\text{m s}^{-1}$ in soft tissue and $1560\,\text{m s}^{-1}$ in the organ.

Answer

$$x = \frac{v\Delta t}{2}$$

For the tissue:

$$x = \frac{1500\,\text{m s}^{-1} \times (75 - 25) \times 10^{-6}\,\text{s}}{2} = 3.8 \times 10^{-2}\,\text{m}$$

For the organ:

$$x = \frac{1560\,\text{m s}^{-1} \times (150 - 75) \times 10^{-6}\,\text{s}}{2} = 5.9 \times 10^{-2}\,\text{m}$$

A–scans can be used to detect a detached retina in the eye. They are also used in echoencephalography to determine accurately the midline of the brain: the midline is the gap between the hemispheres, and any deviation of the midline could indicate the presence of a tumour or haemorrhage on one side of the brain.

B–scans (brightness scans), in addition to detecting the position of the reflecting boundary, give a display where the brightness of the reflection

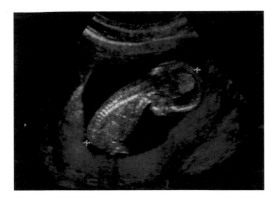

Figure 15.29 Ultrasound scan of a 13-week-old foetus.

represents the fraction of energy reflected. The familiar image from a foetal scan in Figure 15.29 shows clearly the skeletal structure.

Ultrasound images are generally of lower **resolution** than X-ray images – that is, they give much less detail. Resolution depends on wavelength, with the shorter-wave X-rays giving a clearer, more-detailed image. Medical ultrasound uses wavelengths of a few millimetres, whereas the wavelength of the X-rays used in diagnostic imaging is about 1×10^{-11} m. However, X-ray radiation is ionising and can kill or damage cells in the body. A developing foetus is particularly vulnerable, so ultrasound provides a much safer alternative.

The resolution of ultrasound can be improved by increasing the frequency, but the shorter waves are absorbed more readily and the useful range is reduced. The effective range is about 200 wavelengths, so 3 MHz waves are used for more detailed images of regions close to the skin, while the lower frequencies are used to examine deeper organs.

Tip

In the Year 2 book you will study some of these concepts further by looking at the Doppler Effect, and how this relates to 'red shift'. Some preparatory information on this can be found online by using the QR code for this chapter, which can be found in the *Free online resources* section at the end of this book.

Test yourself

17 Give three examples where 'pulse-echo' techniques are used for judging distances.

18 The echo from a gunshot from a distant cliff face is heard 1.2 s after the gun was fired. If the speed of sound in air is 340 m s^{-1}, calculate the distance of the gun from the cliff.

19 What is ultrasound?

20 Give one advantage and one disadvantage of ultrasound imaging in medicine compared with X-ray imaging.

Exam practice questions

1 Blue light is deviated more than red light when it enters a glass block because:

 A it has a longer wavelength

 B it has a lower frequency

 C it travels at a greater speed in glass than red light

 D it travels at a lower speed in glass than red light.

 [Total 1 Mark]

2 Light cannot be polarised by:

 A passing through a narrow slit

 B passing through sheets of Polaroid®

 C reflecting off a glass surface

 D scattering off dust particles in the atmosphere.

 [Total 1 Mark]

3 Ultrasound is preferred to X-rays for some diagnostic images because:

 A it gives a more detailed image

 B it is a longitudinal wave

 C it is less harmful to the patient

 D it penetrates the body more easily.

 [Total 1 Mark]

4 Calculate the critical angle for the interface between the core of an optical fibre with refractive index 1.60 and the cladding with refractive index 1.52. **[Total 2 Marks]**

5 Give one advantage and one disadvantage of using higher frequency ultrasound for diagnostic images in medicine. **[Total 2 Marks]**

6 A camera has a converging lens of focal length 2.0 cm. The image formed on the CCD screen of an object placed about a metre from the camera is

 A real, erect and diminished

 B real, inverted and diminished

 C real, inverted and magnified

 D virtual, inverted and diminished.

 [Total 1 Mark]

7 A recorder at the finish line of a 100-metre race sees the flash of the starting pistol and starts her stopwatch. A second timer fails to see the flash and starts his watch on hearing the bang. The winner's time differed by 0.3 s on the two watches. Explain the likely reason for this and use the difference to estimate the speed of sound in air. **[Total 2 Marks]**

8 In an experiment to determine the speed of sound in air, a loudspeaker is connected to a signal generator and the sound is picked up by a microphone placed a short distance in front of the speaker. The output from the signal generator is connected to the Y_1 input of a double beam oscilloscope (CRO) and the microphone is connected to the Y_2 input. The CRO display is shown in Figure 15.30.

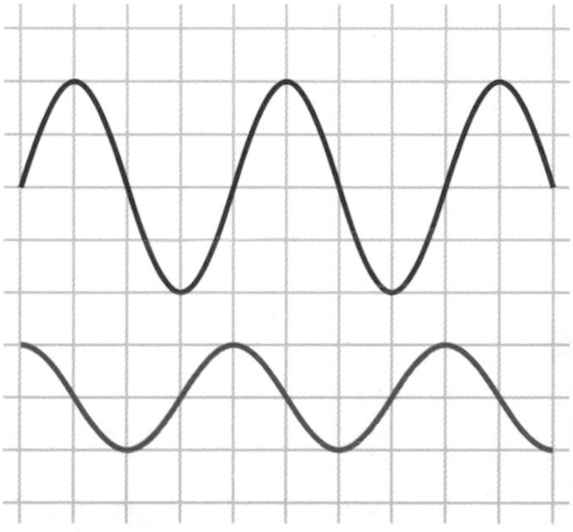

Figure 15.30

The time base setting is $200\,\mu s\,div^{-1}$ and the voltage sensitivities are $0.5\,V\,div^{-1}$ for the Y_1 input and $50\,mV\,div^{-1}$ for the Y_2 input.

a) Determine the values of the frequency of the sound and the peak values of the voltages from the signal generator and the microphone. **[2]**

The base of the microphone is placed onto a metre rule and moved away from the speaker until the traces on the CRO are in antiphase ($180°$ out of phase). The position on the scale is $22.0\,cm$. The microphone is slowly moved further away from the speaker until the two signals are once again in antiphase. The position of the microphone on the scale is now $49.2\,cm$.

b) Sketch the traces on the CRO screen when the signals are in antiphase. **[2]**

c) Calculate the value of the speed of sound in air. **[2]**

d) Explain any adjustments that could be made to the CRO to enable the antiphase positions to be accurately judged. **[2]**

[Total 8 Marks]

9 a) Explain the terms wavefront and wavelength. **[2]**

b) Copy the diagram of light waves travelling from glass to water in Figure 15.31 and continue the passage of the wavefronts into the water. **[2]**

[Total 4 Marks]

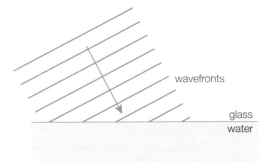

wavefronts

glass
water

Figure 15.31

10 a) A ray of light enters one side of a rectangular glass block at an angle of incidence of $40°$. Calculate the angle of refraction if the refractive index of the glass is 1.55. **[2]**

b) The opposite side of the block is immersed in a clear liquid. The ray makes an angle of $28°$ to the normal when it passes into the liquid. Calculate the refractive index of the liquid. **[2]**

[Total 4 Marks]

11 A trawlerman uses sonar to detect shoals of fish. A strongly reflected pulse is received $1.60\,s$ after it was transmitted. If the speed of sound in water is $1500\,m\,s^{-1}$, how far from the boat are the fish? **[Total 2 Marks]**

12 Find the image position, nature and linear magnification when an object is placed $25\,cm$ from:

a) a converging lens of focal length $20\,cm$ **[2]**

b) a converging lens of power 2.5 dioptres **[2]**

c) a diverging lens of focal length $50\,cm$. **[2]**

[Total 6 Marks]

13 The distance from the centre of the eye lens to the retina is about $25\,mm$ for an average adult. The lens is able to focus on objects from the near point ($250\,mm$) to objects at an infinite distance by changing its shape and hence its power.

a) Calculate the range of power of the eye lens focusing between the far and the near points. **[2]**

15 Transmission and reflection of waves

As people get older the lens loses its flexibility and they are unable to focus on close objects and often need glasses for reading.

b) Calculate the power and type of spectacle lens needed by an older person whose near point has extended to 1.25 m. **[3]**

A short-sighted person has the same accommodation range as someone with normal eyesight but the distance between the lens and the retina is 27 mm.

c) Calculate the power of the lens needed to rectify the problem. **[2]**

[Total 7 Marks]

14 a) One method of measuring the focal length of a converging lens uses the magnification method. If both sides of the lens equation are multiplied by v we get:

$$\frac{v}{f} = \frac{v}{u} + \frac{v}{v} \Rightarrow m = \frac{v}{f} - 1$$

Plan an experiment to find the focal length of a converging lens that uses this equation. You will need to draw a labelled diagram of the equipment you would use, state the measurements you would take and explain how the value of the focal length could be found using a graphical method. **[5]**

b) A second way of measuring f is the two-position method. With the object and screen separated by a distance greater than $4f$, there are two positions for the lens where a sharp image can be obtained, as shown in Figure 15.32.

It can be shown that the focal length is given by the equation

$$f = \frac{d^2 - l^2}{4d}$$

where d is the distance between the object and the screen and l is the distance between the two positions of the lens at which a sharp image is obtained. Explain

i) why the object and image are interchangeable

ii) why the image in one position is enlarged and the other diminished. **[2]**

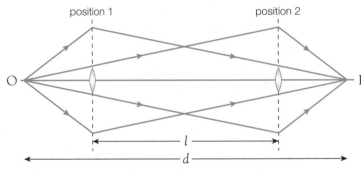

Figure 15.32

c) State one advantage of each method over the standard lens equation method (i.e. plotting $\frac{1}{v}$ against $\frac{1}{u}$). **[2]**

[Total 9 Marks]

15 The intensity of the radiation received by a light meter is directly proportional to the square of the amplitude of the waves.

a) Show that the intensity of polarised light passing through a polarising filter with its plane of polarisation at $\theta°$ to that of the incident light is given by $I = I_0 \cos^2 \theta$. **[2]**

b) Describe an experiment to check the validity of the relationship. **[5]**

c) How may a value for I_0 be obtained from the results? Explain why this may differ from the value read off the meter if the polarising filter were removed. **[2]**

[Total 9 Marks]

16 Polaroid® glasses can be useful when sailing as they reduce the glare from the sea.

a) Explain, with the aid of a diagram, what is meant by polarisation. **[2]**

b) When sunlight is reflected from a water surface, it is partially plane polarised. Explain how Polaroid® glasses reduce the glare from the sea. **[3]**

c) At a particular angle of incidence, 53° in the case of water, the reflected light is completely polarised. The refractive index of water is 1.33.

    i) Calculate the angle of refraction in the water.

    ii) Hence show, with the aid of a diagram, that the reflected and refracted rays are at right angles. Mark all the relevant angles on your diagram. **[6]**

[Total 11 Marks]

Stretch and challenge

17 Figure 15.33 shows a ray of light passing symmetrically through a prism of apex angle A. The emergent ray has been deviated through an angle D.

a) Show that the refractive index of the prism is given by the expression

$$n = \frac{\sin \dfrac{A + D}{2}}{\sin \dfrac{A}{2}}$$
[4]

For small angle prisms ($A \leq 10°$) the angle of deviation can be found using $D = (n - 1)A$.

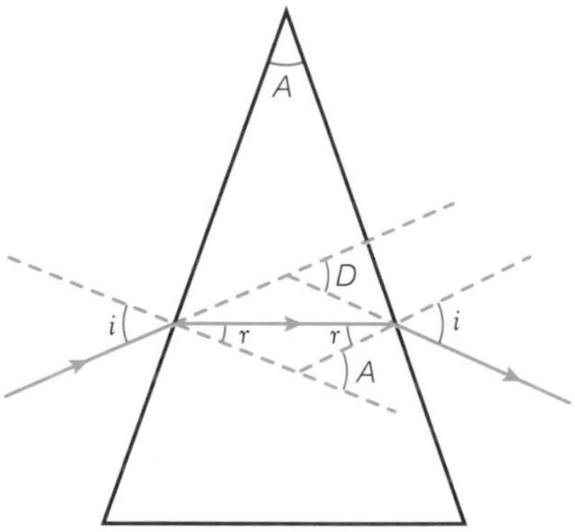

b) Calculate the deviation of a ray of light passing through a prism of apex angle 7.0° and refractive index 1.518. **[2]**

When white light passes through a prism, a spectrum is produced.

c) Sketch an arrangement for the production of a spectrum using a prism and explain how it is created. **[3]**

d) Calculate the angular dispersion (the difference between the deviations for red and violet light) for a 7.0° prism made from crown glass with a refractive index of 1.515 for red light ($\lambda = 656\,\text{nm}$) and 1.523 for violet light ($\lambda = 486\,\text{nm}$). **[2]**

Figure 15.33

Dispersion of white light can cause chromatic aberration on images produced using prisms and lenses. This can be remedied using pairs of prisms or lenses with different refractive indices. This can be achieved for the crown glass prism in part **d** by placing a flint glass prism behind it.

e) Draw an arrangement of the two prisms needed to produce no dispersion. **[2]**

f) Calculate the apex angle of the flint glass prism if the glass has a refractive index for red light of 1.639 and 1.667 for violet light. **[2]**

g) Calculate the mean deviation of the combination by considering a ray of yellow light (refractive index for crown glass = 1.518 and flint glass = 1.654). **[2]**

[Total 17 Marks]

16 Superposition of waves

Prior knowledge

In this chapter you will need to:
→ have an understanding of the modes of transmission of both mechanical and electromagnetic waves
→ be aware that the phase of the oscillations changes with distance from the source
→ understand that if two similar waves meet at a point, the resultant displacement of the particle (field) at that point will equal the vector sum of the displacements of both waves.

The key facts that will be useful are:
→ wave speed = frequency × wavelength
→ amplitude = maximum displacement
→ there is a phase difference of π radians between adjacent points half a wavelength apart
→ π radians is equivalent to $180°$.

Test yourself on prior knowledge

1 a) Change the following angles from radians to degrees:

    i) 2π rad

    ii) $\dfrac{5\pi}{2}$ rad

    iii) 1.2π rad.

  b) Change the following angles from degrees to radians:

    i) $90°$

    ii) $57.3°$

    iii) $150°$.

2 Calculate the resultant displacement of the particles at a point where a sound wave of amplitude 0.20 mm meets another of amplitude 0.30 mm:

  a) in phase

  b) π radians out of phase.

3 Calculate the resultant maximum electric field strength when two microwaves, each having an amplitude E, meet in phase when

  a) both are polarised in the same plane

  b) one is polarised in a plane at $40°$ to the other.

4 A transverse wave travelling along a rope has a wavelength of 1.20 m. State the phase difference between points along the wave that are

  a) 0.60 m apart

  b) 0.90 m apart

  c) 3.00 m apart.

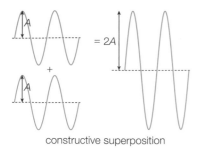

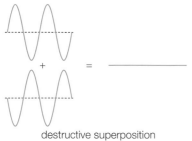

constructive superposition

destructive superposition

Figure 16.1 Constructive and destructive superposition.

16.1 Superposition

When two or more waves of the same type meet at a point, the resultant displacement of the oscillations will be the vector sum of the individual displacements.

Later in this chapter you will see how the **superposition** of waves leads to positions of maximum and minimum amplitude and how coherent sources can create regular regions of high and low intensity called interference patterns. You will also study the production and applications of standing waves, with particular reference to musical instruments, and diffraction, which is the effect of waves being obstructed by objects or apertures.

Figure 16.1 shows the superposition of two waves in phase and in antiphase (π radians out of phase). The resultant amplitudes are $2A$ and zero. When the waves add together to give maximum amplitude, **constructive** superposition occurs; when they combine to produce zero amplitude, **destructive** superposition occurs.

Superposition can only occur for identical wave types. It would not be possible for a sound wave to combine with a light wave, for example. Superposition would also not happen if two transverse waves polarised at right angles were to meet.

Destructive superposition is used in active noise reduction. Sound from a tractor, aeroplane or heavy machinery is picked up using a microphone, electronically processed and transmitted to earphones π radians (180°) out of phase. Destructive superposition takes place and the noise is effectively cancelled out for the operator.

16.2 Interference

In most cases, the waves from two or more sources will overlap at a point in a haphazard fashion, so the resulting amplitude variations will not be noticeable. In some situations, however, two sources can produce a pattern of **maxima** and **minima**, where the waves combine constructively or destructively at fixed positions relative to the sources. This effect is known as **interference**.

Activity 16.1

Interference of sound

Two identical loudspeakers are placed about one metre apart in a large room with little furniture (for example, a school hall) and both are connected to the same signal generator. An observer walking across the room in front of the speakers will pass loud and quiet regions regularly spaced as shown in Figure 16.2. Because the speakers are connected to the same signal generator, they will emit sound waves that have identical frequencies and amplitudes and are in phase.

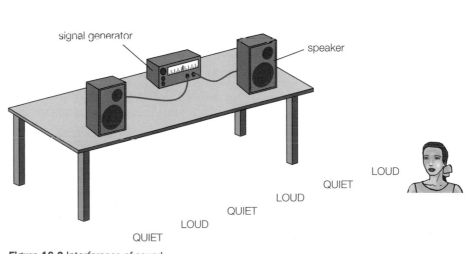

Figure 16.2 Interference of sound.

The positions of the maximum and minimum sounds are marked on the floor of the room and the frequency of the sound is read off the signal generator. The distances of the maxima and minima from each speaker are taken using a tape measure.

In one experiment the first maximum to one side of the central position is 10.23 m from one speaker and 9.10 m from the other speaker and the frequency of the sound is 300 Hz.

Questions

1 Explain why the wavelength of the sound must equal 1.13 m.
2 Determine the speed of sound in the room.

In Figure 16.3, point P_1 is half a wavelength further away from S_2 than S_1. Because the sound from each source is in phase and travels with the same speed, the waves will always be out of phase at P_1, so destructive superposition occurs. Similarly, sound from S_2 will have travelled one whole wavelength further than that arriving at the same time from S_1, so the waves will be in phase at point P_2 and will interfere constructively to give sound of maximum intensity.

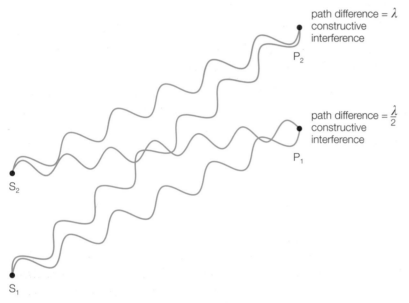

path difference = λ
constructive interference

path difference = $\frac{\lambda}{2}$
constructive interference

Figure 16.3 Path difference.

The difference in distance from each source to a particular point – for example, $(S_2P_1 - S_1P_1)$ – is called the **path difference**. Positions of maximum amplitude occur when the path difference is zero or a whole number of wavelengths, when the waves are always in phase and constructive superposition takes place. When the path difference is an odd half wavelength, the waves are π radians out of phase and the amplitude will be zero.

Stable interference patterns only occur if:

- the waves are the same type
- the sources are coherent (they have the same wavelength and frequency and maintain a constant phase relationship)
- the waves have similar amplitude at the point of superposition.

An interference pattern can be observed using a ripple tank. Circular wavefronts are generated on the surface of the water by two prongs attached to an oscillator. Where a trough from one source meets a trough from the other – or a crest meets a crest – maximum disturbance occurs (Figure 16.4).

Key term

Coherent sources have the same frequency and maintain a constant phase relationship.

Tip

Many candidates lose marks by stating that coherent sources are in phase. Any pair of sources could be in phase for an instant so the candidate needs to write 'always in phase'.

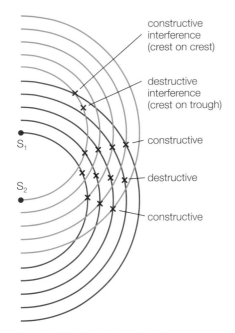

constructive interference (crest on crest)

destructive interference (crest on trough)

constructive

destructive

constructive

S_1

S_2

Figure 16.4 Ripple tank interference.

Calm water indicates the destructive interference where a trough meets a crest. (There are some excellent simulations to be found online. These two websites are to be particularly recommended: www.animations.physics.unsw.edu.au and https://phet.colorado.edu/en/simulations/category/physics.)

It is possible to measure the wavelength of radiation using interference patterns. A simple method for measuring the wavelength of microwaves is shown in Figure 16.5a. Aluminium sheets are used to create a pair of slits about 5 cm apart and are placed with their midpoint perpendicular to an aerial and about 10 cm away from a single source. This double slit allows radiation from the same wavefront to pass through both gaps, which effectively creates a pair of coherent sources. The receiving aerial is placed so that it is in line with the transmitter and the midpoint of the slits and is about half a metre to a metre from the slits.

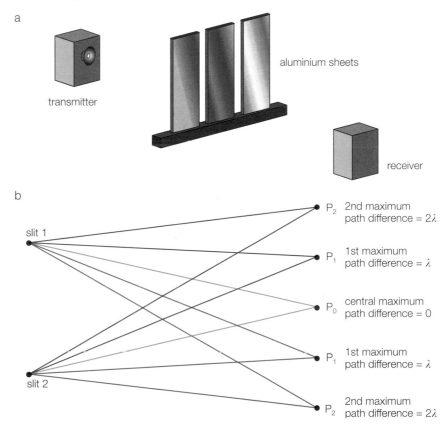

a

transmitter

aluminium sheets

receiver

b

slit 1

slit 2

P_2 2nd maximum path difference = 2λ

P_1 1st maximum path difference = λ

P_0 central maximum path difference = 0

P_1 1st maximum path difference = λ

P_2 2nd maximum path difference = 2λ

Figure 16.5 a) Apparatus for creating double slit interference of microwaves and b) resultant path differences.

The path difference of the radiation from each slit is zero, so a central maximum signal is received. If the receiver is moved to either side of the slits, a series of alternate maxima and minima will be detected. A sheet of paper is placed on the bench and the position of the aerial at the highest available maximum (usually three or four can be detected) is marked on it. The positions of the slits are then pencilled onto the paper and the distance of each slit from the maximum is measured. The wavelength of the microwaves can be obtained using the relationship:

path difference = $S_2P_n - S_1P_n = n\lambda$

where n = order of maximum from the centre.

Interference of light

It is quite tricky to observe interference effects with light sources because of the short wavelengths and the difficulty in providing coherent sources.

In 1800, Thomas Young devised a method to produce coherent light sources from wavefronts generated by passing single wavelength (monochromatic) light through a fine slit and then using a double-slit arrangement (like that used in Figure 16.5). The bright and dark lines observed through a microscope eyepiece gave Young the evidence he needed to support his theory of the wave nature of light. The lines are still referred to as Young's fringes.

It can be shown that

$$\lambda = \frac{Sx}{D}$$

where S is the slit width, x is the fringe separation and D is the distance from the slits to the microscope eyepiece.

Nowadays it is much easier to observe interference fringes. Lasers produce intense beams of light that have the same wavelength, are in phase and are polarised in the same plane. A laser beam passed through a pair of parallel lines scored about 1 mm apart on a blackened glass plate will project a fringe pattern onto a wall several metres away (Figure 16.6). Measurements of the slit separation, the distance of the slits from the wall and the fringe width allow the wavelength of the light to be calculated.

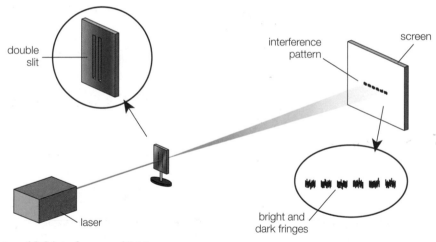

Figure 16.6 Interference of light.

Safety note: The laser should be a Class 2 low power laser.

Interferometers

In precision engineering, and particularly in the optical industry, surfaces need to be ground with tolerances of better than one thousandth of a millimetre. **Interferometers** use the patterns created by the recombination of a laser beam that has been split into two separate beams (Figure 16.7). Small changes in the path difference are detected by a shift in the fringe pattern.

Example

In an experiment using microwaves, the position of the fourth maximum from the centre was 48 cm from one slit and 60 cm from the other. Calculate the wavelength of the waves.

Answer
Path difference $= 12\,\text{cm} = 4\lambda$

$$\lambda = 3.0\,\text{cm}$$

Example

A helium–neon laser produces a beam of light that passes through a double slit and creates a fringe pattern on a screen 4.00 m from the slits. The slit separation is 1.0 mm and the distance across 20 bright fringes is 51 mm. Calculate the wavelength of the laser light.

Answer

$$\lambda = \frac{Sx}{D}$$
$$= \frac{1.0 \times 10^{-3}\,\text{m} \times 2.55 \times 10^{-3}\,\text{m}}{4.00\,\text{m}}$$
$$= 638\,\text{nm}$$

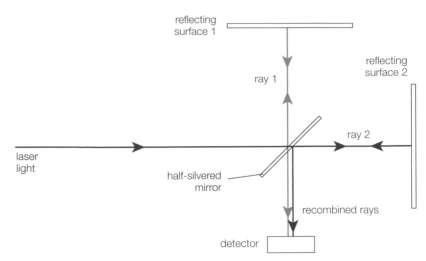

Figure 16.7 Interferometer.

CD players

The information on a CD is in a digital form: the data are stored using binary coding to represent 'bytes' of information in the form of ones and zeros. Electronically these are created by rapidly switching a circuit on and off. Compact discs have a spiral groove with a width of less than 2 μm cut from the edge to the middle of a highly reflective silvered surface. The data are recorded onto the disc as millions of small 'bumps' within the grooves and the surface of the CD has a clear plastic protective covering.

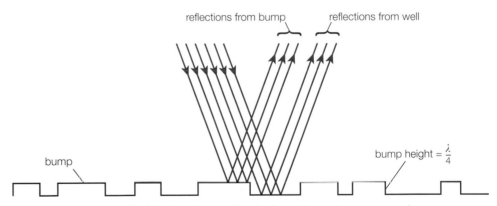

Figure 16.8 Principle of a CD player. In practice the beam is much narrower and perpendicular to the surface of the disc.

A laser beam focused onto the groove is reflected back to a photodiode where the light is converted into an electrical signal (Figure 16.8). Each bump has a height equal to a quarter of the wavelength of the laser (usually about 200 nm), which is a quarter of the wavelength of radiation in the red/infrared region of the electromagnetic spectrum. When the laser illuminates the edge of a bump, the path difference between light reflected from the top of the bump and that from the bottom of the groove is equal to half a wavelength, so the waves interfere destructively and the output of the photodiode is zero (binary 0). When the entire beam is reflected from the upper surface or the gap between the bumps, the intensity of the reflected beam is strong and a high electrical output (binary 1) is generated by the photodiode. The disc rotates rapidly so that the beam follows the groove outwards and the data on the disc are collected as a stream of binary digits.

The principle of the CD player can be demonstrated using a standard microwave kit. Most microwave kits have a wavelength of about 3 cm. The 'disc' is made from a wooden board, with 'bumps' of wood 0.75 cm thick glued along its length. The whole length of the board is covered with a strip of aluminium foil, and the board is mounted onto a pair of dynamics trolleys, as shown in Figure 16.9. When the board is pushed across the transmitter/ receiver, a sequence of high and low signals is observed.

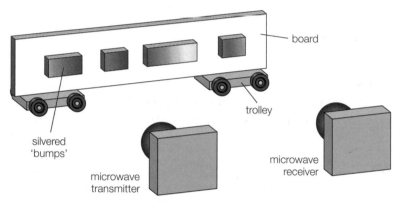

Figure 16.9 Microwave model of a CD player.

Test yourself

1 What is meant by the superposition of waves? Illustrate your answer with sketches showing constructive and destructive superposition.

2 Explain what is meant by coherent sources.

3 Write out an expression for the path difference from the sources to the nth maximum and the nth minimum interference line from the central maximum.

4 A pair of similar loudspeakers are connected to a signal generator set at a frequency of 510 Hz so that the sounds are emitted in phase. A position of the third maximum from the centre of the resulting interference pattern is 4.20 m from one speaker and 6.20 m from the other. Calculate:

   a) the wavelength of the sound

   b) the speed of sound in the room.

5 Calculate the separation of two parallel slits if the interference fringes produced when a laser beam of wavelength 640 nm is shone through them onto a wall 5.40 m from the slits are 2.9 mm apart.

16.3 Standing waves

Standing waves, sometimes called stationary waves, are created by the superposition of two progressive waves of equal frequency and amplitude moving in opposite directions.

If two speakers connected to the same signal generator face each other, a standing wave will exist between them (Figure 16.10). At P, the midpoint between the speakers, the waves, having travelled the same distance at the same speed, will always be in phase and interfere constructively. At A, $\frac{\lambda}{4}$ from P, the distance from speaker S_1 has increased by a quarter of a wavelength, while that from speaker S_2 has decreased by the same distance. The path difference ($S_1A - S_2A$) is therefore half a wavelength, the waves are in antiphase and destructive interference takes place. Similarly at B, half a wavelength from P, the waves will be back in phase and produce a sound of maximum intensity. A person walking along the line between the speakers will detect a series of equally spaced maxima and minima along the standing wave.

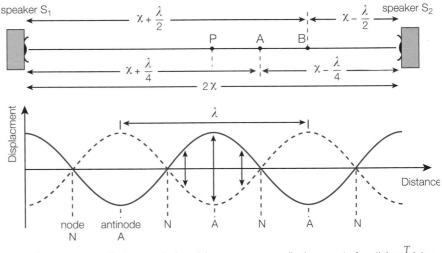

Figure 16.10 Formation of standing wave.

The points of zero amplitude within a standing wave are called **nodes** and the maxima are called **antinodes**. Figure 16.10 shows that the separation of adjacent nodes or antinodes is always half a wavelength.

Activity 16.2

Measuring the wavelength of microwaves and sound

A microwave transmitter is set up in front of an aluminium sheet as in Figure 16.11. Standing waves are formed by the superposition of the reflected wave onto the incident wave. The receiving aerial is moved along the line from the reflector to the transmitter and the distance between ten nodes is measured. This is equal to five wavelengths. The nodes and antinodes are more apparent close to the reflector where the amplitudes of the incident and reflected waves are similar.

Figure 16.11 shows that the same principle can be used to find the speed of sound in the laboratory. The nodes and antinodes are detected using a small microphone placed between a loudspeaker and a reflecting board and connected to a cathode ray oscilloscope (CRO). The height of the trace on the CRO will indicate positions of maximum and minimum intensity. The wavelength of the sound is measured for a range of frequencies and a graph of λ against $\frac{1}{f}$ is plotted.

The gradient of the graph represents the speed of sound ($\lambda = v\frac{1}{f}$).

Questions

1 The distance between 10 maxima in the microwave experiment is 32.0 cm. Calculate
   a) the wavelength of the microwaves
   b) the frequency of the microwaves.
2 The following results were obtained using the standing waves for sound:

Table 16.1 Experimental results.

f/Hz	400	600	800	1000	1200
2λ/m	1.70	1.14	0.86	0.68	0.56

Plot a suitable graph and use it to determine the speed of sound in air.

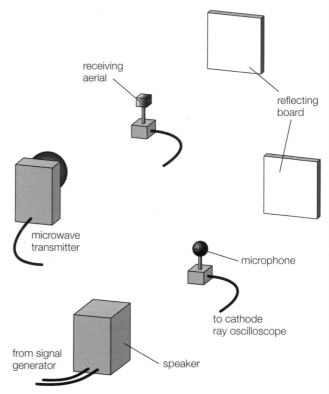

Figure 16.11 Using standing waves to measure the wavelength of microwaves and sound.

Example

A standing wave is set up by reflecting a sound of frequency 1200 Hz from a laboratory wall. The distance between four successive nodes is 42 cm. Calculate the speed of sound in the laboratory.

Answer

The distance between nodes is half a wavelength, so there will be 1.5 wavelengths between the first and the fourth node:

$$\frac{3}{2}\lambda = 42 \text{ cm}$$
$$\lambda = 28 \text{ cm}$$
$$v = f\lambda = 1200 \text{ Hz} \times 0.28 \text{ m} = 340 \text{ m s}^{-1}$$

Standing waves differ from travelling waves in the following ways:

- Standing waves store energy, whereas travelling waves transfer energy from one point to another.
- The amplitude of standing waves varies from zero at the nodes to a maximum at the antinodes, but the amplitude of all the oscillations along a progressive wave is constant.
- The oscillations are all in phase between nodes, but the phase varies continuously along a travelling wave.

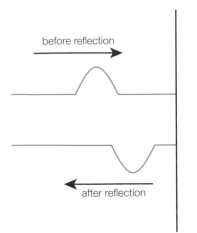

Figure 16.12 Reflection of a wave in a string.

Standing waves in strings

When a pulse is sent along a rope that is fixed at one end, the reflected pulse is out of phase with the incident pulse. A phase change of π radians (180°) takes place at the point of reflection (Figure 16.12). This means that destructive interference will occur and the fixed position will (not surprisingly) be a node.

Standing waves in a string can be investigated using Melde's experiment (Figure 16.13). A thin length of string is attached to an oscillator, passed over a pulley wheel and kept taut by a weight hanging from its end. The frequency of the oscillator is adjusted until nodes and antinodes are clearly visible. A strobe lamp can be used to 'slow down' the motion so that the standing wave can be studied in more detail. The wavelength of the wave is found by measuring the distance between alternate nodes.

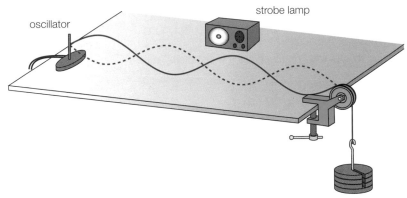

Figure 16.13 Melde's experiment.

Safety note: Frequencies in the range 5–30 Hz Should be avoided as they may trigger epileptic fits in some individuals.

Standing waves of shorter or longer wavelength are observed if the frequency of the oscillator is increased or decreased or the weight on the string is altered. The speed of the incident and reflected waves in the string can be calculated using $v = f\lambda$. An investigation into how this depends on the tension or thickness of the string can be performed.

Stringed instruments

Stringed instruments such as guitars, violins and pianos all produce standing waves on strings stretched between two points. When plucked, bowed or struck, the energy in the standing wave is transferred to the air around it and generates a sound. Because the string interacts with only a small region of air, the sound needs to be amplified – either by a resonating sound box or electronically.

The principle of stringed instruments can be demonstrated using a **sonometer** (Figure 16.14). When the string is plucked at its midpoint, the waves reflected from each end will interfere to set up a standing wave in the string. As both ends are fixed, they must be nodes, so the simplest standing wave will have one antinode between two nodes – that is, the length of the string will be half a wavelength. Using the expression $v = \sqrt{\frac{T}{\mu}}$ and the wave equation, $v = f\lambda$, the frequency of the note emitted by the wire in this mode will be $f = \frac{v}{\lambda} = \frac{1}{2l}\sqrt{\frac{T}{\mu}}$.

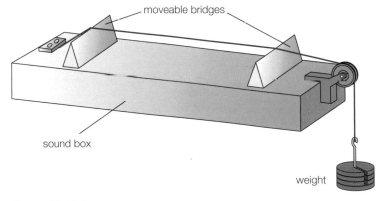

Figure 16.14 Sonometer.

Core practical 7

Investigating the factors that affect the frequency of vibrations on a stretched string

The factors affecting the frequency of the vibrations of a stretched wire can be investigated using wire stretched across two bridges like the sonometer shown in Figure 16.14. The frequency of the sound emitted from the sonometer can be obtained using a microphone connected to a cathode ray oscilloscope.

If a sonometer or cathode ray oscilloscope is not available, the wire can be fixed onto a bench between two triangular blocks held taut by a weight hanger plus masses with the midpoint of the wire between the poles of a pair of magnadur magnets. A standing wave can be produced by connecting the ends of the wire to a signal generator so that the alternating current in the wire interacts with the magnetic field to make the wire vibrate with the frequency of the generator. When the frequency of the alternating current equals the natural frequency of the wire, the wire vibrates with a large amplitude. (This is called resonance and will be studied in detail in Year 2 Student's book.)

The relationships between the frequency and length, tension and the mass per unit length of the wire can be investigated by keeping two quantities constant and varying the third one.

The relationship between the frequency and the length of a particular string is found by moving the bridges and measuring the frequency for a range of different lengths, keeping the same weight on the hanger. A graph of f against $\frac{1}{l}$ is plotted.

A straight line through the origin would indicate that the frequency of the vibrations is inversely proportional to the length of the string.

The relationship between the frequency and the tension of a particular string is investigated by keeping the length of the wire constant and finding the frequency of vibration for a range of weights (i.e. a range of tensions in the wire). A graph of f against $\sqrt{T}$, or f^2 against T is plotted.

A straight line through the origin would indicate that frequency of the vibrations is proportional to the square root of the tension.

If a range of different wires is available, the relationship between the frequency and the mass per unit length can be investigated.

Alternative methods using Melde's experiment (Figure 16.13) can be performed here. The frequency is varied using a signal generator connected to an oscillator to produce a standing wave for the fundamental frequency.

If a metal wire is used, the output from the signal generator can be connected across it and a pair of magnadur magnets either side of the wire at its mid-point. Variations of the current in the wire will make it vibrate with the frequency of the signal generator.

The analysis of the data for the above methods is the same as for the sonometer experiment.

Questions

1 Describe how you would investigate how the frequency of vibrations of a stretched wire depends on the mass per unit length of a number of different piano wires. You should state the instrument used for each measurement and describe how you would analyse the results.

2 A student carried out an investigation on the variation of frequency with length of a piece of piano wire. She used a hanger with weights having a total mass of 5.0 kg and then plotted a graph of frequency against $\frac{1}{l}$. The graph was a straight line with a small intercept on the frequency axis and a gradient of 60 m s^{-1}.

    a) Suggest a reason why the graph did not pass through the origin.

    b) Use the relationship $f = \frac{1}{2l}\sqrt{\frac{T}{\mu}}$ to show that the mass per unit length of the wire is about 3.5 g m^{-1}.

In general, for stringed instruments, the frequency is greater for:

- shorter strings
- strings with greater tension
- strings that have a lower mass per unit length – that is, thinner strings of the same material or strings made from a lower density material.

Overtones and harmonics

You saw earlier that the fixed ends of vibrating strings must be nodes and that the simplest standing wave has a single antinode at the midpoint. The frequency of the note emitted from such a wave is called the **fundamental frequency** of the string. By plucking the string off centre it is possible to create several standing waves on the same string.

Figure 16.15 shows the fundamental mode and two other possible waves. The fundamental vibration has the longest wavelength ($\lambda = 2l$) and the

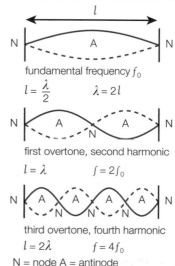

fundamental frequency f_0

$l = \frac{\lambda}{2}$ $\lambda = 2l$

first overtone, second harmonic

$l = \lambda$ $f = 2f_0$

third overtone, fourth harmonic

$l = 2\lambda$ $f = 4f_0$

N = node A = antinode

Figure 16.15 Standing waves.

others reduce in sequence. The notes emitted by vibrations other than the fundamental are called **overtones**. Overtones that have whole number multiples of the fundamental frequency are **harmonics**.

The sounds we hear from a guitar, for example, are a complex mixture of harmonics and are noticeably different from the same tune played on a violin. The property that enables us to distinguish different musical instruments is the **quality**, or **timbre**, of the note.

Activity 16.3

Examining the waveforms of stringed instruments

The waveforms of stringed instruments can be examined as in Figure 16.16, in which notes from different instruments are picked up by a microphone connected to a CRO.

A fuller analysis of wave patterns can be achieved using software such as Multimedia Sound. In addition to displaying the sounds played in the room (Figure 16.17a), such software includes a library of pre-recorded notes and a spectrum analyser. The analyser shows the frequency and relative amplitude of the harmonics (Figure 16.17b).

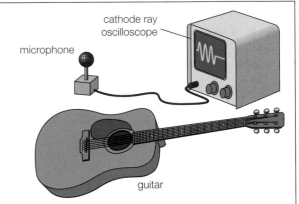

Figure 16.16 Waveforms of a guitar.

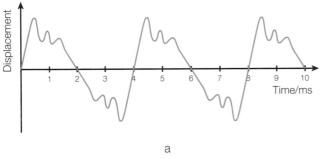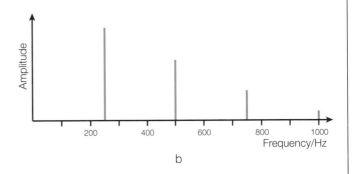

a

b

Figure 16.17 Waveforms of musical instrument.

The CRO trace shows the fundamental frequency to be 250 Hz. The analysis reveals that there are overtones of 500 Hz, 750 Hz and 1000 Hz.

Wind instruments

Wind instruments are basically tubes in which standing air waves are formed from vibrations produced in a mouthpiece. Unlike strings, the wave boundaries can be nodes or antinodes.

A small speaker connected to a signal generator can be used to set up standing waves in tubes of length l open at one or both ends. The speaker is clamped just above an open end and the frequency is slowly increased by adjusting the signal generator. When a standing wave is formed in the tube, the air column resonates and an intense booming sound is heard.

Figure 16.18 shows some of the possible wavelengths in a tube open at each end and a tube closed at one end. The waves are traditionally drawn as displacement–distance variations along the tube, but it is important to remember that sound waves are longitudinal and that the vibration of air molecules is along their length.

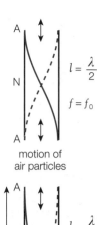

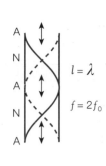

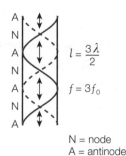

$l = \dfrac{\lambda}{2}$

$f = f_0$

motion of
air particles

$l = \lambda$

$f = 2f_0$

$l = \dfrac{3\lambda}{2}$

$f = 3f_0$

$l = \dfrac{\lambda}{4}$

$f = f_0$

$l = \dfrac{3\lambda}{4}$

$f = 3f_0$

$l = \dfrac{5\lambda}{4}$

$f = 5f_0$

N = node
A = antinode

Figure 16.18 Possible wavelengths in tubes both open and closed.

At the open end, the reflections always create antinodes. At the closed ends, where the particles are unable to oscillate, nodes are formed. The fundamental frequency of the open-ended pipe is therefore twice that of the closed pipe. Figure 16.18 also shows that the first two overtones from the closed tube are the third and fifth harmonics.

Example

A small loudspeaker connected to a signal generator emits a sound of frequency 425 Hz. It is fixed above a long glass tube that is filled with water and has a drain at the bottom so that the water can be slowly released from it. When the level has fallen 20 cm from the top of the tube, a standing wave is formed and the air column resonates.

Calculate:

a) the wavelength of the sound

b) the speed of sound in the air column

c) the distance of the water surface from the top of the tube when the next standing wave is formed.

Answers

a) $\dfrac{\lambda}{4} = 20\,\text{cm}$

$\lambda = 80\,\text{cm}$

b) $v = f\lambda = 425\,\text{Hz} \times 0.80\,\text{m} = 340\,\text{m s}^{-1}$

c) Next standing wave formed when $l = \dfrac{3\lambda}{4} = 60\,\text{cm}$ from the top.

The notes played on a wind instrument are selected by opening or closing holes along its length and blowing into, or across, a mouthpiece. A recorder is one of the more basic instruments and can be used to demonstrate this principle (Figure 16.19). The air is free to move when the holes are open, so, at resonance, there will be antinodes at the holes. With all stops closed, there will be antinodes at the mouthpiece and the end of the recorder and the wavelength of the lowest fundamental will equal twice the length of the instrument. With some holes open, the fundamental frequency will be that of the standing wave, with antinodes at the mouthpiece and the closest open hole.

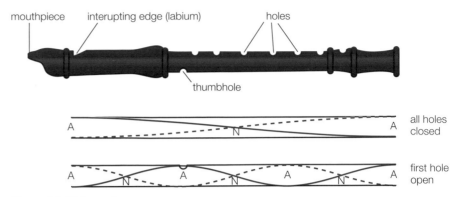

Figure 16.19 Recorder.

Higher harmonics are possible by leaking air into the thumbhole. This effectively divides the air column into two, which means that the wavelength is halved and hence the frequency is doubled. The pitch of the note increases when the frequency rises, so a higher pitched note is emitted.

Test yourself

6 State three differences between standing waves and progressive waves.

7 The distance across six nodes (i.e. five internodal gaps) in the standing waves produced by a microwave reflected from an aluminium surface is 80 mm. Calculate

a) the wavelength of the microwaves

b) the frequency of the microwaves.

8 A wire under tension produces a sound of frequency 480 Hz when plucked. Calculate the frequency of the note when

a) the tension is halved but the length stays the same

b) the length is halved with the original tension

c) both the length and the tension are doubled.

9 A guitar string has a length of 75 cm. Calculate the wavelength of

a) the fundamental note

b) the third harmonic.

10 A small speaker is connected to a signal generator and placed at the open end of a long tube containing water. Water is drained from the tube until the note from the speaker is amplified when the air column resonates. The distance from the surface of the water to the open end is 8.2 cm. Calculate

a) the wavelength of the note

b) the distance of the water surface from the open end of the tube at the next position of resonance.

16.4 Diffraction

When a wave passes through a gap or is partially obstructed by a barrier, the wavefront spreads out into the 'shadow' region (Figure 16.20a). This effect is called **diffraction** and is easily demonstrated using a ripple tank. The dark and light lines represent crests and troughs on the surface of the water, and can therefore be considered as wavefronts.

If the oscillator in a ripple tank is adjusted to a higher frequency, the wavelength shortens and the spreading is reduced. Narrowing the aperture of the gap between two barriers causes even more spreading of the wave (Figure 16.20b). When the width is similar to the wavelength, the wavefronts are almost circular.

Diffraction can be explained using Huygens' construction (shown in Figure 16.21) Huygens considered every point on a wavefront as the source of secondary spherical wavelets which spread out with the wave velocity. The new wavefront is the envelope of these secondary wavelets.

When a wavefront is obstructed by a slit or an obstacle, the secondary wavelets at the edges are transmitted into the geometrical shadow causing the diffraction spreading. In some instances patterns can be seen due to the interference of imagined point sources on the wavefront as it passes through the slit or around the obstruction.

Figures 16.22 (a) and (c) show how the secondary wavelets at S_1 and S_2 spill over into the geometrical shadow. Figure 16.22 (b) illustrates how the wavefront in a very narrow aperture behaves as a single source of a secondary wavelet.

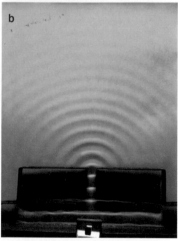

Figure 16.20 Wave diffraction experiment.

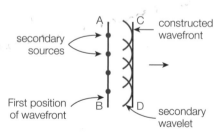

Figure 16.21 Huygens' construction

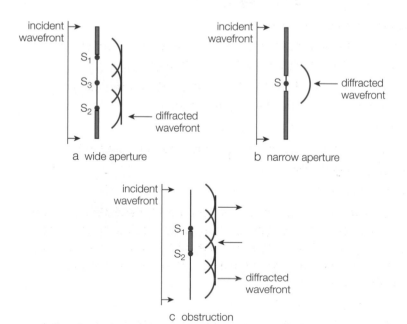

Figure 16.22 Diffraction of wavefronts

Short-wave radio and television signals do not diffract significantly into valleys or around the curvature of the Earth's surface, but military communications radio sends out signals with wavelengths of several kilometres that can be detected by submarines almost halfway around the world.

Close inspection of the water waves on the ripple tank emerging from the narrow aperture in Figure 16.20b reveals regions of constructive and destructive superposition at the edges of the circular wavefront. This effect can be further investigated using a microwave kit (Figure 16.23). The transmitter is placed about 20 cm from a pair of aluminium sheets and directed at a gap of about 5 cm. The receiver is placed facing the aperture and about 50 cm away from the other side of the sheets. The output from the receiver is measured at the midpoint and at 10° intervals as it is moved along a semi-circular path.

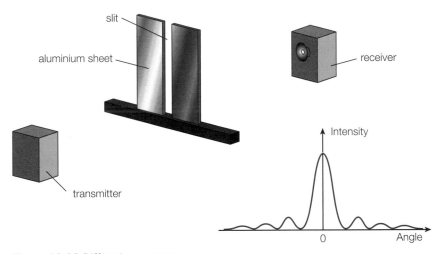

Figure 16.23 Diffraction pattern.

The diffraction pattern shows a central maximum edged by a series of lower intensity maxima and minima as opposed to the regular pattern of interference from a double slit. The central maximum will broaden when the slit width is reduced.

It can be shown that, for a slit of width, a, the angle θ between the central maximum and the first minimum is given by the equation:

$$\sin\theta = \frac{\lambda}{a}$$

In general, to create a diffraction pattern using a slit, the wavelength should be of the same order of magnitude as the width of the slit. Patterns can be seen for slightly wider slits but the value of θ becomes smaller. For angles of less than about 10°, the approximation $\sin\theta = \tan\theta = \theta$ radians can be used.

Diffraction of light

Shadows formed from objects placed in front of a point source of light seem to be sharply defined. There is no noticeable overlap of the light into the shadow region or interference patterns at the edges. The wavelength of visible light is so short (400–700 nm) that such patterns are difficult to detect with the naked eye. However, many everyday observations are due to diffraction. The star-like pattern produced when light from the Sun, or a bright lamp, is viewed through a fine-mesh material like silk and the multicoloured reflections from the narrow grooves on a compact disc are common examples. Single-slit patterns can be seen between your thumb and forefinger when they are almost touching and between the jaws of Vernier calipers.

Diffraction gratings

When light is reflected from a surface with thousands of equally spaced, parallel grooves scored onto each centimetre, or transmitted through thousands of equally spaced, microscopic gaps, a diffraction pattern is produced. Such arrangements are called **diffraction gratings**.

Like the single slit, the width of the spacing must be of the same order of magnitude as the wavelength of the light, but the patterns are different. The maxima occur at specific angles where the small, coherent waves from each groove or slit superimpose constructively producing sharply defined lines, as shown in Figure 16.24.

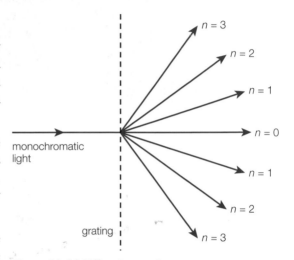

Figure 16.24 Diffraction grating.

If the number of lines per metre on the grating is known, it is possible to determine the wavelength of light transmitted or reflected by the grating by measuring the angles between the central maximum and the diffracted maxima. The relationship between the angles and the wavelength can be shown to be:

$$n\lambda = d \sin\theta$$

where d is the slit separation and n is the order of the maximum.

It follows that for smaller values of d, the value of θ will be bigger for a particular wavelength, and that the maximum number of orders of diffraction will be when $n \leqslant \frac{d}{\lambda}$.

Example

Light from a laser passes through a diffraction grating and a pattern is seen on a screen 2.40 m from the grating. The separation of the second-order maxima is measured and found to be 1.26 m. If the grating has 200 lines per millimetre, calculate:

a) the wavelength of the laser light

b) the maximum number of orders of the diffraction pattern.

Answers

a) The angle between the central maximum and the second maximum,

$$\theta = \tan^{-1}\frac{0.63\text{ m}}{2.40\text{ m}} = 14.7°$$

The slit separation, $d = \dfrac{1}{200 \times 10^3\text{ m}^{-1}} = 5.0 \times 10^{-6}\text{ m}$

$$2\lambda = 5.0 \times 10^{-6}\text{ m} \times \sin 14.7°$$

$$\lambda = 6.34 \times 10^{-7}\text{ m} = 634\text{ nm}$$

b) $n_{max} \leqslant \dfrac{d}{\lambda} \leqslant \dfrac{5.0 \times 10^{-6}\text{ m}}{6.34 \times 10^{-7}\text{ m}} \leqslant 7.9$

Hence $n_{max} = 7$ (must be an integer below 7.9)

Diffraction of light can be studied in more detail using a laser. Shadows of small objects and apertures cast onto a screen several metres away clearly show the patterns (Figure 16.25).

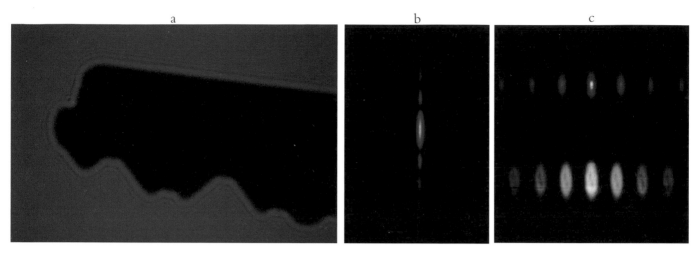

Figure 16.25 a) The shadow of a house key using laser light, b) diffraction pattern of laser beam through a single slit and c) diffraction patterns for red and green laser light using a diffraction grating.

Core practical 8

Diffraction patterns

Diffraction patterns can be explored using lasers. Conventional light sources can produce diffraction patterns, but these are often difficult to observe and too small to take accurate measurements. Lasers have the advantage of having a very high intensity beam so that images can be formed on screens several metres from the slits, making the separation of the maxima and minima much larger and easier to measure.

Some general purpose lasers for school use (often helium–neon lasers) come with a kit containing slits of various widths, apertures and gratings.

When using a laser in the laboratory strict **safety rules** must be observed. Looking directly at a laser beam can cause permanent eye damage, even blindness, so whenever possible, operators and students should be behind the laser with the beam pointing away from them. When the position of images needs to be marked on a screen it should be done with the students' backs to the laser and, if possible, wearing protective glasses.

Most of the experiments can be carried out using a laser pointer. These use a laser diode and are restricted to 5 mW beams and are therefore quite safe. They are, however, powerful enough to project patterns onto walls several metres from the source in subdued light.

If no kit is available with the laser, the following two experiments are recommended.

Experiment 1: Diffraction patterns for a single slit

The laser is positioned several metres from a screen (a piece of paper stuck to a wall will do), so that the beam gives a clear spot image on it. The open jaws of a set of Vernier calipers are placed so that the beam passes through the gap between them. The jaws are slowly closed until there is a narrow slit between them and a diffraction pattern similar to that shown in Figure 16.25 is seen on the screen.

Questions

1 Describe the appearance of the central maximum as the gap between the jaws is narrowed and then widened.
2 Calculate the width of the gap if the distance between the central maximum and the first minimum is 2.2 cm, the calipers are 3.40 m from the screen and the wavelength of the laser beam is 633 mm. (Hint: use $\sin \theta = \tan \theta$ as the angle is small.)

Experiment 2: Diffraction grating

The grating is placed close to the laser between the source of the beam and a screen. The beam is passed through the grating and the pattern is observed on the screen. The distance of the grating to the wall is measured and the positions of the maxima are marked on the screen. Measure the distance, x, of each maximum from the central maximum and find the angle θ, using $\tan\theta = \frac{x}{D}$. The wavelength of the light is found using $n\lambda = d\sin\theta$.

Questions

1 Describe the appearance of the diffraction pattern produced by the grating.
2 Calculate the number of lines per cm on the grating if the position of the first-order maximum from the central maximum is 25.5 cm, the screen is 2.00 m from the grating and the wavelength of the laser beam is 633 nm.

3 It is suggested that the accuracy of the answer to **2** could be improved by using the position of the fourth-order spectrum. Explain why this is true and give one other method by which the uncertainty in the final answer could have been reduced.

If no lasers are available the experiments can be adapted using a ray box with a series of coloured filters. Because the intensity of the light is much less than that of a laser, the experiments must take place in a darkened room with the screen much closer to the source. This will mean that accurate data are more difficult to measure.

Diffraction gratings are commonly used to determine the wavelengths of unknown sources, and are particularly useful for identification of elements by measuring the wavelengths of line spectra. Details of the production of line spectra will be given in Chapter 17.

Resolution

In Chapter 15 you saw that medical ultrasound images showed much less detail than X-rays. The reason given for this was that the X-rays have a much shorter wavelength than the ultrasound. Although this is valid, it is only part of the explanation. Image resolution is limited by the diffraction that occurs at the aperture of the receiver and so depends not only on the wavelength but also on its diameter.

Figure 16.26 shows the positions of images on the retina from two distant objects. The resolving power of the eye is the smallest angular separation of the objects for which I_1 and I_2 can be distinguished separately. The diffraction patterns formed by the light passing through the pupil of the eye are shown for the images of the objects as their separation decreases. As the images converge, the diffraction patterns begin to overlap on the retina. When the central maximum of one image coincides with the first minimum of the other, they can just be resolved as separate entities, but for smaller angles they merge into a single image. The diameter of the objective lens is much bigger than the pupil, so if the two objects were viewed using a telescope, the central maxima would be narrower and a greater resolution would be possible ($a \approx \frac{\lambda}{D}$, where D is the diameter of the aperture).

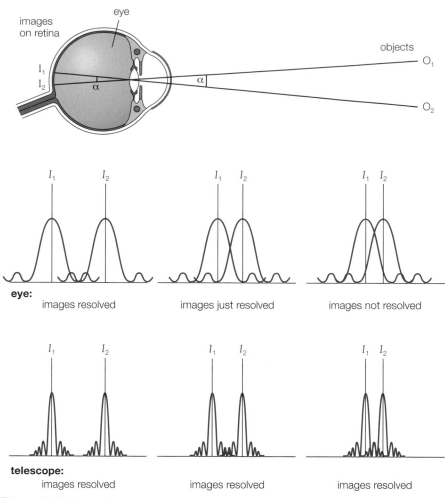

Figure 16.26 Resolution.

Activity 16.4

Estimating the resolving power of your eye

Two small black dots are marked 1 mm apart on a piece of white paper. The paper is fixed to a wall and you walk away from it until the dots just appear as one. Your distance from the wall, x, is measured.

The resolving power,

$$\alpha - \frac{1 \times 10^{-3} \text{ m}}{x} \text{ radians}$$

Electron diffraction

In 1912, Max von Laue suggested that if X-rays had a wavelength similar to atomic separations, they should produce diffraction patterns when fired through single crystal materials. He earned a Nobel Prize for his work, and X-ray diffraction techniques have developed so that they now are able to explore the structures of the most complex molecules.

In von Laue's time, electrons had been shown to be particles of mass 9.1×10^{-31} kg carrying a charge of 1.6×10^{-19} C. The idea that particles like electrons could also behave as waves was proposed in 1924 by Louis de Broglie. In Chapter 17, you will use the expression:

$$\lambda = \frac{h}{mv}$$

where h is Planck's constant (6.6×10^{-34} J s). This is often referred to as the 'de Broglie wavelength' of the electron.

This equation links the wavelength to the momentum, (*mv*), of the 'particle'. The relationship was subsequently verified using electron diffraction and can be demonstrated using the vacuum tube shown in Figure 16.27. High-velocity electrons are fired from the electron gun at the graphite crystal. A voltage of 1 kV gives the electron a speed of about $2 \times 10^7 \, \text{m s}^{-1}$ and hence a de Broglie wavelength of around $4 \times 10^{-11} \, \text{m}$. The thin graphite crystal has a regular hexagonal structure with atomic separations similar to this wavelength. It behaves a bit like a three-dimensional diffraction grating and creates a diffraction pattern of concentric rings on the fluorescent screen. When the voltage across the electron gun is increased, the faster moving electrons have a shorter wavelength so the diffraction is less and the diameter of the rings is reduced.

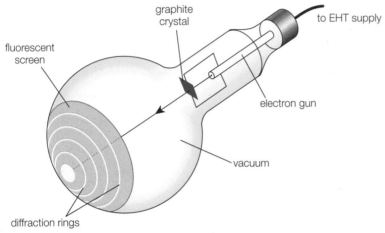

Figure 16.27 Electron diffraction.

Test yourself

11 What is meant by diffraction of waves?

12 Figure 16.28 shows the shape of ripples that have passed through a gap when the width of the gap is about three times the wavelength of the ripples.

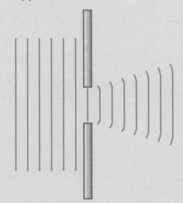

Figure 16.28

Make sketches to show the shape of the ripples when their wavelength is

a) about three times greater than the wavelength shown in the diagram

b) about one-third as much as the wavelength shown in the diagram.

13. Explain why the signal for a mobile phone (cell phone) may be lost in a valley but a radio signal from a transmitter in the same vicinity as the phone aerial can be received.

14 Why is the detail of an image produced using an optical telescope with an objective lens of diameter 10 cm greater than that of

a) an optical telescope with a 5 cm diameter lens

b) a radio telescope with a disc of diameter 10 m?

15 Why are X-rays used to study the atomic structure of crystalline materials?

16 Calculate the angular separation of the third maximum from the central maximum of the diffraction pattern for light of wavelength 599 nm passing through a diffraction grating with 250 lines per mm.

Exam practice questions

1 Coherent sources always have the same:

   A   amplitude   C   intensity

   B   frequency   D   phase.   **[Total 1 Mark]**

2 To form a minimum on an interference pattern, the path difference
  from two coherent sources that are in phase could be:

   A $\dfrac{\lambda}{4}$ C $\dfrac{3\lambda}{4}$

   B $\dfrac{\lambda}{2}$ D λ **[Total 1 Mark]**

3 A guitar string has a length of 0.62 m from the bridge to the first fret.
  The wavelength of the standing wave in the fundamental mode of
  vibration is:

   A   0.31 m   C   0.93 m

   B   0.62 m   D   1.24 m.   **[Total 1 Mark]**

4 The fundamental frequency of the G string on the guitar is 196 Hz.
  The second harmonic has a frequency of:

   A   98 Hz   C   294 Hz

   B   147 Hz   D   392 Hz.   **[Total 1 Mark]**

5 When a violin is played, a wave in the string and a sound wave
  are produced. Which of the following statements is true?

   A   The sound wave is longitudinal and stationary.

   B   The sound wave is transverse and progressive.

   C   The wave in the string is longitudinal and progressive.

   D   The wave in the string is transverse and stationary.   **[Total 1 Mark]**

6 A student connects two loudspeakers to the same terminals of a signal
  generator. She walks along a line parallel to the speakers and hears a
  series of maxima and minima, with a minimum when she is equidistant
  from the speakers. This is because:

   A   there is a no path difference between her and the two speakers

   B   there is a path difference of half a wavelength between her and the two
       speakers

   C   one speaker is much louder than the other

   D   she has connected the speakers in antiphase.   **[Total 1 Mark]**

7 The central maximum of a single-slit diffraction pattern can be made
  wider by increasing the:

   A   frequency   C   slit width

   B   intensity   D   wavelength.   **[Total 1 Mark]**

8 Describe how the principle of superposition is used to reduce the engine noise heard by a fighter pilot. Why are communication signals unaffected by this process? **[Total 3 Marks]**

9 A child dips a wire frame into soap solution to blow some bubbles. She notices that there are multi-coloured patterns on the film when it is held up to the light.

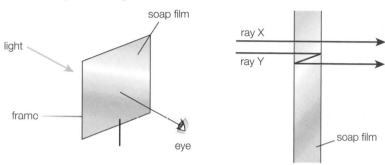

Figure 16.29

The second diagram in Figure 16.29 shows a cross-section of the soap film with two adjacent rays passing through a certain region. Ray X passes straight through the film, but ray Y undergoes two reflections at the inner surfaces before emerging. In this region the thickness of the film is about 450 nm. The wavelength of blue light in the soap solution is about 300 nm.

a) Explain why this region appears blue to the child. **[3]**

b) The soap film increases in thickness from top to bottom. Suggest a reason for this. **[1]**

c) When a red lamp is viewed through the film, a series of bright and dark horizontal lines is seen. Explain this effect, and determine the minimum thickness of the film for a dark line to appear if the wavelength of red light in the solution is 500 nm. **[4]**

[Total 8 Marks]

10 Figure 16.30 shows a simplified cross-section of part of a DVD.

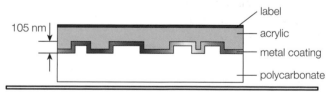

Figure 16.30

a) Light of wavelength 640 nm from a laser passes through the plastic coating, which has a refractive index of 1.52. The light is reflected by the flat section and is scattered in all directions if it strikes a bump. Calculate:

   i)   the frequency of the light from the laser

  ii)   the speed of this light in the plastic coating. **[3]**

b) Destructive interference occurs between the light scattered by a bump and that reflected from a flat section.

i) Show that the wavelength of the laser light in the plastic coating is approximately 420 nm.

ii) Hence explain how destructive interference can occur in this situation. **[3]**

[Total 6 Marks]

11 A recorder is basically an open pipe in which a standing wave is produced by blowing into the mouthpiece.

a) i) Draw the positions of the nodes and antinodes for a standing wave in an open-ended pipe in the fundamental mode and for the first overtone.

ii) Express the wavelength of the note in terms of the length of the tube in each case. **[4]**

b) The effective length of a recorder is from the open end at the mouthpiece to the first open hole along its length (Figure 16.31).

Figure 16.31

In an experiment to determine the speed of sound in air, the notes played for a range of finger positions were recorded onto a spectrum analyser. The frequencies for each length of tube are given in the table.

l/m	f/Hz	$\frac{1}{l}$/m^{-1}
0.485	349	
0.435	392	
0.385	440	
0.345	494	
0.325	523	
0.245	698	
0.215	784	

i) Copy and complete the table and plot a graph of f against $\frac{1}{l}$.

ii) Show that $f = \frac{v}{2l}$, where v is the speed of sound in air.

iii) Measure the gradient of your graph and hence calculate a value for v. **[8]**

c) A recorder is said to play sharp (higher pitched) at high air temperatures and flat (lower pitched) at low temperatures. Give a reason for these variations. **[2]**

[Total 14 Marks]

12 In microwave ovens, the microwaves reflect off the metal walls. In basic models with no revolving tray or rotating reflectors, direct and reflected rays can interfere to produce hot and cold spots, leading to uneven cooking.

a) Explain how these hot and cold spots are produced.　　　**[2]**

b) A thin slice of cheese cooked in the oven was found to have small molten regions about six centimetres apart. Estimate the wavelength of the microwaves and show that the microwave frequency is approximately 2.5 GHz.　　　**[2]**

Figure 16.32 shows a diagram of the microwave oven, in which two waves reach the point X having travelled different paths. The source and X can be assumed to be at the midpoints of the upper and lower surface.

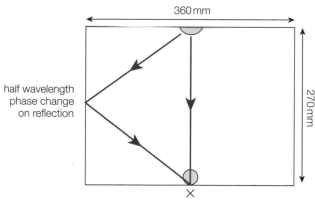

Figure 16.32

c) Explain why there is a hot spot at X.　　　**[3]**

[Total 7 Marks]

13 The fundamental frequency of the sound emitted by a stretched wire is given by the equation

$$f = \frac{1}{2l} \sqrt{\frac{T}{\mu}}$$

where T is the tension in the wire, l is the length of the wire and μ is its mass per unit length.

a) Write a plan for experiments to check the validity of the equation for a particular piece of wire. Your plan should include:

　i) a list of all materials to be used

　ii) a fully labelled diagram of the apparatus

　iii) a description of all measurements to be taken and the appropriate instrument to be used for each reading

　iv) comments on how the experiment will be safely performed

　v) a discussion on how the data collected will be used to verify the relationship.　　　**[8]**

b) Explain how you would use the results of one experiment to determine the value of μ of the wire.　　　**[2]**

[Total 10 Marks]

14 A single filament (festoon) lamp is viewed through a fine slit as shown in Figure 16.33a.

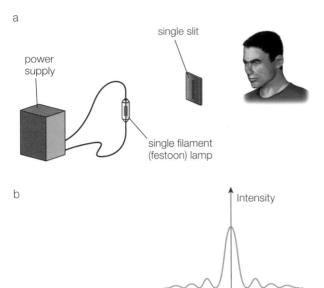

a

single slit

power supply

single filament (festoon) lamp

b

Intensity

Figure 16.33

The intensity profile for green light is given in Figure 16.33b.

a) Using the same scales, sketch the diffraction patterns for:

   i) a narrower slit

   ii) red light with the same slit width

   iii) blue light with the same slit width. **[3]**

b) A second lamp is placed close to and parallel with the first, and the pair are viewed through the slit with a red filter, then a green filter and finally a blue filter. If the green filaments are just distinguishable as separate images, use your sketches to explain the appearance of the red and blue images. **[3]**

[Total 6 Marks]

15 A ray box is set up in a dark room so that a single ray of white light is projected onto a screen 1.00 m from the front of the box. A diffraction grating with 500 lines per mm is placed just in front of the ray box so that a diffraction pattern is observed on the screen.

a) Describe the appearance of the central maximum and the first-order spectra. **[3]**

b) Explain what measurements you take from the first-order spectra in order to determine the range of wavelengths of the visible spectrum and how you would make the appropriate calculations from these measurements. **[3]**

The range of visibility of light from a ray box is about 400 nm–700 nm.

c) Calculate the approximate values of the angular separation between the central maximum and:

16 Superposition of waves

i) the red and violet positions on the first-order spectrum

ii) the red end of the second-order spectrum

iii) the violet end of the third-order spectrum. **[4]**

A student suggested that it would improve the accuracy of the values of the wavelength if measurements were taken from the second-order or third-order spectra.

d) Explain why the student's suggestion may be valid, but why it is not feasible using the grating in the above experiment. **[2]**

[Total 12 Marks]

Stretch and challenge

● 16 A Young's double-slit experiment is set up to measure the wavelength of light emitted from a sodium vapour lamp. The apparatus used consists of a single slit aligned with a double slit and a scale viewed through a microscope eyepiece, as shown in Figure 16.34.

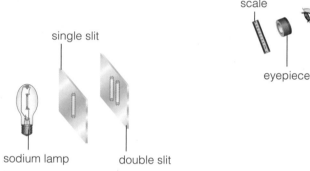

Figure 16.34

a) Describe how diffraction of the light at the slits leads to the production of interference fringes on the scale. **[3]**

b) For a particular experiment the slit separation is 0.80 mm and the distance from the slits to the scale is 50.0 cm. The distance across ten fringes is measured as 3.7 mm. Calculate the wavelength of sodium light. **[2]**

c) A thin piece of clear glass is placed over one slit. Describe and explain the effect this has on the interference pattern. **[3]**

d) The relative intensity of the fringes varies on each side of the central (maximum intensity) fringe. Sketch a graph showing how the relative intensity of the lines changes with distance from the central maximum value. **[2]**

e) It is observed that there is a gap in the interference pattern where the fourth fringe on each side of the central maximum should appear. Determine the width of the single slit. **[3]**

f) If the sodium lamp were replaced by a white light source, describe the pattern obtained and compare it with the original pattern. **[3]**

[Total 16 Marks]

17

Particle nature of light

Prior knowledge

In this chapter you will need to:
→ recall earlier work on waves and, in particular, that the interference, diffraction and polarisation of light suggest that light is an electromagnetic wave
→ understand that light forms part of what we call the electromagnetic spectrum, which stretches from γ-rays of wavelength as short as 10^{-16} m to radio waves having wavelengths as long as 10^3 m
→ be familiar with the concept of a simple atomic model in which the nucleus (protons and neutrons) is surrounded by electrons (of equal number to the protons in the nucleus).

The key facts that will be useful are:
→ speed of waves: $v = f\lambda$
→ speed of electromagnetic waves in a vacuum: $c = 3.0 \times 10^8$ m s^{-1}
→ work done on a charge: $W = QV$.

Test yourself on prior knowledge

1 Calculate the wavelength of an electromagnetic wave of frequency 909 kHz. What type of wave is this?

2 What is the frequency of an electromagnetic wave of wavelength 12.25 cm? In which part of the electromagnetic spectrum does such a wave occur?

3 What is the wavelength of an X-ray of frequency 7.5×10^{18} Hz? Explain why such X-rays can be diffracted using a graphite crystal.

4 X-rays are produced by firing electrons at a tungsten target. The X-rays in Question 3 were produced by accelerating electrons through a potential difference of 1.0 kV. Calculate the kinetic energy gained by the electrons and hence the speed of these electrons when they hit the target (electron charge = 1.6×10^{-19} C; electron mass = 9.1×10^{-31} kg).

17.1 Some early theories

Sir Isaac Newton was undoubtedly one of the world's greatest scientists. In a letter to Robert Hooke in 1676, he rather modestly wrote, 'If I have seen further it is by standing on ye shoulders of Giants.'

It is important that you understand how science evolves – or what is called 'scientific method'. This consists of the investigation of phenomena and acquisition of new knowledge, or the correction and integration of previous knowledge. It is based on the collection of data through observation and experimentation and the formulation and testing of hypotheses.

The very nature of light has been the subject of discussion and debate since the time of the Greek philosophers some 2500 years ago. The study of how theories of light have evolved is therefore a good way to get an understanding of how we analyse and evaluate scientific knowledge and processes.

'And God said, "Let there be light"; and there was light.' These words appear right at the very beginning of the Bible and reflect the importance that light has in all religions. During Diwali – the 'festival of lights' – Hindus celebrate the victory of good over evil and of light over darkness. The Koran says that 'God is the light of Heaven and Earth'. Evidence that very early religions worshipped light is provided by great monuments such as the Pyramids and Stonehenge.

Without light we would not even exist. Life depends on three things: long-chain carbon molecules, water and light. The Earth had all three, and so within the oceans a rich organic soup that ultimately bore life began to form. It is no wonder, then, that scientists have been fascinated by light and many have devoted much of their lives in trying to unravel its mysteries.

Some 2500 years ago, the Greek philosophers were divided in their opinions as to the nature of light. Pythagoras suggested that visible objects emitted a stream of particles that bombarded the eye, while Plato reasoned that light originated from the eye and then reflected off objects to enable us to see them. Euclid first put forward the concept of light as rays travelling in straight lines, while Aristotle thought of light as waves and tried to relate colour to music (as did Newton some 2000 years later).

When Western science, like art and literature, emerged from the dark ages (no pun intended!), the scientists of the day once more debated the nature of light. In the seventeenth century, great men such as Galileo, Kepler, Huygens and Newton pushed back the barriers of physics and astronomy with their experiments and theories. Galileo attempted, unsuccessfully, to measure the speed of light and concluded that if it was not infinite (as was thought to be the case by many), it was certainly very fast. It was left to the Danish astronomer Olaus Roemer to make the first determination of the speed of light in 1676, 34 years after Galileo's death. His value of $200\,\text{million m s}^{-1}$ was remarkably close to today's defined value of $299\,792\,458\,\text{m s}^{-1}$. Yes, we *define* the speed of light in a vacuum to nine significant figures as it is such a fundamental constant within our Universe! Since 1983, the *metre* has been defined as being the distance travelled by light in a vacuum in $1/299\,792\,458\,\text{s}$. This means that the speed of light c is now a defined quantity and distances are measured accurately using laser pulses and atomic clocks.

Meanwhile, Newton and Huygens were fiercely debating the nature of light. Huygens, like Newton, also contributed much to the development of mechanics and calculus. Newton proposed the **corpuscular theory**, in which he thought of light as being made up of corpuscles (or tiny particles). Using this theory, he was able to give a mathematical interpretation of the laws of reflection and refraction. He imagined the light particles to bounce off surfaces, like a ball bouncing off a wall, to explain reflection and suggested that the light particles changed speed when they moved from one material to another to explain refraction.

Huygens put forward a completely different idea, however. He maintained that light consisted of waves, like ripples spreading out when a stone is dropped in a pond. His **wave theory** was also able to explain the laws of reflection and refraction, with one fundamental difference from Newton – the wave theory required the speed of light to be *less* in glass or water than in air, whereas in the particle theory the opposite was the case.

Thomas Young provided further evidence for the wave theory around 1800. His famous two-slit experiment demonstrated the phenomenon of interference (see Chapter 16), which could be satisfactorily explained at the time only by considering light to be in the form of waves.

Test yourself

1 The photograph shown in Figure 17.1 is of the interference fringes produced by shining a laser through a double slit.

   The slit separation $s = 0.25\,mm$. The fringes were displayed on a screen placed at a distance $D = 1.95\,m$ from the slits.

   a) Use the photograph, which is full scale, to determine the fringe width x, (i.e. the distance between the centres of successive bright fringes). Explain what you did in order to get as precise a value as possible.

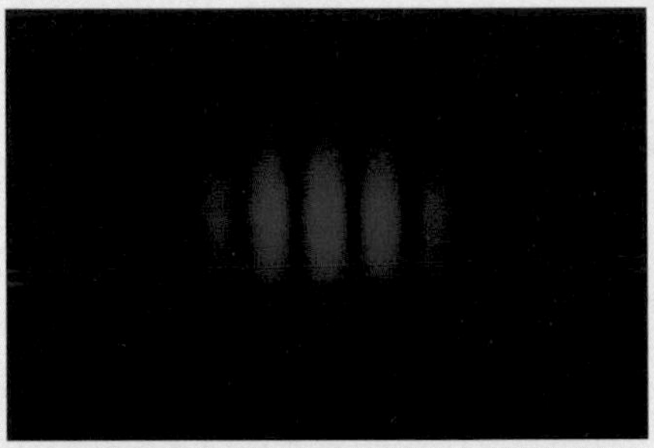

   b) In an experimental arrangement such as this, the wavelength, λ, of the laser light can be found using the formula $\lambda = \dfrac{xs}{D}$. Show that the wavelength is about 630 nm.

Figure 17.1 Interference fringes from a school laboratory Young's slits experiment using a laser.

   c) With reference to the photograph, suggest whether this value for the wavelength is reasonable.

Tip

Warning! The equation $\lambda = \dfrac{xs}{D}$ can only be used when $x \ll D$, which in most cases is true when the wavelength is very small, as it is here. In all other interference problems you must work from first principles, using the geometry of the set up.

There was only one way to resolve the argument – to measure the speed of light in air and then in glass or water!

That's just what Foucault did. In 1850, he showed qualitatively that the speed of light in water was indeed slower than it was in air. Further, more accurate, measurements confirmed that the ratio of the speed of light in air to that in water was equal to the refractive index of water – exactly as predicted by the wave theory.

About the same time, James Clerk Maxwell developed a model for an electromagnetic wave based on the mathematics of electric and magnetic fields (which are considered in the A level course). From Maxwell's equation for an electromagnetic wave, it was possible to calculate the speed of the wave from the electric and magnetic field constants. This gave a value for the speed of light that agreed exactly with that found experimentally.

The wave theory proposed the existence of a fluid medium, called the aether, necessary for transmission of the waves. A celebrated experiment performed by Michelson and Morley in 1887 failed to detect the motion of the Earth through the aether, and it was left to the genius of Albert Einstein to resolve the matter. In 1905, he took the Michelson–Morley experiment as the starting point of his theory of relativity. This theory, which concluded that all motion was relative, denied the existence of any stationary medium and so effectively abolished the aether. Game, set and match to the wave theory … or was it?

17.2 Intensity of light

In everyday usage, the word 'intensity' means the strength or level of something. In physics, it has a more precise meaning. We define **intensity** as the power per unit area. The units of intensity are therefore $W\,m^{-2}$.

When we talk about the radiation from the Sun falling on the Earth, we sometimes use the phrase **radiation flux density** instead of intensity. You need to be familiar with both terms.

Key terms

Intensity is the power per unit area and has units of $W\,m^{-2}$.

Radiation flux density = intensity = $\dfrac{power}{area}$

Example

A solar heating panel has an area of $2.8\,m^2$. How much energy falls on the panel per second if the solar radiation flux density is $300\,W\,m^{-2}$ when:

a) the panel is normal to the Sun's radiation

b) the radiation makes an angle of 55° with the panel, as in Figure 17.2.

Answers

a) Power = radiation flux density × area = $300\,W\,m^{-2} \times 2.8\,m^2 = 840\,J\,s^{-1}$

b) If the angle between the radiation and the panel is 55°, we need to consider the component of the radiation that falls at right angles (that is, 'normal') to the panel, so the energy reduces to: $840\,\sin 55° = 690\,J\,s^{-1}$.

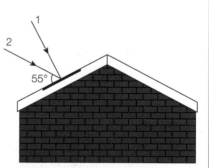

Figure 17.2

If we are a distance r from a point source of radiation, the radiation will have spread out in all directions over a sphere of surface area $4\pi r^2$ by the time it reaches us. The intensity will therefore be:

$$\text{intensity} = \frac{power}{4\pi r^2}$$

Example

The solar constant (the power of the radiation from the Sun falling normally on $1\,m^2$ of the outer surface of the Earth's atmosphere) is approximately $1.4\,kW\,m^{-2}$. Estimate the power radiated by the Sun, assuming that it is 150 million kilometres away.

Answer

$$\text{Intensity} = \frac{power}{4\pi r^2}$$

$$\Rightarrow power = \text{intensity} \times 4\pi r^2$$

$$= 1.4 \times 10^3\,W\,m^{-2} \times 4\pi \times (150 \times 10^6 \times 10^3\,m)^2$$

$$= 4.0 \times 10^{26}\,W$$

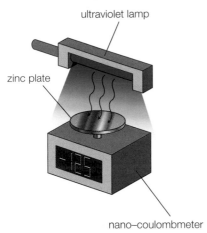

ultraviolet lamp

zinc plate

nano–coulombmeter

Figure 17.3

17.3 Photoelectric effect

The **photoelectric effect** can be demonstrated by means of a negatively charged zinc plate, as shown in Figure 17.3.

A freshly cleaned zinc plate is given a negative charge (for example, with a polythene rod – see Figure 7.1). The nano-coulombmeter indicates that in visible light, even from a 60 W lamp shining directly onto the zinc plate, the zinc does not discharge (or at least only very slowly if the atmosphere is humid). However, if even a weak ultraviolet light is shone onto the zinc plate, it begins to discharge immediately.

Safety note: Care must be taken to minimise the exposure to ultraviolet radiation and not to direct the source towards observers or irradiate the hands.

On the other hand, if the zinc plate is given a *positive* charge, *nothing at all* is observed as any photoelectrons transmitted would immediately be attracted back again by the positive charge.

This phenomenon is called the photoelectric effect – light ('photo') causing the emission of electrons ('electric'). Energy from the light is given to the electrons in the zinc, and some electrons near the surface of the zinc gain enough energy to escape from the attraction of the positive charge on the nucleus. These electrons are called **photoelectrons**. As the visible light is much more intense (and therefore has more energy) than the ultraviolet light, how can we explain why this happens with the ultraviolet and not the visible light?

Once more it is Einstein to the rescue! In 1905, he explained the photoelectric effect in terms of a **quantum theory**. The word 'quantum' means a fixed indivisible amount. For example, we say that charge is quantised because all electric charges are made up of a whole number multiple of 'fixed' electron charges (that is, 1.60×10^{-19} C). Five years earlier, Max Planck had proposed mathematically that electromagnetic radiation did not exist as continuous waves but as discrete bundles (quanta) of energy, which we now call **photons**. Each photon has a discrete amount of energy, hf, where f is the frequency of the radiation and h is a universal constant, now called the Planck constant ($h = 6.63 \times 10^{-34}$ J s). This theory enabled Planck to explain successfully the distribution of energy with wavelength for the radiation from a hot body. Einstein applied Planck's theory to the photoelectric effect.

Tip

Remember that *one* photon (if it has enough energy) will release *one* photo electron.

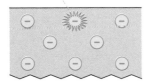

photon of
visible light

loosely bound
electrons near
zinc surface

a) *hf* for visible light is less than the energy needed for an electron to escape–nothing happens

Figure 17.4 Photoelectric emission.

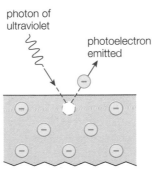

photon of
ultraviolet

photoelectron
emitted

b) *hf* is much greater for the higher frequency ultraviolet and so the electron is given enough energy to escape

Instead of the energy from the light being gradually absorbed by the electrons near the surface of the zinc until they had enough energy to escape, Einstein reasoned that the electron would be emitted only if a **single quantum** of the light had enough energy for the electron to escape. The frequency of visible light is not high enough for a photon (energy = hf) to provide the necessary energy for an electron to escape; however, ultraviolet light has a higher frequency than visible light. Its quanta each have sufficient energy to release an electron, so the zinc plate begins to discharge immediately in the presence of ultraviolet light.

The intensity of a light source depends on both the number of quanta and the energy associated with each quantum. However intense a visible light source may be, the energy of each of its quanta will be insufficient to liberate an electron from the zinc, so the zinc plate will not discharge in the presence of visible light.

Example

An advertisement for a green laser pointer gives the data in the table.

Wavelength	532 nm
Power	5 mW
Beam diameter at source	1.5 mm
Current	300 mA
Power supply	2 × 1.5 V AAA cells

1 What is the efficiency of the laser in converting electrical energy to light energy?

2 What is the intensity (radiation flux density) of the light beam close to the laser?

3 How much energy is in one photon of the light?

4 How many photons per second are emitted by the laser?

5 The manufacturer claims that this laser is 'much brighter than a 5 mW red laser'. Suggest why this statement may be justified.

Answers

1 Electrical power consumed
$$P = IV$$
$$= (300 \times 10^{-3})\,A \times 3.0\,V$$
$$= 0.90\,W \text{ (or } 900\,mW)$$

$$\text{Efficiency} = \frac{\text{output}}{\text{input}} \times 100\% = \frac{5\,mW}{900\,mW} \times 100\% = 0.5\%$$

2 $\text{Intensity} = \dfrac{\text{power}}{\text{area}} = \dfrac{5 \times 10^{-3}\,W}{\pi(0.5 \times 1.5 \times 10^{-3}\,m)^2} = 2.8\,kW\,m^{-2}$

3 Energy of one photon, $E = hf$, where f is given by $c = f\lambda$
$$f = \frac{c}{\lambda} = \frac{3.00 \times 10^8\,m\,s^{-1}}{532 \times 10^{-9}\,m} = 5.64 \times 10^{14}\,Hz$$
$$E = hf = 6.63 \times 10^{-34}\,J\,s \times 5.64 \times 10^{14}\,s^{-1} = 3.7 \times 10^{-19}\,J$$

4 Energy emitted per second $= 5\,mW = 5 \times 10^{-3}\,J\,s^{-1}$

$\text{Number of photons emitted per second} = \dfrac{5 \times 10^{-3}\,J\,s^{-1}}{3.7 \times 10^{-19}\,J}$
$$= 1.4 \times 10^{16}\,s^{-1}$$

5 Green light (wavelength 532 nm) is in the middle of the visible spectrum (400–700 nm). As the human eye is most sensitive to light in this region, the green laser will seem much brighter than a red laser of comparable power, so there may be some justification in the manufacturer's claim.

Test yourself

2 A 60 W filament lamp is suspended from the ceiling in the middle of a room so that it is 2.4 m above the floor. The lamp is 5% efficient.

a) What does 'the lamp is 5% efficient' mean?

b) State what is meant by the intensity, or radiation flux density, of light.

c) The intensity at a distance r from a point source of radiation is given by the equation, $\text{intensity} = \frac{\text{power}}{4\pi r^2}$. Estimate the intensity of the light shining on the floor. What assumptions have you had to make in arriving at your answer?

3 The energy emitted by a typical laptop screen is 5.0 W.

a) Assuming that the average wavelength of the radiation is 600 nm, how many photons does the screen emit per second?

b) If the screen has a resolution of 1024 × 768 pixels, how many photons, on average, are emitted per second by each pixel? Explain why this is an average value.

Key term

An **electron-volt** is the work done on (or the energy gained by) an electron when it moves through a potential difference of 1 volt.

Tip

Remember:

$1\,\text{eV} = 1.6 \times 10^{-19}\,\text{J}$

17.4 Electron-volt

In the question about a green laser, we found that the energy of a photon of green light was $3.7 \times 10^{-19}\,\text{J}$. This is a very small amount of energy, so rather than keep having to include powers of 10^{-19} when considering photon energy, we often use the **electron-volt** (eV) as a convenient unit of energy.

As an electron-volt is the work done on an electron when it moves through a potential difference of 1 volt, from $W = QV$ we get:

$$1\ \text{electron-volt} = 1.6 \times 10^{-19}\,\text{C} \times 1\,\text{V} = 1.6 \times 10^{-19}\,\text{J}$$

As the electron-volt is a very small unit of energy, we often use keV ($10^3\,\text{eV}$) and MeV ($10^6\,\text{eV}$). For example, typical X-rays have energy of $120\,\text{keV}$ and alpha particles from americium-241 (commonly found in smoke detectors) have energy of $5.6\,\text{MeV}$.

Example

Calculate:

a) the energy, in electron-volts, of a photon of light from a red laser pointer with a wavelength of 650 nm.

b) the wavelength of a 120 keV X-ray.

Answers

a) $E = hf = \dfrac{hc}{\lambda} = \dfrac{6.63 \times 10^{-34}\,\text{J s} \times 3.00 \times 10^{8}\,\text{m s}^{-1}}{650 \times 10^{-9}\,\text{m}}$

$\quad = 3.06 \times 10^{-19}\,\text{J}$

$\quad = \dfrac{3.06 \times 10^{-19}\,\text{J}}{1.6 \times 10^{-19}\,\text{J eV}^{-1}} = 1.9\,\text{eV}$

b) $E = 120\,\text{keV} = 120 \times 10^3\,\text{eV} \times 1.6 \times 10^{-19}\,\text{J eV}^{-1} = 1.92 \times 10^{-14}\,\text{J}$

$\quad E = hf = \dfrac{hc}{\lambda}$

$\quad \rightarrow \lambda = \dfrac{hc}{E} = \dfrac{6.63 \times 10^{-34}\,\text{J s} \times 3.00 \times 10^{8}\,\text{m s}^{-1}}{1.92 \times 10^{-14}\,\text{J}} = 1.0 \times 10^{-11}\,\text{m}$

17.5 Einstein's photoelectric equation

If a photon of energy hf has more than the bare minimum energy needed to just remove an electron from the surface of a metal (called the **work function**, symbol ϕ), the remaining energy is given to the electron as kinetic energy, $\frac{1}{2}mv^2$.

Applying the conservation of energy, the maximum kinetic energy $\frac{1}{2}mv_{\text{max}}^2$ that an electron can have is given by:

$$hf = \phi + \frac{1}{2}mv_{\text{max}}^2$$

This is known as Einstein's photoelectric equation. Electrons emitted from further inside the metal will need more than the work function to escape and so will have less than this maximum kinetic energy. Therefore, electrons are emitted with a range of kinetic energies up to the maximum defined by the equation.

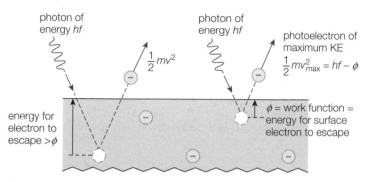

Figure 17.5 Einstein's photoelectric equation.

Einstein's equation indicates that there will be no photoelectric emission unless $hf > \phi$. The frequency that is just large enough to liberate electrons, f_0, is called the **threshold frequency**, so $\phi = hf_0$.

Tips

Note that two different approaches are used in the calculation below. In the first part, the photon energy is compared with the work function, while in the second part the photon frequency is compared with the threshold frequency. Either method is perfectly acceptable – the method that you choose to use may depend on the way in which the data are provided.

You are expected to be familiar with the electromagnetic spectrum and to know the approximate wavelengths of the visible region. A fully **quantitative** answer (i.e. giving numerical values to justify your answer) would therefore be expected in order to get full marks for this question.

Example

The work function for zinc is 4.3 eV. Explain why photoelectric emission is observed when ultraviolet light of wavelength in the order of 200 nm is shone onto a zinc plate but not when a 60 W filament lamp is used.

Answer

A photon of the ultraviolet light has energy:

$$E = hf = \frac{hc}{\lambda} = \frac{6.63 \times 10^{-34} \text{ J s} \times 3.00 \times 10^8 \text{ m s}^{-1}}{200 \times 10^{-9} \text{ m}}$$

$$= 9.95 \times 10^{-19} \text{ J}$$

$$= \frac{9.95 \times 10^{-19} \text{ J}}{1.6 \times 10^{-19} \text{ J eV}^{-1}} = 6.2 \text{ eV}$$

This is greater than the 4.3 eV work function for zinc. This means that each photon will have sufficient energy to remove an electron from the surface of the zinc and photoemission will occur.

The shortest wavelength (highest frequency) visible light is 400 nm (at the blue end of the visible spectrum). The threshold frequency, f_0, for zinc is given by:

$$\phi = hf_0 \Rightarrow f_0 = \frac{\phi}{h} = \frac{4.3 \text{ eV} \times 1.6 \times 10^{-19} \text{ J eV}^{-1}}{6.63 \times 10^{-34} \text{ J s}} = 1.0 \times 10^{15} \text{ Hz}$$

The frequency of the blue end of the visible spectrum is:

$$f = \frac{c}{\lambda} = \frac{3.00 \times 10^8 \text{ m s}^{-1}}{400 \times 10^{-9} \text{ m}} = 7.5 \times 10^{14} \text{ Hz (that is, } < 1.0 \times 10^{15} \text{ Hz)}$$

Photoemission therefore will not take place.

17.6 Phototube

A phototube is the name given to a particular type of **photocell** that generates photoelectrons when light falls on a specially coated metal cathode. The other types of photocells are **photovoltaic** photocells, in which an e.m.f. is generated by the presence of light across the boundary of two semiconducting materials, and **photoconductive** cells, or **light-dependent resistors** (LDRs). As we saw in section 11.12 (on page 169), an LDR is a semiconductor whose resistance decreases (that is, it becomes a better conductor) when it is exposed to electromagnetic radiation. This is because the photon energy release more electrons to act as charge carriers: n increases in $I = nAvq$.

Example

A phototube can be used to investigate the photoelectric effect by connecting a variable potential difference across it and measuring the current in it with a very sensitive ammeter, as shown in Figure 17.6.

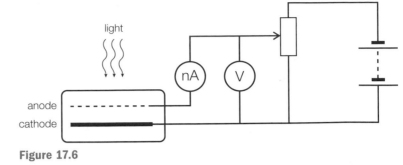

Figure 17.6

If the power supply is connected so that the anode is *negative*, the reverse potential difference does work slowing down the photoelectrons. When the work done is equal to the maximum kinetic energy of the photoelectrons, they will not reach the anode and the current will be zero. The reverse potential to just do this is called the **stopping potential**, V_s. It is given by:

$$eV_s = \frac{1}{2}mv_{max}^2$$

Measuring the stopping potential enables us to determine the maximum kinetic energy of the photoelectrons.

The graph in Figure 17.7 shows how the maximum kinetic energy $\frac{1}{2}mv_{max}^2$ of photoelectrons emitted from the surface of caesium varies with the frequency, f, of the incident electromagnetic radiation.

Use the graph to find:

a) the value for the Planck constant

b) the work function for caesium.

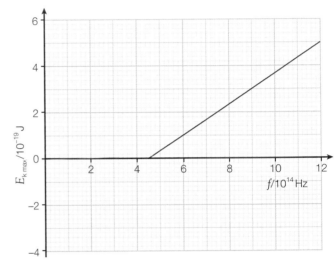

Figure 17.7

Answers

a) Rearranging $hf = \phi + \frac{1}{2}mv_{max}^2$ gives:

$$\frac{1}{2}mv_{max} = hf - \phi$$

Comparing with $y = mx + c$, h will be the gradient of the graph and $-\phi$ will be the intercept:

$$h = \frac{(5.0 - 0.0) \times 10^{-19}\,\text{J}}{(12.0 - 4.5) \times 10^{-14}\,\text{s}^{-1}} = 6.7 \times 10^{-34}\,\text{J s}$$

b) Extrapolating the graph to the y-axis gives $-\phi = -3.0 \times 10^{-19}\,\text{J}$

$$\phi = \frac{3.0 \times 10^{-19}\,\text{J}}{1.6 \times 10^{-19}\,\text{J eV}^{-1}} = 1.9\,\text{eV}$$

Test yourself

4 The table gives some data for a blue, a green and a red laser.

Colour	Frequency/Hz	Wavelength/nm	Quantum energy/eV
1	a)	633	b)
2	6.55×10^{14}	c)	d)
3	e)	f)	2.34

Copy and complete the table by adding answers for a)–f) and hence the three colours 1–3.

5 Laser 3 in Question 4 is shone on a sodium surface for which the work function is 2.28 eV.

a) Explain why photoelectrons will be emitted.

b) Calculate the maximum kinetic energy of these photoelectrons in joules.

c) What is meant by the threshold frequency of a surface? Calculate the value of the threshold frequency for the sodium surface.

d) What reverse voltage would have to be applied to prevent photoelectrons being emitted when laser 2 is shone on the sodium surface?

e) Explain why no photoelectrons are emitted at all when laser 1 is shone on the surface.

17.7 Atomic spectra

A flame can be coloured by holding the salts of certain metals in it: a yellow flame is produced by sodium, a lilac flame by potassium and a green flame by barium. If the flame is viewed through a diffraction grating, a series of bright lines is observed. Sodium gives two yellow lines close together, potassium a red line and a violet line and barium several lines running from red to violet – the brightest being green. Such spectra are called **emission spectra** as light is being emitted, or given out. We will define this later when we see what causes the emission.

In the middle of the nineteenth century, Kirchhoff established that the spectra were characteristic of the atoms and molecules that produce them. In his researches Kirchhoff collaborated with the chemist Bunsen, who provided him with chemicals of the highest purity and the burner that now bears his name, in order to provide a suitable flame.

Spectral lines can also be produced by applying a large potential difference between the ends of a tube filled with gas at low pressure. Such devices are called gas discharge tubes, and they can be seen all around us in the familiar form of street lights. These appear yellow as they contain sodium, which predominantly emits deep yellow wavelengths at 589.0 nm and 589.6 nm. Such light is considered safer to drive under than white light, as its monochromatic output improves the perception of contrast and allows the light to penetrate fog and rain with the minimum of dispersion.

Activity 17.1

Observing emission spectra from a gas discharge tube using a diffraction grating

The arrangement shown in Figure 17.8a is set up.

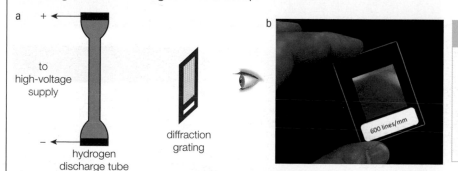

Safety note

WARNING! As the gas discharge tube used in Activity 17.1 has a very high voltage connected to it, the utmost care must be taken and the activity should only be undertaken by students under direct supervision.

a to high-voltage supply hydrogen discharge tube diffraction grating b 600 lines/mm

Figure 17.8 a) Experimental arrangement, b) diffraction grating.

The discharge tube is switched on and viewed through a diffraction grating as shown in Figure 17.8b. Line spectra either side of the central maximum can be seen as shown in Figure 17.9. The colours of the lines observed will depend on the gas inside the discharge tube. The experiment can be repeated using tubes having different gases inside, such as sodium, helium and neon.

Questions

1 Typically a high-voltage supply will have a 50 MΩ resistance connected in series. Explain the reason for this.
2 Explain why the central maximum is the same colour as the tube as seen without the diffraction grating.
3 Explain, using the diffraction grating formula $n\lambda = d\sin\theta$, why
   a) the blue lines of each order spectra are closer to the centre than the red lines
4 Show that the value of the grating spacing shown in Figure 17.8b is about 1.7×10^{-6} m.
5 Calculate the angular difference, in degrees, between the first order blue and red lines that would be observed using the grating shown in Figure 17.8b. Take the wavelengths to be 400 nm and 660 nm, respectively.

Figure 17.9 Spectra from hydrogen gas discharge tube seen through a diffraction grating.

When a hydrogen-filled tube is observed in detail through a diffraction grating, using an instrument called a spectrometer, four lines can be identified: two blue lines, a bluish-green line and a red line. Balmer analysed the frequency of these lines and discovered in 1885 that the frequency, f, of the hydrogen lines was given by a simple mathematical formula. Balmer's equation was modified by Rydberg, who wrote it in the form:

$$f = R\left(\frac{1}{2^2} - \frac{1}{m^2}\right)$$

where R is a constant (now called the Rydberg constant) and m had the values 3, 4, 5 and 6 for each of the lines in turn.

Figure 17.10a shows the visible spectrum for white light. Figure 17.10b shows the **emission spectrum** of hydrogen and Figure 17.10c shows the **absorption spectrum** for hydrogen. Absorption spectra are discussed on page 287.

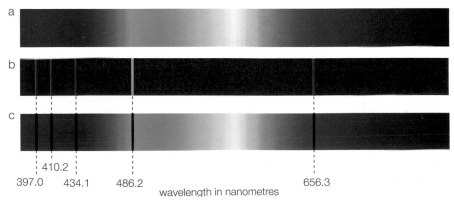

a

b

c

397.0　410.2　434.1　486.2　656.3

wavelength in nanometres

Figure 17.10 Hydrogen spectra.

Balmer performed some elegant mathematical analysis without the aid of a calculator, let alone a computer, but he was unable to explain why the frequencies were given by a mathematical series formed by the difference of two discrete terms. The mystery was solved by Bohr in 1913. He gave a simple and remarkably exact explanation of the hydrogen spectrum by applying quantum theory to the model of the nuclear atom proposed two years earlier by Rutherford, according to which the hydrogen atom consists of a single electron revolving around a proton. The Rutherford model is discussed in more detail in Unit 4 of the Year 2 Student's book.

Bohr proposed that the electron orbits the proton like a satellite orbits the Earth, with the necessary force being provided by the attractive nature of the electrical charges of opposite sign on the electron (negative) and proton (positive). From the value of this attractive electrical force he was able to calculate the radius of orbit of the electron, and also its energy, when it was most stable, or in its ground state.

Just as satellites can orbit at different distances from the Earth, Bohr reasoned that the electron could be given energy (or **excited**) and therefore exist in less-stable, higher-energy orbits. If sufficient energy is supplied, the electron can escape completely from its atom. This is called ionisation, and the energy needed for an electron to escape from its lowest energy level is called the ionisation energy of the atom.

By applying quantum theory to his model, Bohr concluded that the electron could only exist in certain **discrete** or **quantised** orbits. This means orbits having distinct, fixed amounts of energy.

As a particular value of energy is associated with each orbit, we say that the electrons have discrete or quantised **energy levels**. When an electron drops from a higher energy level to a lower level, it gives out the energy difference in the form of one quantum of radiation, hf.

Thus, when an electron drops from energy level E_2 to a level E_1:

$$hf = E_2 - E_1$$

and we get an **emission spectrum**.

Key terms

An electron is in its **ground state** when it is at its lowest possible energy level.

Ionisation energy is the energy that must be supplied for an electron in the lowest energy level to just escape from the atom.

A **quantum** is a fixed, or discrete, amount of energy, $E = hf$.

The **emission spectrum** of a chemical element or compound is the spectrum of frequencies of electromagnetic radiation emitted when the electrons in an atom make a transition from a high-energy state to a lower-energy state.

Example

Use the data in Figure 17.11 to answer the following questions.

1 Calculate the energy in joules that would have to be supplied for an electron at the lowest energy level to escape from the atom (i.e. the ionisation energy of the atom).

2 What is the wavelength of the spectral line emitted when an electron falls from the −1.51 eV level to the −3.41 eV level? Suggest what colour this line would be.

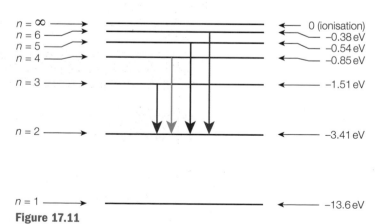

Figure 17.11

3 Between which energy levels must an electron fall to emit a blue line of wavelength 434 nm?

4 Without doing any detailed calculations, explain why the radiation emitted when an electron falls to the lowest energy level cannot be seen.

Answers

1 Ionisation energy $= +13.6\,\text{eV} = 13.6\,\text{eV} \times 1.6 \times 10^{-19}\,\text{J eV}^{-1} = 2.2 \times 10^{-18}\,\text{J}$

2 $hf = E_2 - E_1 = (-1.51\,\text{eV}) - (-3.41\,\text{eV}) = 1.90\,\text{eV}$

$$= 1.90\,\text{eV} \times 1.6 \times 10^{-19}\,\text{J eV}^{-1} = 3.04 \times 10^{-19}\,\text{J}$$

$$f = \frac{3.04 \times 10^{-19}\,\text{J}}{6.63 \times 10^{-34}\,\text{J s}} = 4.59 \times 10^{14}\,\text{Hz}$$

$$c = f\lambda \Rightarrow \lambda = \frac{c}{f} = \frac{3.00 \times 10^8\,\text{m s}^{-1}}{4.59 \times 10^{14}\,\text{s}^{-1}} = 6.54 \times 10^{-7}\,\text{m} = 654\,\text{nm}$$

This is visible red light.

3 $f = \dfrac{c}{\lambda} = \dfrac{3.00 \times 10^8\,\text{m s}^{-1}}{434 \times 10^{-9}\,\text{m}} = 6.91 \times 10^{14}\,\text{Hz}$

$$E_2 - E_1 = hf = 6.63 \times 10^{-34}\,\text{J s} \times 6.91 \times 10^{14}\,\text{s}^{-1} = 4.58 \times 10^{-19}\,\text{J}$$

$$= \frac{4.58 \times 10^{-19}\,\text{J}}{1.6 \times 10^{-19}\,\text{J eV}^{-1}} = 2.86\,\text{eV}$$

This could arise from an electron falling from the $-0.54\,\text{eV}$ level to the $-3.41\,\text{eV}$ level (allowing for rounding differences).

4 The smallest possible energy difference when an electron falls to the lowest energy level is:

$$\Delta E = (-3.41\,\text{eV}) - (-13.6\,\text{eV}) = 10.19\,\text{eV}$$

This is much greater than the 2.86 eV in Question 3, which gives a blue line, so the frequency of the emitted radiation must be much greater than that for blue light. This means it would be in the ultraviolet region of the spectrum and so would not be visible.

The answer to Question 4 of the example above explains why the equation discovered by Balmer contains the term:

$$\frac{1}{2^2}$$

The '2' represents an electron falling to the second energy level. As the lines produced by electrons falling to the lowest energy level were in the ultraviolet range, they could not be seen. By substituting '1' into the first term, the wavelengths of spectral lines in the ultraviolet range could be predicted. It was not long before the existence of these lines was established experimentally by Lyman.

This is a good example of how science develops. Scientists had observed spectral lines for some time, and then Balmer derived a mathematical equation to fit these observations. Bohr proposed a model, based on the quantum theory, which gave a physical explanation of Balmer's equation. Bohr's model predicted further lines in the ultraviolet spectrum, which were subsequently found by Lyman.

The development of the quantum theory led to wide-ranging experiments on solid materials ('solid-state physics') in the 1920s and 1930s. After the Second World War, during which the attention of physicists was diverted elsewhere, Bardeen and Brattain discovered the transistor effect in 1947. With Shockley, they were jointly awarded the Nobel Prize for physics in 1956 for *'their researches on semiconductors and their discovery of the transistor effect'*. Their work on semiconductors led to the development of silicon chips, without which we would not have today's computers or your iPod!

It is worth mentioning, in the context of the Second World War, that in 1935 Robert Watson-Watt – a Scottish physicist – was asked by the Air Ministry to investigate the possibility of developing a 'death-ray' weapon using radio waves. Watson-Watt did not create a 'death-ray' weapon, but he did find that his radio transmitters could create an echo from an aircraft that was more than 200 miles away. Hence evolved **radar**, an acronym for **ra**dio **d**etection **a**nd **r**anging (appropriately, a palindrome – a word that reads the same backwards as forwards!). Today, worldwide travel would be virtually impossible without radar. (What if the *Titanic* had had radar?) We would not have DVD players, mobile phones or microwave ovens (or police speed traps come to that!).

Another application of quantum emission is in the **laser**. This is another acronym, standing for **l**ight **a**mplification by **s**timulated **e**mission of **r**adiation. In a laser, a flash of light is injected into a suitable material, now usually a semiconductor. This excites some of the electrons in the semiconductor atoms into higher unstable energy levels. When the electrons return to their lower energy levels, they emit photons. As these photons pass through the semiconductor, they stimulate emission in other atoms, causing an avalanche effect. The light intensity is rapidly amplified and a beam of monochromatic light is emitted; the colour of this depends on the laser material. This beam is **coherent** (see Chapter 16, page 248). In other words, the emitted waves are in phase with one another and are so nearly parallel that they can travel long distances. Originally developed for scientific research, lasers are now widely used in industry, medicine and about the home – for example, in DVD players and laser printers.

In 1802, Woolaston observed a number of dark lines in the spectrum of the Sun. These lines were accurately mapped by von Fraunhofer in 1814 and now bear his name. Figure 17.13 depicts von Fraunhofer's original drawing of the absorption spectrum of the Sun.

Figure 17.12 a) A police officer using a radar speed gun to check the speed of a car. b) A radar screen on the bridge of a chemical tanker approaching the Dutch coast near Rotterdam. The coast is represented by the bright yellow area bottom left of the radar image. The yellow line moving from the centre (the ship's position) towards bottom right indicates the ship's heading.

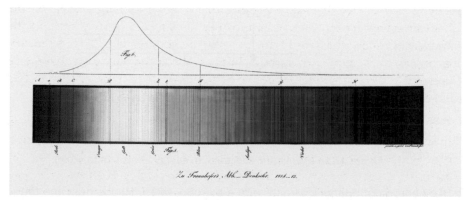

Figure 17.13 Joseph von Fraunhofer's absorption spectrum of the Sun.

Such a spectrum is called an **absorption** spectrum, because the gases surrounding the Sun selectively absorb certain wavelengths of the Sun's radiation. Electrons in the gas absorb quanta (that is, precise amounts) of energy, *hf,* that are exactly sufficient to excite the electrons into higher energy levels. When the electrons drop down to lower energy levels, they re-emit the quanta of radiation randomly in all directions, so the corresponding wavelengths are not observed. This selective absorption of particular wavelengths is an example of **resonance**.

Key term

In an **absorption spectrum**, radiation is absorbed by a substance at frequencies that match the energy difference between two quantum states of electrons in an atom or molecule of the substance.

The frequencies of the absorption lines due to a particular element are exactly the same as the frequencies of the emission spectrum of that element. This is shown in Figure 17.10 on page 284. This means we can determine the elements present in the gases surrounding the Sun and other stars and galaxies.

Indeed, the element helium was first discovered in 1868 by observation of a strong yellow emission line in the Sun's spectrum (hence its name, derived from the Greek word for Sun, *helios*). A few years earlier the French philosopher Comte had stated: 'There are some things of which the human race must remain in ignorance, for example the chemical composition of the heavenly bodies.' This shows the rashness of prophecy!

Test yourself

6 Figure 17.14a shows the emission spectrum for a mercury vapour lamp. Figure 17.14b is a simplified diagram of the energy levels for electrons in a mercury atom.

a) Define the term ionisation energy and determine the ionisation energy, in joules, for a mercury atom.

b) What is meant by an emission spectrum? Illustrate your answer with reference to Figure 17.14a and Figure 17.14b.

c) Show, by calculating its wavelength in nm, that transition Y in Figure 17.14b corresponds to the green line shown in Figure 17.14a.

d) Without doing any calculation, explain which lines in the spectrum correspond to transitions W, X and Z.

e) Mercury emits a very strong line of wavelength 254 nm. Explain why this would not appear in the spectrum shown in Figure 17.14a even if the wavelength scale was extended.

f) Calculate the energy, in eV, of a quantum of radiation of wavelength 254 nm. Hence determine the transition corresponding to the 254 nm line.

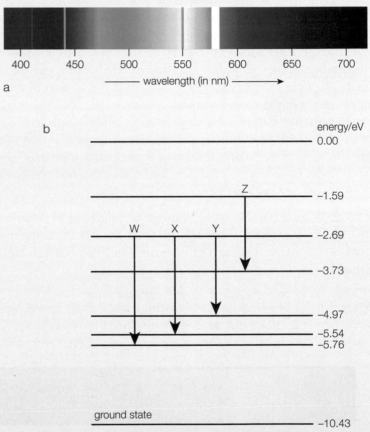

Figure 17.14 a) Emission spectrum, b) energy levels (not to scale).

g) Explain what happens when a 2.14 eV photon of energy interacts with a mercury atom. Hence describe the absorption spectrum produced when light from a filament lamp passes though mercury vapour.

17.8 Wave–particle duality

In earlier chapters we saw that light (and, indeed, *all* electromagnetic radiation) has properties associated with waves – for example, polarisation, diffraction and interference. In this section, we have seen that electromagnetic radiation sometimes behaves like particles – for example, the photoelectric effect and atomic spectra. So, is light made of particles or waves? The answer is both!

At the end of Chapter 16 we saw that electrons, which we think of as particles, could be diffracted and therefore could behave like a wave. Therefore, particles can sometimes behave like waves and waves can sometimes behave like particles! These ideas emerged from the theory of quantum mechanics, which was developed in the early twentieth century. In 1924, de Broglie linked waves and particles by the formula:

$$\lambda = \frac{h}{p}$$

where λ is the wavelength corresponding to a particle of momentum p (i.e. a particle of mass m travelling with a velocity v). Thus was born the concept of **wave–particle duality**.

Examples

1 Calculate the wavelength of a football of mass 440 g travelling at 30 m s^{-1} and hence explain why it does not show wave-like properties.

2 a) Show that an electron of energy 15 keV has a wavelength of about 10^{-11} m.

  b) In which region of the electromagnetic spectrum is this wavelength?

Answers

1 $\lambda = \frac{h}{p}$ gives $\lambda = \frac{h}{mv} = \frac{6.63 \times 10^{-34}\,\text{J s}}{0.44\,\text{kg} \times 30\,\text{m s}^{-1}} = 5.0 \times 10^{-35}\,\text{m}$

This wavelength is far too short for us to observe any wave properties of the football.

2 a) $15\,\text{keV} = 15 \times 10^3\,\text{eV} \times 1.6 \times 10^{-19}\,\text{J eV}^{-1} = 2.4 \times 10^{-15}\,\text{J}$

$$\frac{1}{2}mv^2 = 2.4 \times 10^{-15}\,\text{J}$$

$$\Rightarrow v^2 = \frac{2 \times 2.4 \times 10^{-15}\,\text{J}}{9.11 \times 10^{-31}\,\text{kg}}$$

$$\Rightarrow v = 7.26 \times 10^7\,\text{m s}^{-1}$$

$$\lambda = \frac{h}{mv} = \frac{6.63 \times 10^{-34}\,\text{J s}}{9.11 \times 10^{-31}\,\text{kg} \times 7.26 \times 10^7\,\text{m s}^{-1}}$$

$$= 1.00 \times 10^{-11}\,\text{m}$$

  b) This wavelength is in the X-ray region of the electromagnetic spectrum.

The answer to part b) of the worked example explains why electrons can be diffracted by crystals in which the planes of atoms are in the order of 10^{-10} m apart. Wave–particle duality and the de Broglie equation are considered in more detail in the Year 2 Student's book in the chapter on particle physics.

Three hundred years after Newton and Huygens argued about whether light consisted of particles or waves, we see that both were right. When considering some phenomena, it is more meaningful to think of light as waves and when considering other phenomena it is more meaningful to think of light as particles. This principle of complementarity may be one of the most profound contributions of twentieth-century physics.

Let the last word rest with Newton who, in order to explain the interference effects that he observed, speculated that 'When a Ray of Light falls upon the surface of any pellucid [transparent] Body, and is there refracted or reflected, may not Waves or Vibrations, or tremors be excited in the refracting or reflecting Medium [substance] at the point of Incidence?'

Key term

The **complementarity principle** says that sometimes electrons have the properties of particles and sometimes the properties of waves, but never both together.

7 A scanning electron microscope uses electrons of energy 30 keV.

   a) Show that keV has units of energy.

   b) Calculate the energy of a 30 keV electron in joules.

   c) Hence show that the velocity of the electron is about $1 \times 10^8 \, \text{m s}^{-1}$. (Take the mass of an electron to be $9.11 \times 10^{-31} \, \text{kg}$).

   d) Calculate the momentum of the electron.

   e) Hence show that the wavelength corresponding to a 30 keV electron is about 0.007 nm.

   f) In which part of the electromagnetic spectrum is radiation of this wavelength?

   g) Suggest why electron microscopes rather than optical microscopes are used in the study of the structure of materials.

8 A sodium street lamp emits yellow light of frequency $5.09 \times 10^{14} \, \text{Hz}$.

   a) Calculate the wavelength of this radiation.

   b) Use the de Broglie equation to show that the momentum of a quantum of this radiation is about $1 \times 10^{-27} \, \text{kg m s}^{-1}$.

   c) The mass of a sodium atom is 23 u, where u = unified mass unit = $1.66 \times 10^{-27} \, \text{kg}$. Calculate the recoil velocity of a sodium atom when it emits a quantum of yellow light.

Exam practice questions

1 A 60 W reading lamp gives out only 5% of its power as visible light. The visible light intensity at a distance of 1.5 m from the lamp is approximately:

    A $0.11\,W\,m^{-2}$ **C** $0.42\,W\,m^{-2}$

    B $0.16\,W\,m^{-2}$ **D** $2.02\,W\,m^{-2}$ **[Total 1 Mark]**

Questions 2–4: When blue light of wavelength 447 nm is shone onto a sodium surface, photoelectrons are emitted. The work function of sodium is 2.28 eV.

2 The threshold frequency of sodium is:

    A $2.9 \times 10^{-34}\,Hz$ **C** $5.5 \times 10^{14}\,Hz$

    B $1.8 \times 10^{-15}\,Hz$ **D** $3.4 \times 10^{33}\,Hz$ **[Total 1 Mark]**

3 The energy of a photon of the blue light is:

    A 1.74 eV   **C** 2.78 eV

    B 2.28 eV   **D** 2.96 eV   **[Total 1 Mark]**

4 The potential difference that would have to be applied to just prevent photoemission would be:

    A 0.50 V   **C** 2.3 V

    B 0.68 V   **D** 5.1 V   **[Total 1 Mark]**

5 When a photon of sunlight is incident on a voltaic cell, an electron in the cell gains sufficient energy to move through a potential difference of 0.48 V.

    a) What is a photon?   **[2]**

    b) Show that the energy to move an electron through a potential difference of 0.48 V is about $8 \times 10^{-20}\,J$.   **[2]**

    c) Photons of sunlight typically have energy $4.0 \times 10^{-19}\,J$. Calculate the efficiency of conversion of the energy of the photon.   **[2]**

    d) What is the wavelength corresponding to a photon of this energy?   **[3]**

    [Total 9 Marks]

6 A laser pen, emitting green light of wavelength 532 nm, is shone onto a caesium surface that has a work function of 1.9 eV.

    a) What is meant by the term work function?   **[2]**

    b) Show that the energy of photons from the laser is about 2.3 eV.   **[4]**

    c) What is the maximum kinetic energy of the photoelectrons emitted from the caesium surface? Give your answer in joules.   **[2]**

    d) No photoelectrons are emitted when a red laser pen is shone onto the caesium surface. Suggest the reason for this.   **[2]**

    [Total 10 Marks]

7 A lighting manufacturer supplies red, green and yellow LEDs.

a) Explain which of these will emit photons having the highest energy. **[2]**

b) The manufacturer's catalogue gives the following data for the red LED:

Wavelength of maximum intensity   630 nm

Visible light emitted   18 mW

Power consumption   120 mW

   i) Show that the photons at the stated wavelength each have energy of about 3×10^{-19} J.

  ii) What is this energy in electron volts? **[4]**

c) i) Assuming that the LED acts as a point source and radiates equally in all directions, show that the intensity at a distance of 30 cm away is about $16\,\text{mW}\,\text{m}^{-2}$.

  ii) Assuming that the diameter of your pupil is 5 mm, approximately how many photons would enter your eye per second at this distance?

 iii) Explain, in terms of photons, why the intensity of the LED gets less as you move further away. **[7]**

d) What is the efficiency of the LED in converting electrical energy to light energy? **[2]**

e) Transport departments are beginning to replace filament bulbs in traffic lights with red, green and yellow LEDs. Discuss the advantages of this. **[2]**

[Total 17 Marks]

8 a) Most physicists believe that light can behave as both a wave and a particle. Name a property of light that shows it can behave as a wave. **[1]**

b) In 1916, Millikan published the results of an experiment on the photoelectric effect, which proved that light also behaves as particles. He shone monochromatic light onto plates made of different metals. What is meant by the term monochromatic? **[2]**

c) When light hit the plates, photoelectrons were produced. Millikan found the potential difference that was just large enough to stop these electrons being released. He also investigated how this stopping voltage varied with the frequency of the light used. **[3]**

The table shows the results of an experiment like Millikan's, in which sodium was used as the metal plate. Plot a graph of V_s against f.

Stopping voltage V_s/V	Frequency of light f/10^{14} Hz
0.26	5.01
0.43	5.49
0.75	6.28
1.00	6.91
1.18	7.41

d) The following photoelectric effect equation applies to this experiment:

$hf = \phi + eV_s$, where ϕ is the work function of the metal.

What information about the photoelectrons does the value of the term eV_s give? **[2]**

e) i) Use your graph to determine the threshold frequency of sodium.

  ii) Hence calculate the work function, in eV, for sodium. **[4]**

f) Explain why no electrons are emitted below the threshold frequency. **[2]**

[Total 14 Marks]

9 Figure 17.15, which is not to scale, shows some of the energy levels of a tungsten atom.

a) An excited electron falls from the −1.8 keV level to the −69.6 keV level. Show that the wavelength of the emitted radiation is approximately 0.02 nm. **[4]**

b) To which part of the electromagnetic spectrum does this radiation belong? **[1]**

[Total 5 Marks]

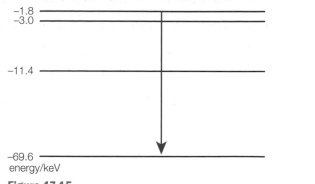

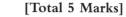

Figure 17.15

10 Figure 17.16 shows some of the main components of one type of fluorescent tube.

a) When the tube is switched on, a charge flows between the electrodes and the mercury atoms become excited. The mercury atoms then emit radiation.

   i) Explain the meaning of the word 'excited' as used above.

   ii) Explain how the excited mercury atoms emit radiation.

   iii) Explain why only certain wavelengths of radiation are emitted. **[6]**

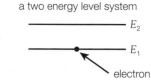

Figure 17.16

b) Some of the radiation emitted is in the ultraviolet part of the spectrum. Humans cannot see ultraviolet radiation, so the inside of the tube is coated with phosphor. The atoms of phosphor absorb the ultraviolet radiation and then emit visible light.

   i) Suggest why the phosphor emits wavelengths different from the mercury.

   ii) A typical classroom fluorescent tube takes a current of 200 mA. Calculate the amount of charge that flows during a lesson that lasts $1\frac{1}{2}$ hours.

   iii) Explain one advantage and one disadvantage of using a tube such as this rather than a tungsten filament light. **[8]**

[Total 14 Marks]

11 The diagrams shown in Figure 17.17 are taken from an explanation of how a laser works. Each diagram illustrates some aspect of a 'two-energy level system'. The system consists of an electron in an isolated atom.

a) What is meant by energy level? **[2]**

b) What is a photon? **[2]**

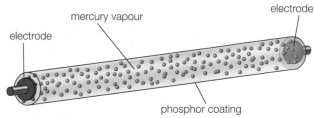

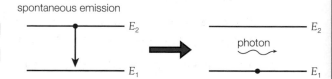

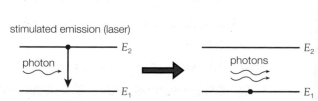

Figure 17.17

c) Write down a formula in terms of E_1 and E_2 for the energy of the photon in the absorption diagram. [1]

d) The laser light emitted by the stimulated emission process must have the same wavelength as the photon in the spontaneous emission diagram. Explain this. [2]

e) The laser light is said to be coherent. Explain the meaning of coherent. [2]

[Total 9 Marks]

12 A diffraction grating, having 50.0 lines per mm, is used to produce a diffraction pattern for light from a green laser pen on a screen placed 330 ± 1 mm from the grating. Figure 17.18 is a full-scale photograph of the diffraction pattern.

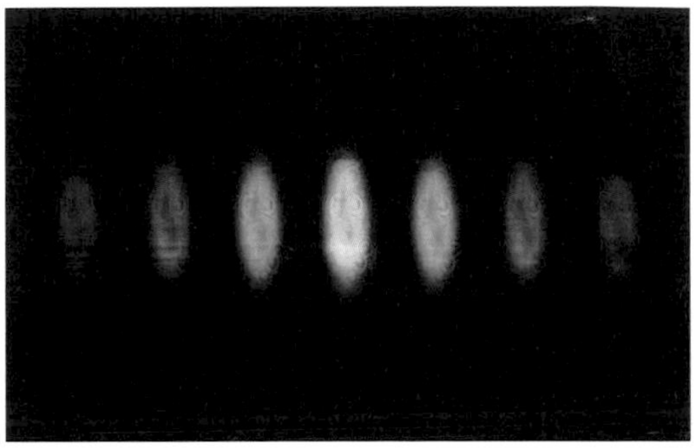

a) Take measurements from the photograph to find the angle, θ, of the first-order maximum and hence the wavelength, λ, of the laser light using the formula $n\lambda = d \sin\theta$. [6]

b) Estimate the uncertainty in your value for the wavelength. You may assume that the uncertainty in the value of the grating spacing is negligible. [6]

Figure 17.18

c) Discuss the extent to which your value for the wavelength agrees with the stated value of 543.4 nm. [2]

d) Explain two ways in which the uncertainty in the measurement of the wavelength could be reduced. [4]

[Total 18 Marks]

13 Diffraction patterns can be observed for light using a diffraction grating and for electrons using a crystal.

a) Show that the relationship between the momentum, p, of a particle and its kinetic energy, E_k, is given by $p = \sqrt{2mE_k}$. [3]

b) Hence show that the wavelength associated with a 15 keV electron is about 0.01 nm. [3]

c) Explain why a crystal can be used to produce electron diffraction with such electrons. [3]

[Total 9 Marks]

Stretch and challenge

14 An article about solar cells states that:

The silicon semiconductors in a solar cell are geared toward taking infrared light and converting it directly to electricity. Meanwhile, the visible spectrum is lost as heat and longer wavelengths pass through unexploited. This limits the efficiency of a solar cell to typically 20%. Photons (light particles) produce an electrical current as they strike the surface of the thin silicon wafers and a typical 12 volt panel about 120 cm by 60 cm will contain 36 cells wired in series to produce about 17 volts peak output. When under load (charging batteries for example), this voltage drops to 12 to 14 volts resulting in 75 to 100 watts for a panel of this size.

a) If the solar radiation flux density incident normally on the panel is $600 \, \mathrm{W \, m^{-2}}$, how much energy will fall on the panel per second? [2]

b) If the efficiency of the panel is 20%, how much electrical energy will be produced per second? Does this correspond with the data in the article? [2]

c) Suggest why, when under load, the voltage drops from about 17 V to 12–14 V. [3]

d) Use the data in the question to estimate the e.m.f. and internal resistance of a single solar cell. Explain how you chose the data to use in your estimation. [9]

e) Assuming that the average wavelength of the photons striking the panel surface is 550 nm:

   i) Calculate an approximate value for the number of photons striking the panel per second.

   ii) In which part of the electromagnetic spectrum is 550 nm?

   iii) Suggest why an average of 550 nm is assumed and why your answer is only very approximate. [8]

[Total 24 Marks]

15 a) Show that if the atoms in carbon had a cubic structure, the atoms would be about 0.2 nm apart. You will need to know that 6.02×10^{26} carbon atoms have a mass of 12 kg and that the density of carbon is $2.27 \times 10^3 \, \text{kg m}^{-3}$. [5]

b) Show that the de Broglie wavelength of a 4.0 kV electron is about 0.02 nm. (Electron mass = 9.11×10^{-31} kg, Planck constant = 6.63×10^{-34} J s). [5]

c) With reference to your answers to **a** and **b**, explain why 4.0 kV electrons could be diffracted by a carbon crystal. [2]

d) Figure 17.19 is a photograph (to half scale) of the diffraction pattern produced using the tube shown in Figure 16.27 on page 267. The accelerating voltage V is 4.0 kV.

Graphite, which is a form of carbon having a hexagonal structure, is used to diffract the electrons. The first two bright rings are the result of diffraction by two different planes of graphite atoms having a different lattice spacing, d. To a good approximation, the diameter, D, of the first two bright rings is given by the formula

$$D = \frac{a}{d \sqrt{V}} \text{ where } a \text{ (a constant)} = 3.32 \times 10^{-10} \, \text{m}^2 \, \text{V}^{\frac{1}{2}}.$$

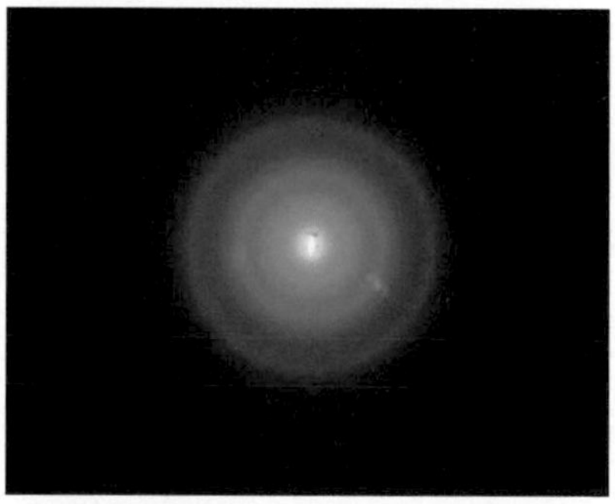

Figure 17.19

   i) Measure the diameters of the first two bright rings and hence determine the two different lattice spacings for graphite.

   ii) To what extent do your results confirm that the ratio of the two lattice spacings is $\sqrt{3}$?

   iii) Comment on your values of d compared with your answer to part **a**. [12]

[Total 24 Marks]

18 Maths in physics

The mathematical requirements for the AS and A level are given in Appendix 6 of the Edexcel specifications. Most of these have been used throughout the book as *Examples*, *Test yourself questions* and *Exam practice questions*.

In this chapter the mathematics is presented in sections covering a range of skills. In each section there are practice calculations to test these skills.

Bold text indicates that the requirement is for the **A level examination only**.

18.1 Numbers

The Test yourself questions in this section will need you to:

- be aware that the value of a physical quantity may be expressed to a certain number of significant figures and understand that the number of significant figures in a result should not be more than the smallest number of significant figures in the data
- express numbers in standard form and add, subtract, multiply and divide numbers in standard form with and without using a calculator
- use a calculator to raise numbers to powers, and to take square and cube roots of numbers
- remember the meanings of the prefixes p (pico-), n (nano-), μ (micro-), m (milli-), k (kilo-), M (mega-) and G (giga-).

Test yourself

1 Express the following numbers in standard form, to two significant figures:

  a) 0.0834   d) 171.3

  b) 0.003582   e) 824

  c) 0.612

2 In each of the following calculations give your answer to the appropriate number of significant figures.

  a) $7.32 \times 1.23 \times 2.2$

  b) $3.33 \times (2.41)^2 \times 1.234$

  c) $54 \times 1.87 \times 0.020$

3 *Without using a calculator*, find a value of

a) $10^3 \times 10^9$

b) $10^6 \times 10^2$

c) $10^{-3} \times 10^{-6}$

d) $10^6 \div 10^3$

e) $10^3 \div 10^{-6}$

f) $\dfrac{(10^3 \times 10^{-6})}{10^{-5}}$

4 *Using a calculator* find a value of

a) $10^6 \times 10^{-6}$

b) $10^6 \times 10^{-8}$

c) $10^2 \div 10^3$

d) $10^{-6} \div 10^3$

e) $1 \div 10^{-4}$

5 Do the following calculations, giving your answers in standard form, to the appropriate number of significant figures.

a) $\pi(21.5)^2$

b) $4\pi^2(1.75 \times 10^{-3})^2$

c) $\sqrt{\dfrac{0.12}{\pi}}$

d) $\sqrt[3]{(4.09 \times 10^8)}$

6 Using a calculator, determine the values of the following expressions.

a) $4^{3.1}$

b) 2^5

c) $10^{4.3}$

d) $\sqrt{(10^4)}$

e) $\sqrt[3]{(27)^2}$

f) $\dfrac{1}{0.0786}$

7 What is the percentage increase in a quantity if it is

a) doubled

b) trebled?

8 a) If 25% of X is 25, what is X?

b) If 13% of X is 28.8, what is X?

c) If 80% of X is 0.144, what is X?

9 The tolerance of the thickness of aluminium sheets produced in a factory is ±0.2%. The thicknesses of eight batches were tested and gave the following values

t/mm= 3.22, 3.13, 3.25, 3.16, 3.20, 3.24, 3.19, 3.27.

The required thickness is 3.20 mm. Which of the batches were out of tolerance?

10 Calculate the value of X in the following

a) $1.2\,\text{mA} \times 330\,\text{k}\Omega = X\,\text{V}$

b) $9.3\,\text{GHz} \times 32\,\text{mm} = X\,\text{m s}^{-1}$

c) $47\,\mu\text{F} \times 2.2\,\text{k}\Omega = X\,\text{s}$

18.2 Units and equations

The Test yourself questions in this section will need you to:

- remember that the base units in the SI system are the metre, kilogram, second, ampere and kelvin
- rearrange equations so that different quantities are the subject of the equation
- use equations to show how a derived unit may be expressed in terms of base units
- be able to check whether an equation is homogenous with regard to units.

Test yourself

1 Express the following quantities in the appropriate base units, in standard form to two significant figures.

a) 23.2 cm

b) 1.8 mm

c) 446 nm

d) 1.228 km

e) 38.2 g

f) 417 mg

g) 12 µg

h) 30 minutes

i) 20 ns

j) 12 hours

k) 303 mA

l) 1.5 litres

2 Express the following areas and volumes in the appropriate base units.

a) 2.3 cm²

b) 8.4 mm²

c) 0.023 cm²

d) 3.3 cm²

e) 91 mm²

3 Calculate the areas of the circles with the following diameters. In each case give your answer in m² and to two significant figures.

a) 1.25 m

b) 48 cm

c) 328 mm

d) 12.5 mm

4 The force, F, exerted on an area, A, by a pressure, p, is given by the equation $F = pA$. Calculate F, giving your answer to the appropriate number of significant figures, when $p = 1.01 \times 10^5 \, \text{N m}^{-2}$ and $A = 1.50 \, \text{m}^2$.

5 The volume of a cylinder is given by the equation $V = \pi r^2 h$. Calculate V, giving your answer in the base unit and to the appropriate number of significant figures, when $r = 2.44 \, \text{cm}$ and $h = 0.20 \, \text{m}$.

6 Rearrange the following equations so that the quantity shown in brackets after each equation is the subject of the equation.

a) $s = vt$; (v)

b) $A = \pi r^2$; (r)

c) $V = \pi r^2 h$; (h)

d) $v = u + at$; (t)

e) $s = ut + \frac{1}{2}at^2$; (a)

f) $\eta = 1 - \frac{T_1}{T_2}$; (T_2)

g) $P = I^2 R$; (I)

h) $V = \varepsilon - Ir$; (r)

i) $T = 2\pi\sqrt{\left(\frac{\ell}{g}\right)}$; (ℓ)

7 Use the equation $c = f\lambda$ to calculate f when $c = 3.00 \times 10^8 \, \text{m s}^{-1}$ and $\lambda = 598 \, \text{nm}$.

8 Use the equation $V = \frac{4}{3}\pi r^3$ to calculate r when $V = 2.6 \times 10^{-3} \, \text{m}^3$.

9 Use the equation $E = F\frac{\ell}{eA}$ to calculate e when $F = 12 \, \text{N}$, $\ell = 2.2 \, \text{m}$, $A = 5.3 \times 10^{-8} \, \text{m}^2$ and $E = 120 \, \text{GN m}^{-2}$.

10 **Use the equation $GM = \frac{4\pi^2 r^3}{T^2}$ to calculate T when $G = 6.67 \times 10^{-11} \, \text{N m}^2 \, \text{kg}^{-2}$, $M = 5.97 \times 10^{24} \, \text{kg}$ and $r = 6.37 \times 10^3 \, \text{km}$.**

11 Use the equation $F = ma$, where m is mass and a is acceleration, to express the unit of force, the newton, in base units.

12 Use the equation $p = \frac{F}{A}$, where F is force and A is area, to express the unit of pressure, the pascal, in base units.

13 Use the equation $W = Fs$, where F is force and s is distance, to express the unit of work, the joule, in base units.

14 Use the equation $P = \frac{W}{t}$, where W is energy and t is time, to express the unit of power, the watt, in base units.

15 Use the equation $I = \frac{Q}{t}$, where I is current and t is time, to express the unit of charge, the coulomb, in base units.

16 Use the equation $V = \frac{W}{Q}$, where W is the energy transferred and Q is electric charge, to express the unit of potential difference, the volt, in base units.

17 Use the equation $T = \frac{1}{f}$, where T is the time period, to express the unit of frequency, the hertz, in base units.

18 Use the equation $C = \frac{Q}{V}$, where Q is electric charge and V is potential difference, to express the unit of capacitance in base units.

19 Calculate the following, giving the results as numbers in standard form with a single unit.

   a) $25\,\text{m s}^{-1} + (5\,\text{m s}^{-2})(4.0\,\text{s})$

   b) $\frac{1}{2}(40\,\text{N m}^{-1})(0.05\,\text{m})^2$

   c) $(1.29\,\text{kg m}^{-3})(9.81\,\text{N kg}^{-1})(2.0\,\text{m})$

   d) $\frac{4}{3}\pi(0.12\,\text{m})^3(8900\,\text{kg m}^{-3})$

20 Check the homogeneity of the following equations with respect to units.

   a) $\Delta W = mg\Delta h$, where ΔW is the change in gravitational potential energy, m is mass and Δh is the change in height.

   b) $\Delta W = \frac{1}{2}\sigma\varepsilon$, where W is the change in elastic potential energy per unit volume, σ is stress and ε is strain.

   c) $\Delta p = \rho g\Delta h$, where Δp is the change in pressure, ρ is the density of a fluid and Δh is the change in height.

   d) $T = 2\pi\sqrt{\left(\frac{\ell}{g}\right)}$, where T is the period, ℓ is the length and g is the gravitational acceleration at the surface of the Earth.

   e) $E = c^2\Delta m$, where E is the energy equivalent of a mass m and c is the speed of light.

18.3 Relationships between quantities and graphs

The Test yourself questions in this section will need you to:

- understand the statement that one quantity is proportional to another
- be aware that the equation $y = mx + c$ represents a linear relationship and a graph of y against x gives a straight line of gradient m with an intercept c on the y-axis
- rearrange equations so that the quantities plotted are of the form $y = mx + c$
- calculate or measure the gradient of a straight-line graph
- calculate the gradient at a point on a non-linear graph by drawing a tangent to the curve at that point
- understand that the area between a graph and the horizontal axis may represent a physical quantity and be able to calculate a value of the quantity by measuring this area.

1 State the meaning of

   a) $x > 12\,\text{cm}$

   b) $10\,\text{kg} < m \leqslant 20\,\text{kg}$

   c) $V \propto r^3$

   d) $\Delta x = 25\,\text{mm}$

2 Write the following statements as equations using the appropriate mathematical symbols.

   a) The sine of the angle, θ_{m}, subtended by the first minimum positions of a single-slit diffraction pattern, is directly proportional to the wavelength, λ, of the light and inversely proportional to the slit width, d.

   b) **The pressure exerted by an ideal gas is directly proportional to the product of the mass, m, of the molecules and the mean square of their speeds, $<c^2>$, and inversely proportional to the volume, V, occupied by the gas.**

3 State which of the following statements are true and which are false.

   a) $F = mg$; so F is proportional to g if m is constant.

   b) $F = mg$; so F is proportional to m if g is constant.

   c) $P = \dfrac{W}{t}$; so P is proportional to W if t is constant.

   d) $P = \dfrac{W}{t}$; so P is proportional to t if W is constant.

   e) $P = \dfrac{W}{t}$; so P is proportional to $\dfrac{1}{t}$ if t is constant.

   f) $E_k = \dfrac{1}{2}mv^2$; so E is proportional to v if m is constant.

   g) $E = \dfrac{1}{2}mv^2$; so E is proportional to v^2 if m is constant.

   h) $E = \dfrac{1}{2}mv^2$; so E is proportional to m if v is constant.

4 Figure 18.1 shows three graphs representing varying quantities.

   a) Which of the graphs show a linear relationship between y and x?

   b) Which of the graphs show that y is proportional to x?

5 Calculate the gradients of the graphs shown in Figure 18.2.

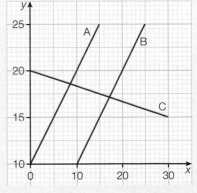

Figure 18.1

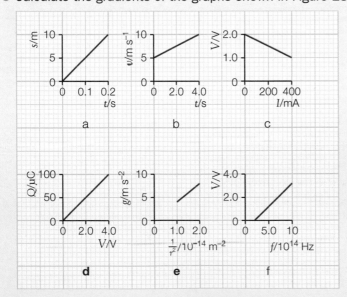

Figure 18.2

In each case express the unit of the gradient in its simplest form.

6 Calculate the value of the physical quantity represented by the area under the graphs drawn in Figure 18.3. In each case give a unit for the quantity.

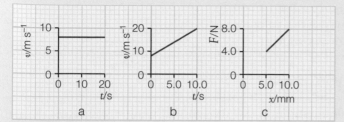

a   b   c

Figure 18.3

7 Estimate the value of the physical quantity represented by the area beneath the graphs drawn in Figure 18.4. In each case give a unit for the quantity.

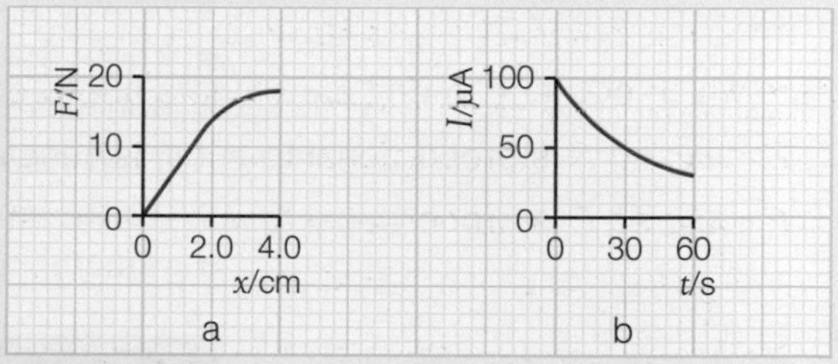

8 For the following relationships state the quantities that need to be plotted to obtain a straight-line graph. In each case state what is represented by the gradient of the graph.

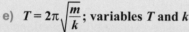

Figure 18.4

a) $E_k = \frac{1}{2}mv^2$; variables E and v

b) $c = f\lambda$; variables f and λ

c) $F = \dfrac{Gm_1m_2}{r^2}$; variables F and r

d) $f = \dfrac{1}{2\pi}\sqrt{\dfrac{1}{g}}$; variables f and l

e) $T = 2\pi\sqrt{\dfrac{m}{k}}$; variables T and k

f) $eV_s = hf - \varphi$; variables f and V_s

9 A variable resistor R is connected across a cell of emf, ε. Values of the potential difference, V, and the current, I, are measured for a range of values of R, and the following results are obtained:

I/mA	100	200	300	400	500
V/V	1.45	1.38	1.30	1.23	1.16

a) Plot a graph of V against I.

b) Is V proportional to I?

c) Calculate the gradient of the graph.

d) The relationship between V and I for the cell is in the form $V = mI + c$. Use the graph to determine the values of m and c.

10 The following table shows how the values of the intensity I of the light received from a small intense light source of power P varies with the distance x of the detector from the source.

x/m	0.40	0.50	0.60	0.70	0.80
I/W m⁻²	3.12	2.04	1.34	0.98	0.80

a) Plot graphs of

    i) I against x ii) I against $\dfrac{1}{x^2}$

b) What can you deduce from these graphs about the relationship between the intensity of light and the distance of the detector from the source?

c) It is suggested that the intensity of light at a distance x from a source of power P is given by the equation $I = \dfrac{P}{4\pi x^2}$. Assuming this to be correct, use graph **a ii** to calculate a value for the power of the source.

18.4 Angles

The Test yourself questions in this section will require you to:

- use angles measured in both degrees and radians
- be aware that neither radian nor degree is a proper SI unit
- know that π rad (radians) is equivalent to $180°$ (degrees)
- understand that angles can also be measured in degrees, °, minutes, ′, and seconds, ″.
- use the word 'subtend' when referring to angles subtended by an arc of a circle
- remember that for small angles, $\sin\theta \approx \tan\theta$ and that each is $\approx \theta$ (in radians).

Test yourself

1 Express the following angles in degrees to three significant figures.

  a) 0.20 rad   **b)** 2.00 rad   **c)** $\frac{\pi}{2}$ rad

2 Express the following angles in radians to three significant figures.

  a) 25°   **b)** 40°   **c)** 65°

3 What are the values of the following angles expressed as fractions or multiples of π radians?

  a) 90°   **b)** 360°   **c)** 30°

4 a) How many minutes of arc are there in one degree?

  b) How many seconds of arc are there in one degree?

5 a) How is the radian defined?

  b) Sketch a circle and two radii separated by an angle of 1.5 rad.

6 If the radius of the circle you have drawn in response to Question **5b** is 40 mm, how long is the arc subtended by the 1.5 rad angle?

7 A bicycle wheel of radius 35.7 cm picks up some mud and makes a dirty mark on the road that is exactly 1.00 m long. Calculate, to two significant figures, the angle subtended by the mud on the bicycle wheel

  a) in radians   **b)** in degrees.

8 The angle between the top and bottom of a telegraph pole is seen to be 6.0° when observed from a distance at ground level of 120 m.

  a) What is the height of the telegraph pole?

  b) What assumption, if any, have you made?

9 In the Year 2 Student's book, a unit of length called the parsec is mentioned. The parsec is defined as the distance from Earth at which the radius of the Earth's orbit around the Sun (1.5×10^{11} m) subtends an angle of exactly 1 second of arc. (1.5×10^{11} m is sometimes called an astronomical unit or 1 AU.) Calculate the size of 1 parsec in metres.

18.5 Trigonometry

The Test yourself questions in this section will need you to:

- remember that the ratios of the sides of right-angled triangles are called sine, cosine and tangent and that these relate to the size of the angle between two of the sides
- remember how each function relates to the ratio between the sides that are opposite the angle, adjacent to the angle and the hypotenuse of the triangle
- remember the values of the sine and cosine of commonly used angles such as $0°$, $30°$, $60°$, $90°$ and $180°$ (π radians)
- use an acute angle of a right-angled triangle with the length of one side to determine the lengths of the other two sides
- understand how to calculate the resolved components of a vector
- use Pythagoras' theorem to calculate the length of the third side of a right-angled triangle.

Test yourself

1 Use your calculator to find the values of

   a) $\sin\theta$ b) $\cos\theta$ c) $\tan\theta$, when $\theta = 35.4°$.

   Give your answers to three significant figures.

2 Without using your calculator, state the values of

   a) i) $\sin\theta$ ii) $\cos\theta$, when $\theta = 0°$

   b) i) $\sin\theta$ ii) $\cos\theta$, when $\theta = 90°\left(\dfrac{\pi}{2}\text{rad}\right)$

   c) i) $\sin\theta$ ii) $\cos\theta$ and $\tan\theta$, when $\theta = 45°$

   d) $\sin 30°$ and $\cos 60°$.

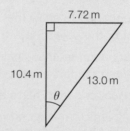

Figure 18.5

3 a) Use the triangle in Figure 18.5 to find the value of $\sin\theta$.

   b) Use the $\sin^{-1}$ function on your calculator to find the value of θ to three significant figures.

4 Check the value of θ in Figure 18.5 by

   a) finding $\cos\theta$ and then using the $\cos^{-1}$ function

   b) finding $\tan\theta$ and then using the $\tan^{-1}$ function.

5 In Figure 18.6 calculate

   a) the length AB

   b) the perpendicular distance from C to the line AB.

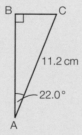

Figure 18.6

6 Figure 18.7 shows several vectors.

   Using only the cosine function, calculate the resolved components of the vector in the x- and y-directions.

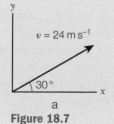 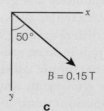

a   b   c

Figure 18.7

7 Use Pythagoras' theorem to check whether the three values given could be the lengths of the sides of a right-angled triangle.

a) 3 cm, 4 cm, 5 cm

b) 10 m, 20 m, 25 m

c) 5 m, 12 m, 13 m

d) 21 cm, 28 cm, 35 cm

e) 9 cm, 12 cm, 13 cm

f) 20 m, 48 m, 52 m

g) 9 m, 40 m, 41 m

8 Find the third side of a right-angled triangle if

a) the two short sides are 3.2 cm and 4.8 cm

b) the shortest side is 2.6 cm and the longest side is 6.2 cm

c) the shortest side is 0.462 cm and the longest side is 0.884 cm.

9 Figure 18.8 shows two forces of 20 N and 12 N acting at a point. The resultant force is represented by the diagonal of the rectangle. Calculate

a) the size of the resultant force

b) the direction of the resultant relative to the 20 N force.

Figure 18.8

10 Figure 18.9 shows an object of weight 10 N attached to a string. The object is held in equilibrium by a horizontal force of 8.0 N so that the string is at an angle θ to the vertical.

a) Write an equation to show that the vertical component of the tension and the weight of the object have the same size.

b) Write an equation to show that the horizontal component of the tension in the string and the force applied to the object have the same size.

c) Use the equations from parts **a)** and **b)** to show that θ is about 40°.

d) If the string is 50 cm long, how far has the object been displaced horizontally from its initial position vertically below the point of suspension?

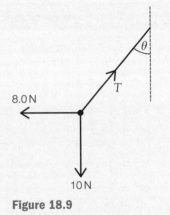

Figure 18.9

18.6 Sinusoidal oscillations

The Test yourself questions in this section will require you to

- **recognise the shape of the sine and cosine functions for an angle θ from 0 to 2π radians (from 0 to 360°)**
- **know what is meant by the amplitude (x_0), frequency (f) and period (T) of a sinusoidally varying quantity**
- **use expressions of the form $x = x_0 \sin(2\pi f t)$ or $x = x_0 \cos(2\pi f t)$ that represent a quantity x which varies sinusoidally with time**
- **understand that the maximum velocity of a sinusoidally varying quantity x is given by $v_{max} = 2\pi f x_0$**
- **understand that the maximum acceleration of a sinusoidally varying quantity x is given by $a_{max} = (2\pi f)^2 x_0$**
- **recognise that the above two statements for v_{max} and a_{max} are the result of differentiating the expression for x with respect to t**
- **use ω, which is often used in place of $2\pi f$, e.g. $x = x_0 \sin \omega t$ represents a simple harmonic motion (s.h.m.)**
- **recognise that in an undamped s.h.m. the shape of the sine and cosine functions continue for increasing values of the angle θ**

- remember that in damped s.h.m. the frequency and period of the motion continue to be fixed
- recognise that equations such as $a = \frac{d^2 x_2}{dt} = -(2\pi f)^2 x$ and $a = \frac{d^2 x}{dt} = -\omega^2 x$ can be used to define simple harmonic motions
- understand that in situations involving resonance there is a maximum transfer of energy between the driving force and the oscillating system.

Test yourself

1 What are the values of $\sin\theta$ and $\cos\theta$ for

   a) $\theta = 0°$

   b) $\theta = 90°$ or $\frac{\pi}{2}$ rad?

2 Describe the difference between a graph of $\sin\theta$ against θ and a graph of $\cos\theta$ against θ.

3 At what angles does a graph of $\sin\theta$ against θ vary most rapidly?

4 At what angles does a graph of $\cos\theta$ against θ have a slope of zero?

5 a) What is the unit of ω and $2\pi f$?

   b) What is the unit of ωt and $2\pi f t$?

6 Calculate

   a) the size of the maximum speed

   b) the size of the maximum acceleration.

   during a sinusoidal oscillation of amplitude $0.60\,\mu m$ and frequency $5.0\,kHz$.

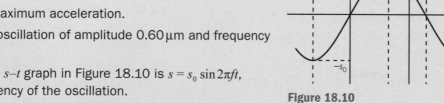

Figure 18.10

7 The equation for the s–t graph in Figure 18.10 is $s = s_0 \sin 2\pi f t$, where f is the frequency of the oscillation.

   a) The speed v at any instant is related to s by the equation $v = 2\pi s_0 \sqrt{(s_0^2 - s^2)}$. What is the value of v

   i) when $s = s_0$

   ii) when $s = 0$?

   b) State a mathematical relationship between the variation of s and a. Sketch a graph of your relationship and label the maximum value of a.

8 In the expression $x_0 \sin(2\pi f t)$, the angle is expressed in radians. Calculate a value for $x_0 \sin(2\pi f t)$, when x_0 is $2.0\,m$, f is $50\,Hz$ and $t = 13\,ms$. (Hint: make sure your calculator is dealing with angles in radians and not in degrees.)

9 Repeat the previous question for $\cos(2\pi f t)$ rather than $\sin(2\pi f t)$.

10 For a sinusoidal oscillation of frequency $f = 50\,Hz$, calculate a value for ω.

11 The tides move water up and down a beach about twice a day.

   a) What is this frequency in hertz?

   b) At a place where the amplitude of the tide is $35\,m$ as measured along a smoothly sloping beach, what (assuming the variation is sinusoidal) is the maximum rate at which the tide moves up the beach?

12 Sketch a graph of x against t for a *damped* s.h.m. of initial amplitude $100\,cm$. Label the time axis in oscillations and assume that the peak of the oscillation decreases by a factor of about 0.5 after each complete oscillation.

18.7 Exponential change

The Test yourself questions in this section will require you to:

- know that the logarithm of a number is *not* a physical quantity
- use logarithms both to base 10 ($\log_{10} x$ or lg x) and to base e ($\log_e x$ or ln x)
- understand the nature of exponentially varying quantities
- use the general expressions for exponential decrease or increase with time t, i.e. $X = X_0 e^{-kt}$ or $X = X_0 e^{kt}$, where kt is a number
- be able to take logarithms of both sides of these equations, giving $\ln X = -kt + \ln X_0$ or $\ln X = kt + \ln X_0$
- use the relationship between the half-life of a sinusoidally varying quantity and k or ω.

Test yourself

1 Use a calculator to find, to three significant figures

  a) lg 30   b) lg 300   c) lg 3000

2 Use a calculator to find, to three significant figures

  a) ln 30   b) ln 300   c) ln 3000

3 What pattern is there to your answers for the first question?

4 What are, to three significant figures

  a) i) lg 10   ii) lg 100   iii) lg 1000

  b) i) ln e   ii) ln e^2 iii) ln e^3?

5 Why is it *not possible* to find lg x or ln x when $x = 1.2\,\text{C s}^{-1}$?

6 Calculate the values of e^{kt} where

  a) $k = 0.68\,\text{s}^{-1}$ and $t = 2.5\,\text{s}$

  b) $k = 0.68\,\text{min}^{-1}$ and $t = 2.5\,\text{min}$.

7 Calculate values for (a) and (b) in the previous question, but with k being a negative quantity.

8 What is the value of e$^{-\lambda t}$ when $\lambda = 0.775$ minutes^{-1} and $t = 2.60$ hours?

9 Why might you expect the rate of growth of each of the following to be exponential initially but to stop being exponential after a time?

  a) The number of people who were ill with Spanish 'flu after the First World War.

  b) The number of people who currently own an iPad.

10 The number of people in a Facebook group is found to be 200 and to be increasing exponentially at a rate of 16 per month. What is the rate of increase when the number in the group is

  a) 300

  b) 1000?

11 The charge on a capacitor was initially 32 µC and 30 s later was found to be 24 µC. Assuming the capacitor is losing charge exponentially, what will be the charge on the capacitor after

  a) 60 s

  b) 90 s?

12 What is the half-life period for the decay of a radioactive source that decays according to the equation $N = N_0 e^{-\lambda t}$, where $\lambda = 1.21 \times 10^{-4}\,\text{y}^{-1}$?

13 A specimen of radioactive material contains 2.0×10^{10} atoms at $t = 0$. This specimen decays exponentially. When $t = 10\,\text{s}$, $N = 1.6 \times 10^{10}$.

  a) Draw a graph of N against t from $t = 0$ to $t = 50\,\text{s}$.

  b) Use your graph to read off the half-life period of the specimen.

14 The current I in an RC circuit varies with time t as $I = I_0 e^{-t/RC}$, where the initial current $I_0 = 100\,\text{mA}$.

  a) What is the unit of the product RC if its size is 0.33?

  b) After how long will the current have fallen to 50 mA?

15 The activity of a radioactive source is 5600 s^{-1} when it is first measured. The half-life period of the source is 3.0 h.

  a) What is the source's activity after 6.0 h?

  b) After how many half-life periods will the activity of the source have fallen to less than 1% of its initial value?

16 Atmospheric pressure decreases with height h above the Earth's surface. The equation for

the pressure is $p = p_0 e^{-kh}$, where the constant $k = 1.25 \times 10^{-4}\,\text{m}^{-1}$.

   a) Explain why h must be measured in metres.

   b) Calculate the values of p at
      i) 5 km
      ii) 10 km, given that $p_0 = 101\,\text{kPa}$.

17 The following are readings from a graph showing the exponential growth of bacteria in an opened tin of meat at room temperature. The number of bacteria, N, are given every 6 hours.

(0 h, 1000) (6 h, 2000) (12 h, 4000) (18 h, 8000) (24 h, 16000)

   a) Plot a graph of $\ln N$ against time t.

   b) Explain how your graph shows that the growth is/is not exponential.

   c) How else could you use these numbers to check that the growth was/was not exponential?

   d) Which method, (b) or (c), is the easier to use in this case?

18.8 Calculus

The Test yourself questions in this section will require you to:

- understand what is meant by 'rate of change' of a physical quantity, e.g. $v = \frac{ds}{dt}$ and $a = \frac{dv}{dt}$
- recognise that the gradient of the tangent to a graph of y against x at a point is equal to $\frac{dy}{dx}$
- understand that the physical significance of the area between a curve and the x-axis, for example, the integral of $I\,dt$, is equal to the charge Q flowing during the time integrated
- use equations representing exponential growth and decay, e.g. for the radioactive decay of an isotope: $N = N_0 e^{-\lambda t}$
- use the results of integrating functions, such as $\frac{1}{r^2}$, that are found when studying electric and gravitational fields.

Test yourself

1 The (sinusoidal) motion of an object is given by the equation $x = x_0 \sin(2\pi f t)$.

   a) The frequency, f, of the oscillation is known to be 300 Hz. Calculate the maximum speed of the object if its maximum displacement is 0.25 m.

   b) What is the size of the maximum acceleration of the object?

   c) Why are the speed in a) and the size of the acceleration in b) asked for?

2 Differentiate both sides of $Q = Q_0 \cos kt$. What does $\frac{dQ}{dt}$ represent?

3 The graph in Figure 18.11 shows the relationship between the current in a circuit and the time for which the current is switched 'on'.

How much charge has passed each point in the circuit during the 12 s for which the current was on?

4 Explain what the area between the graph line and the time axis represents when the vertical axis is the speed of an object.

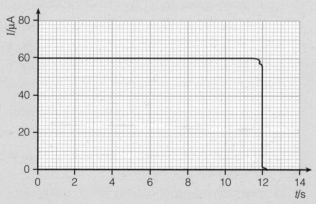

Figure 18.11

5 A graph is drawn of the p.d., V, across a resistor (y-axis) and the current, I, in the resistor (x-axis). Explain what $\frac{dV}{dI}$ represents. Does your answer depend upon the material of the resistor?

6 The graph in Figure 18.12 shows how the horizontal force F on a jogger's foot varies with time t during a single stride.

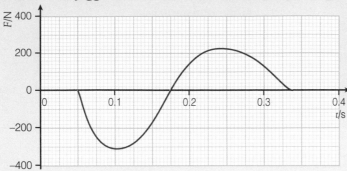

Figure 18.12

a) What is the rate of change of F with t at
   i) $t = 0.105\,s$ ii) $t = 0.175\,s$?
   Explain your answers.

b) Estimate the area between the graph line and the x-axis between $t = 0.050\,s$ and $t = 0.175\,s$.

7 The kinetic energy, E_k, of a body moving at a low speed is given by $\frac{1}{2}mv^2$.
   a) Differentiate E_k with respect to v, i.e. calculate $\frac{dE_k}{dv}$.
   b) Explain what $\frac{dE_k}{dv}$ represents and express this in words.

8 Explain which of the following two statements of Newton's second law of motion is the more general:
$$F = \frac{mdv}{dt} \text{ or } F = \frac{d(mv)}{dt}$$

9 The activity of a radioactive source is defined as being the number of disintegrations the source undergoes per second. Mathematically this can be written as $A = -\frac{dN}{dt}$.
   a) Explain why the − sign appears in this mathematical statement.
   b) A can be written as λN, where λ is called the decay constant for the radioactive decay.
      i) What is the equation $-\frac{dN}{dt} = \lambda N$ telling you?
      ii) How do we write the solution to this equation?

10 Figure 18.13 shows a logarithmic graph of the voltage across a resistor in the first few milliseconds as the current in the resistor rises exponentially. (For example, when a filament light is first switched on.)

   a) Is lg (V/V) proportional to the time t? Explain your answer.
   b) i) Make a table of lg (V/V) and (t/ms) for at least five points on the graph.
      ii) Use your calculator to add a line to the table giving the corresponding values of V in volts.
   c) V is growing exponentially with t. Using values from your table, how else would you know that V was growing exponentially with t?

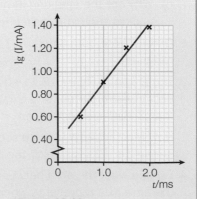

Figure 18.13

11 The number of bacteria in a population increases with time as shown in the table below.

t/days	0	1	2	3	4
N	50	370	2730	20170	149000

a) What difficulty would you find were you to plot these numbers on a graph?

b) How would you try and show that the number N is increasing exponentially with time?

12 What is

a) the result of differentiating

 i) $\frac{1}{r^3}$ ii) $\frac{1}{r^2}$ iii) $\frac{1}{r}$

b) the result of integrating

 i) $\frac{1}{r^3}$ ii) $\frac{1}{r^2}$ iii) $\frac{1}{r}$

in each case with respect to r?

13 The excess pressure p in a leaky cylinder of oxygen gas is measured every day at 1000 hours. A graph of $\ln(p/\text{MPa})$ is plotted for each daily reading (Figure 18.14).

a) The graph can be described by the relation $y = mx + c$.

 i) Calculate a value for m in this case.

 ii) Write down a value for c in this case.

b) Write an equation describing the graph.

c) Using your knowledge of physics, how is the excess pressure of the gas changing with time?

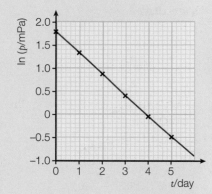

Figure 18.14

19 Preparing for the exams

Preparing for the exams

19.1 The A-level and AS examinations

Both AS and A-level physics will be assessed by written tests at the end of the one- or two-year course. All the AS content will also be included in the A-level examination, together with the further material given in the A-level specification. While the topics in this book are designed for the first year of the A-level course, they also cover all the topics needed for the AS examination.

For the AS there are two papers, each lasting 1 hour 30 minutes. Both papers will consist of two sections, A and B. Section A will have 56–60 marks and Section B 20–24 marks of the 80 available marks.

Paper 1: core physics I covers the core physics I topics (Mechanics and Electric Circuits) in Section A, while Section B will include a data analysis question, possibly with an experimental context, and will draw on topics from the whole specification.

Paper 2: core physics II covers the core physics II topics (Materials, Waves and the Particle Nature of Light) in Section A, while Section B will include a short article with questions on topics from the whole specification.

Both papers may include multiple-choice, short open and open response questions, together with calculations and extended writing questions. Both papers will also examine 'Working as a Physicist'. Briefly this means students

- working scientifically, developing competence in manipulating quantities and their units, including making estimates
- experiencing a wide variety of practical work, developing practical and investigative skills by planning, carrying out and evaluating experiments and becoming knowledgeable of the ways in which scientific ideas are used
- developing the ability to communicate their knowledge and understanding of physics
- acquiring these skills through examples and applications from the entire course.

This book has been specially written to help you develop these skills. The main text is illustrated with numerous applications of physics, with a strong emphasis on practical work. The many examples and questions throughout the book are designed to give you plenty of practice in the sort of questions you will come across in the examination.

For A level, there are three papers. **Paper 1** includes the topics from Core Physics 1 plus those from Advanced Physics I (Further Mechanics, Electric and Magnetic Fields and Particle Physics) and **Paper 2** includes the topics from Core Physics 2 plus those from Advanced Physics II (Thermodynamics, Space, Nuclear Radiation, Gravitational Fields and Oscillations). **Paper 3** covers the General and Practical Physics, and contains synoptic questions that may draw on two or more different topics and include questions on the understanding of experimental methods.

This chapter is mainly aimed at the AS examination. More details of the A level examination will be given in the Year 2 Student's book.

19.2 Learning and revising

Learning

It is not within the scope of this book to consider in any great detail the process of learning and revising. Indeed, there are whole books devoted to this subject, some of them as long as, if not longer, than this book! This is a topic you might wish to investigate further for yourself. There are, however, one or two basic principles that you should put into practice when using this book, and other resources, to prepare for your examinations.

You have clearly been successful at GCSE or you wouldn't be doing AS or A level. However, A level is a quantum leap from GCSE. As one student put it, 'In GCSE you only need to know the facts and remember them, but at A level you need to know much more and be able to think much more deeply.' At A level there is a greater emphasis on learning, rather than just being told. You also need to think for yourself, understand the basic principles and apply your knowledge, sometimes to unfamiliar situations.

A famous golfer is attributed as saying, 'the more I practise, the luckier I get' when an onlooker suggested that he may have been lucky by holing three bunker shots in a row while practising in Texas. Translating this into your A level physics, the more you practise, the more likely you will be able to answer the questions in the exam paper.

Another sporting icon, Jonny Wilkinson, famously won the rugby World Cup final for England in 2003 with a drop kick in the last minute. His mother is quoted as saying, 'I remember hearing him say that all the practice he puts in was so that if he had a kick in the World Cup final then he wanted to be in a position where he wouldn't worry about it.' Relating this to your exams, the more you practise the more confidence you will have in your ability to succeed.

Practice, therefore, is emphasised throughout this book. Working through the *Examples* in the text, and then checking the answers that follow, will help you consolidate each topic of work as you go along. There are also questions at the end of each *Activity* and *Core practical* which are designed to give you practice in answering the type of question on practical work that will be asked in the examination. At the end of each section, the *Test yourself questions* give you further practice. As a student put it 'Regular tests, however tiresome, are a good idea as they let you know what you do or don't understand.'

Your teacher may set the *Test yourself questions* for homework – that's why the answers are not in the text. If you are using this book as a personal resource, you must be strong-minded and work through all the questions yourself. You can then check out how you are doing by looking up the answers on line.

An important principle of the learning process is regular review. This means a continuous, planned revision programme, particularly as your A level exams will test you on two years' work. This book helps you do just this. Having done the *Examples* and tried the *Test yourself questions*, attempting the *Exam practice questions* at the end of each chapter will further consolidate the learning process.

Revising

While each individual has their own particular preference for how they revise, there are nevertheless some basic principles that you need to follow. If you look up '*top 10 tips for revision*' on the internet, you will probably find nearer 100 tips between the numerous sites!

The absolutely essential principle underlying all revision tips is *being active*. Just sitting down and reading your notes or a book will not bring much success. A century of research has shown that *repeated testing* is what works. In other words, spreading your revision over a period of time (i.e. not leaving it to the last minute!) and continually testing yourself will get results. You should therefore:

- plan ahead
- create resources to test yourself
- perhaps study with a friend and test each other
- think positive!

There are some specific points worth mentioning that will help you when you are revising for your physics examination.

- Always choose manageable portions for revision.
- Familiarise yourself with what you need to know — ask your teacher and look through the specification.
- Make sure you have a good set of notes — you can't revise properly from a textbook.
- Learn all the equations indicated in the specification and become familiar with the formulae that will be provided in the examination (at the end of each question paper) so that you can find them quickly and use them correctly.
- Make sure that you learn key terms thoroughly and in detail. For example, the principle of conservation of linear momentum states that in any interaction between bodies, the linear momentum is conserved, *provided no resultant external force acts on the bodies*. The proviso in italics is vital and is often missed out by candidates.
- Be able to describe (with clearly *labelled* diagrams) the basic experiments referred to in the specification.
- Make revision *active* by writing out equations and key terms, drawing diagrams, describing experiments and performing lots of calculations.
- Closer to the examination, it is very useful to attempt as many past papers as you can and, if possible, use the published mark scheme to mark them yourself, or get your teacher to mark them.

One creative resource a lot of students find useful is the construction of *mind maps*. If you haven't come across mind maps before, you can find out all about them by searching the internet where you will find many useful examples.

You could try drawing a mind map of a particular topic, get a friend to draw one for the same topic and then compare your efforts with each other. Just to get you started, here is one for 'resistance'.

> **Tip**
>
> Make revision *active*!

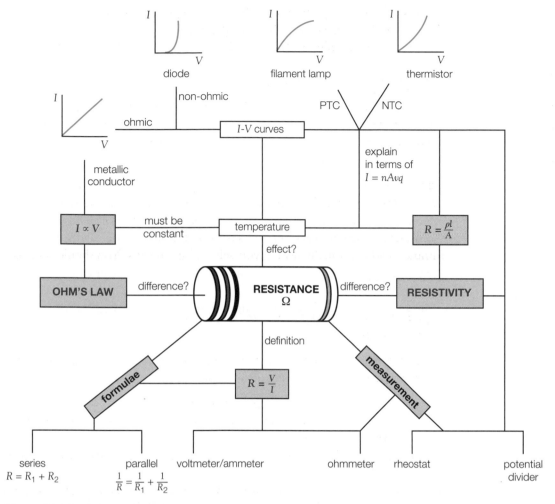

Figure 19.1 A mind map for resistance.

This book has been written as a resource to provide you with plenty of practice, so **no excuses** – just remember Jonny Wilkinson!

19.3 Command terms

Examiners use certain words that require you to respond in a particular way. You must be able to distinguish between these terms and understand exactly what each requires you to do.

The following command terms are used by Edexcel:

- **Add/Label** – Requires the addition or labelling to a stimulus material given in the question, for example labelling a diagram or adding units to a table.
- **Assess** – Give careful consideration to all the factors or events that apply and identify which are the most important or relevant. Make a judgement on the importance of something, and come to a conclusion where needed.
- **Calculate** – Obtain a numerical answer, showing relevant working. If the answer has a unit, this must be included.
- **Comment on** – Requires the synthesis of a number of variables from data/information to form a judgement.

Tip

Remember to show the *units* at every stage of calculations. You are strongly advised to use *quantity algebra* as this helps avoid unit errors.

- **Compare and contrast** –Looking for the similarities and differences of two (or more) things. Should not require the drawing of a conclusion. The answer must relate to both (or all) things mentioned in the question. The answer must include at least one similarity and one difference.
- **Complete** – Requires the completion of a table/diagram.
- **Criticise** – Inspect a set of data, an experimental plan or a scientific statement and consider the elements. Look at the merits and/or faults of the information presented and back judgements made.
- **Deduce** – Draw/reach conclusion(s) from the information provided.
- **Derive** – Combine two or more equations or principles to develop a new equation.
- **Describe** – To give an account of something. Statements in the response need to be developed as they are often linked but do not need to include a justification or reason.
- **Determine** – The answer must have an element which is quantitative from the stimulus provided, or must show how the answer can be reached quantitatively.
- **Devise** – Plan or invent a procedure from existing principles/ideas
- **Discuss** –
  - Identify the issue/situation/problem/argument that is being assessed within the question.
  - Explore all aspects of an issue/situation/problem/argument.
  - Investigate the issue/situation etc by reasoning or argument.
- **Draw** – Produce a diagram either using a ruler or using freehand.
- **Evaluate** – Review information then bring it together to form a conclusion, drawing on evidence including strengths, weaknesses, alternative actions, relevant data or information. Come to a supported judgement of a subject's qualities and relation to its context.
- **Explain** – An explanation requires a justification/exemplification of a point. The answer must contain some element of reasoning/justification, this can include mathematical explanations.
- **Give/state/name** – All of these command words are really synonyms. They generally all require recall of one or more pieces of information.
- **Give a reason/reasons** – When a statement has been made and the requirement is only to give the reasons why.
- **Identify** – Usually requires some key information to be selected from a given stimulus/resource.
- **Justify** – Give evidence to support (either the statement given in the question or an earlier answer).
- **Plot** – Produce a graph by marking points accurately on a grid from data that is provided and then drawing a line of best fit through these points. A suitable scale and appropriately labelled axes must be included if these are not provided in the question.
- **Predict** – Give an expected result.
- **Show that** – Prove that a numerical figure is as stated in the question. The answer must be to at least 1 more significant figure than the numerical figure in the question.
- **Sketch** – Produce a freehand drawing. For a graph, this would require a line and labelled axis with important features indicated, the axes are not scaled.
- **State what is meant by** – When the meaning of a term is expected but there are different ways of how these can be described.
- **Write** – When the questions ask for an equation.

> **Tip**
>
> You should pay particular attention to diagrams, graph sketching and calculations. Candidates often lose marks by failing to label diagrams properly, by not giving essential numerical data on sketch graphs, and by not showing all the working or by omitting units in calculations.

19.4 Examination techniques

An examiner can only mark what is seen on the paper. In physics you need to be sure that your explanations are clearly stated and definitions are accurate. A few tips for examinations are given below.

- Always read the question carefully. Many candidates lose marks by producing a full answer on a related topic that doesn't answer the question, or just gives a description when an explanation is asked for.
- Make best use of the available time. There are 80 marks to be gained in a time of 90 minutes. Don't spend 10 minutes on a 2-mark question! Try to leave enough time to go back and check numerical calculations.
- Try to make descriptions clear and concise. The quality of written communication is important, but a number of clearly stated phrases presented in bullet points can be used and often makes it easier for the examiner to spot the relevant data.
- Always show all your working. You can gain marks for getting part of the way through a calculation. Also, if you make a mistake in calculating a value early on in a calculation, the examiner will still give you credit for using this value correctly in the rest of the calculation – this is called 'error carried forward'.
- Always give the appropriate SI unit for all physical quantities calculated from an equation.

> **Tip**
>
> Remember, make your work clear and concise by using bullet points.

19.5 Questions and answers

In this section there are four questions similar to those that might appear on an AS or A level examination paper and a synoptic question of the type that might be asked in Paper 3 of the A level. Some tips have been included within the question and a typical mark scheme is given for each question. In this book the questions are restricted to those topics that are examined in both the AS and A-level papers.

Mechanics

1 A car of mass 1500 kg tows a trailer of mass 1000 kg along a level road. The driving force on the car is 2000 N and the total resistive forces on the car and trailer are 500 N and 300 N, respectively.

   a) Calculate the resultant force acting on the combination. *(1)*

   b) Determine the acceleration of the car and trailer. *(3)*

   c) How long will it take the car and trailer to accelerate from $10\,\mathrm{m\,s^{-1}}$ to $20\,\mathrm{m\,s^{-1}}$ if the forces remain the same during this period? *(2)*

   d) Draw a diagram of the trailer showing the forces acting upon it in the horizontal direction. *(2)*

   e) Calculate the force the car exerts on the trailer through the coupling. *(3)*

   (Total: 11 marks)

> **Tip**
>
> This question examines the application of Newton's second law of motion relating to a body of fixed mass.

> **Tip**
>
> Many candidates lose marks on Newton's second law questions by failing to use the resultant force, and just including the driving force in the equation $F = ma$. You should also be aware that this force acts on the total mass of the combination.

Mark scheme

a)　Resultant force = 2000 N − (500 N + 300 N) = 1200 N　✓

b)　By Newton's second law, for a fixed mass, $\Sigma F = ma$　✓
　　1200 N = (1500 + 1000)kg × a　✓
　　$a = 0.48\,\text{m s}^{-2}$　✓✓

c)　Use $v = u + at$: $20\,\text{m s}^{-1} = 10\,\text{m s}^{-1} + 0.48\,\text{m s}^{-2} \times t$　✓
　　$t = 21\,\text{s}$　✓

d)

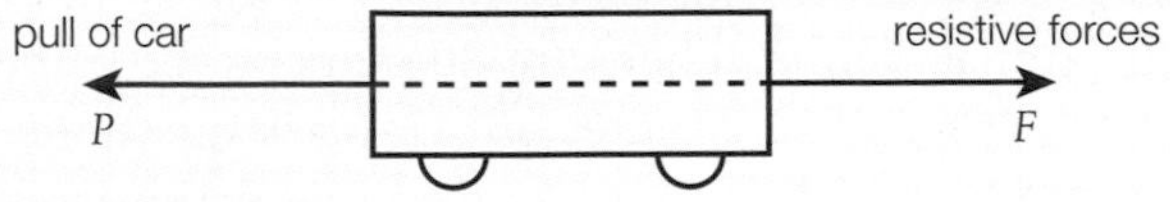

　✓✓

Figure 19.2

Pull of car, P
Resistive forces, F
(Note: Any **extra** forces shown will be penalised.)

e)　Resultant force = $P − F = P − 300\,\text{N}$　✓
　　　　　　　　= 1000 kg × 0.48 m s^{-2}　✓
　　　　　$P = (480 + 300)\text{N} = 780\,\text{N}$　✓✓

(Note: In all of the above numerical answers the appropriate *unit* is needed to gain the mark.)

Tip

Once again, the key is to find an expression for the resultant force acting on the trailer alone. Since the acceleration of the trailer is the same as that of the combination, the value of the resultant force can be calculated and the pull on the trailer can be found.

Tip

This question relates to one of the *Core practicals*. The question needs details of experimental techniques as well as theoretical knowledge.

Tip

The choice of instrument requires a statement on how the resolution is appropriate for the size of measurement. It is worth commenting upon the percentage uncertainty of the measurements.

Electric circuits

2　A student is asked to determine the material used in a metal wire by measuring its resistivity. She plans an experiment to measure the resistance of a range of lengths of the wire. She knows that resistivity is given by the equation $\rho = \dfrac{RA}{l}$ and decides to plot a graph of R against l and calculate the resistivity by multiplying the gradient of the graph by the area of cross-section of the wire. The sample she is given is almost one metre in length and has a diameter in the region of half a millimetre.

a)　State which instruments would be suitable for measuring the length and diameter of the wire. Justify your choice and any techniques that may be used to improve the accuracy of the readings.　*(4)*

b)　The student used an ammeter and a voltmeter to measure current and potential difference and calculated $R = \dfrac{V}{I}$. The length was adjusted by moving crocodile clips to make the connections at the end of the wire. The resistance against length graph drawn by the student is shown in Figure 19.3.

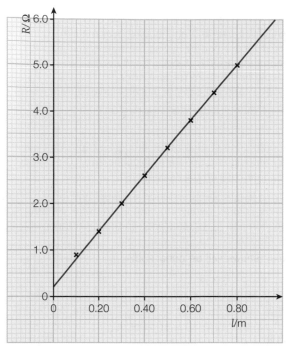

Figure 19.3

The graph does not pass through the origin and there is an anomalous value at the lowest value of length.

  i) Suggest possible reasons for these anomalies.

  ii) Explain how the experiment could be modified to minimise these effects. *(4)*

c) The student was given a list of values of the resistivities of some common metals as shown in the table below:

Resistivity	iron	manganin	nichrome	constantan	copper	steel
$\rho/\times10^{-8}\,\Omega\,m$	9.7	48	100	49	1.8	18

The average value of the diameter of the wire was 0.46 mm. Use this value and the resistance–length graph to find the most likely material of the wire. *(4)*

(Total: 12 marks)

Tip

These are some of the common errors made by students doing this experiment. In addition many people force the line through the origin when, like this one, the line of best fit clearly misses it.

Mark scheme

a) A metre rule is used to measure the length. ✓
It has a range of up to one metre and a resolution of one millimetre.

For a length of 50 cm the % uncertainty will be $\pm \dfrac{0.1\,cm}{50.0\,cm} \times 100\% = \pm0.2\%$

(The uncertainty in the scale reading of a metre rule is ±0.5 mm, but judging the position of the contacts makes ±1 mm more realistic.)

A micrometer is a suitable instrument for measuring the diameter of ✓
the wire as it has a range of several millimetres and a resolution of 0.01 mm.

For a thickness of 0.50 mm the % uncertainty will be $\pm \dfrac{0.005\,mm}{0.5\,mm} \times 100\%$
= ±1%

Tip

In descriptions of this experiment, candidates often lose a mark by stating that the micrometer is used to measure the cross-sectional area of the wire.

- When measuring the length of the wire it should be kept taut (e.g. by adding a weight to one end and looping over a pulley).
- Parallax errors should be avoided by having the wire in contact with the rule, or using a set square if it is not.
- The micrometer should be checked for zero error.
- The diameter should be taken at several places along the wire, with two readings at right angles at each point.

Any two. ✓✓

b) i) There is likely to be contact resistance at the crocodile clips. This will ✓ be the same for every reading and so is a systematic error. The effect is that there will be an intercept on the *R*-axis.

For the shortest value of length the resistance is lowest and it is ✓ possible that the current is large enough to heat up the wire. Heating the wire will cause an increase in its resistivity, so the lowest plot will have a slightly higher resistance than if it were cold.

ii) Contact resistance can be reduced by using knife-edge contacts rather ✓ than crocodile clips.

A fixed resistor in series with the wire will ensure that the current is ✓ kept low for all readings, including those made for short lengths of wire.

c) Cross-sectional area, $A = \dfrac{\pi d^2}{4} = \dfrac{\pi(0.46 \times 10^{-3}\,\text{m})^2}{4} = 1.66 \times 10^{-7}\,\text{m}^2$ ✓

Gradient of graph $= \dfrac{(5.0 - 0.2)\,\Omega}{0.80\,\text{m}} = 6.0\,\Omega\,\text{m}^{-1}$ ✓

Resistivity, ρ = gradient × A = $6.0\,\Omega\,\text{m}^{-1} \times 1.66 \times 10^{-7}\,\text{m}^2$
$= 1.0 \times 10^{-6}\,\Omega\,\text{m or } 100 \times 10^{-8}\,\Omega\,\text{m}$ ✓
(Note: In all of the above answers the appropriate unit is needed to gain the mark.)

It is likely that the wire is made from nichrome. ✓

Tip

If an ohmmeter is used rather than an ammeter and voltmeter, the current in the wire is always very small and heating effects are negligible. However, the precision of most ohmmeters is usually poor compared with ammeters and voltmeters.

Tip

This question relates to a *Core practical* for this unit. You are expected to know what factors affect the terminal velocity of a sphere falling through a fluid, and how the accuracy of the final results depends on how the measurements are taken.

Materials

3 The viscosity of a liquid can be found by measuring the terminal velocity of a sphere falling through the liquid.

a) Draw a free-body force diagram of a sphere falling through a fluid at its terminal velocity. Label your diagram, naming all the forces acting on the sphere. *(3)*

b) With reference to the diagram, explain the meaning of terminal velocity. *(2)*

c) A student carried out an investigation into the variation of the viscosity of syrup with its temperature. The terminal velocity of a 4.0 mm diameter ball bearing was found by timing the ball as it moved between two marks drawn 10.0 cm apart on the side of a test tube. This was repeated for a range of temperatures and the corresponding values of the viscosity of the syrup were calculated.

In her conclusion the student stated, 'The accuracy of the experiment, particularly at the higher temperatures, could be improved by using 2 mm diameter balls and a longer test tube.'

Comment on this conclusion, explaining how the terminal velocity measurements would be affected by these changes. *(5)*

(Total: 10 marks)

Mark scheme

a)

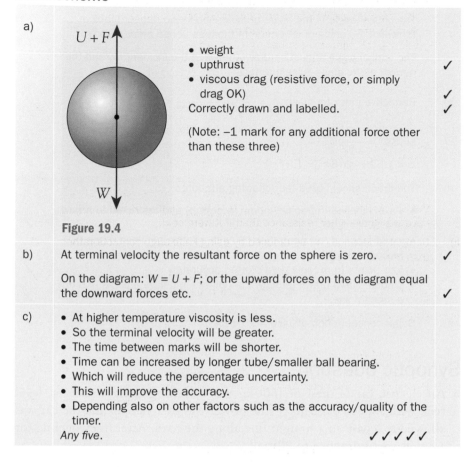

$U + F$

- weight ✓
- upthrust ✓
- viscous drag (resistive force, or simply drag OK) ✓

Correctly drawn and labelled. ✓

(Note: –1 mark for any additional force other than these three)

W

Figure 19.4

b) At terminal velocity the resultant force on the sphere is zero. ✓

On the diagram: $W = U + F$; or the upward forces on the diagram equal the downward forces etc. ✓

c)
- At higher temperature viscosity is less.
- So the terminal velocity will be greater.
- The time between marks will be shorter.
- Time can be increased by longer tube/smaller ball bearing.
- Which will reduce the percentage uncertainty.
- This will improve the accuracy.
- Depending also on other factors such as the accuracy/quality of the timer.

Any five. ✓✓✓✓✓

Tip

To gain all three marks in part a), a candidate must draw all forces along a vertical line through the centre of the sphere. The forces must meet at the centre, or be touching the surface of the sphere.

Tips

To gain the second mark in part b), some reference must be made to the free-body force diagram.

An A-grade candidate would answer this question by referring to the percentage uncertainties in the readings. Such a candidate may also realise that the terminal velocity depends on the square of the radius, and state that a sphere of half the radius will travel four times more slowly.

Waves and the properties of light

4 A refractometer is used in the food industry to determine the sugar concentration in fruit juices by measuring the refractive index of the juices and comparing these with the values for standard sugar solutions. A simple form of refractometer is shown in Figure 19.5.

A ray of light is passed through the prism as shown, and the critical angle is measured using the position of the light–dark boundary on the scale.

a) Explain the meaning of critical angle, and mark its position on a sketch of the refractometer. *(2)*

b) Before the juice is added, the critical angle for the glass–air boundary is measured. If this value is 41.0°, calculate the refractive index of the glass. *(1)*

c) When fruit juice is used, the critical angle at the glass–juice interface is 64.0°. Determine the refractive index of the juice. *(2)*

d) The instrument has been calibrated using sugar solutions of different concentrations. The results are given in the table.

Plot a graph of these results. *(4)*

e) Use the graph to determine the sugar concentration of the fruit juice. *(1)*

(Total: 10 marks)

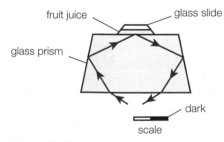

fruit juice — glass slide

glass prism

dark

scale

Figure 19.5

Tip

You will need to use the expressions linking the critical angle to the refractive index of the media on each side of the boundary (page 230).

Concentration of sugar solution/%	Refractive index of sugar solution
0	1.33
20	1.36
40	1.40
60	1.45

Tip

Marks are awarded for correctly labelled axes, appropriate choice of scale, accurate plotting ($\pm\frac{1}{2}$ division) and drawing the line of best fit. In this case a candidate would lose two marks if both scales started at the origin (so that the plots only covered a small region of the graph) and the % was omitted when labelling the concentration axis.

Tip

An A-grade candidate should realise that the graph is a gentle curve; a C-grade candidate may try to force a straight line through the points, and hence lose a mark in part d). The mark for part e) could still be gained if the estimated concentration is within the range specified.

Tip

This question covers material from three of the topics covered in this book. Such questions will not appear in Section A of the AS papers or in **Paper 1** and **Paper 2** of the A level exams. They are typical of the questions that may be asked in Section B of the AS papers and **Paper 3** of the A level.

Tip

You will need to determine the electrical power developed by the motor.

Tip

You should be aware that the kWh is a unit of energy.

Mark scheme

a) The critical angle is the angle of incidence in the denser medium for which the angle of refraction in the less dense medium is 90°. ✓

The critical angle should be drawn between the incident ray and the normal at the glass–juice interface. ✓

b) Refractive index for glass, $n_g = \dfrac{1}{\sin C} = \dfrac{1}{\sin 41.0°} = 1.52$ ✓

c) $\sin 64° = \dfrac{n_j}{1.52}$ ✓

$n_j = 1.52 \times \sin 64° = 1.37$ ✓

d) The graph should have the following attributes:

- y-axis labelled 'refractive index' (accept n) and restricted to a range of 1.30 to 1.50 ✓
- x-axis labelled 'concentration/%' with a range between 0 and 60% (accept 100%) ✓
- four points from table data plotted accurately ✓
- smooth curve drawn through the points – must be a curve and not a straight line. ✓

e) Sugar concentration of juice = 23 ±1% ✓

Synoptic question

5 An electric car of mass of 1600 kg was travelling along a straight, level road with a speed of 30 m s⁻¹. The power was switched off and the car was allowed to 'coast' in a straight line along the road. After freewheeling for 1 km its speed dropped to 20 m s⁻¹.

a) Calculate the average acceleration of the car during this time. *(1)*

b) Calculate the average resistive force experienced by the car. *(1)*

The car is fitted with a massive 144 V lithium–ion battery that powers an electric motor. When driven at a constant speed of 25 m s⁻¹ along the same stretch of road, the current in the motor is 80 A.

c) Calculate the power needed for the car to maintain this speed. *(1)*

d) Determine the efficiency of the motor at this speed. *(2)*

The manufacturer quotes the average energy consumption of the car as 20 kWh per 100 km.

e) i) Convert 1 kWh into the appropriate SI unit. *(1)*

ii) Show that the energy consumption when the car is driven along a level road at 25 m s⁻¹ is only about 13 kWh per 100 km. *(2)*

iii) Give a reason why the calculated value is less than the quoted value. *(1)*

Electric cars are able to improve their efficiency by using regenerative braking. The motor function is reversed so that it becomes a generator and recharges the battery.

f) i) Describe the energy transformations that take place during regenerative braking. *(3)*

   ii) Calculate the maximum possible energy transfer to the battery when the motor is used to slow the car down from 25 m s⁻¹ to 15 m s⁻¹. *(2)*

Electric cars like this are described as having 'zero emissions' as they do not emit petrol or diesel fumes like conventional cars.

g) Suggest two reasons why such cars may not be quite as environmentally friendly as it might seem. *(2)*

At high speeds air resistance has a significant effect on a vehicle's motion. Figure 19.6 shows a large van and a sports car travelling at the same speed along a motorway.

Figure 19.6

The resistive forces are lower if the airflow relative to the vehicle is laminar, and greater if the flow is turbulent.

h) i) Explain what is meant by laminar flow and turbulent flow. *(2)*

   ii) Copy Figure 19.6 and draw lines around the surfaces of the van and the car to represent the airflow, labelling the regions of laminar and turbulent flow. *(2)*

(Total: 20 marks)

Mark scheme

a)	$v^2 = u^2 + 2as \Rightarrow a = \dfrac{v^2 - u^2}{2s} = \dfrac{(20\,\mathrm{m\,s^{-1}})^2 - (30\,\mathrm{m\,s^{-1}})^2}{2 \times 1000\,\mathrm{m}}$ $a = -0.25\,\mathrm{m\,s^{-2}}$	✓
b)	$F = ma = 1600\,\mathrm{kg} \times -0.25\,\mathrm{m\,s^{-2}} = -400\,\mathrm{N}$	✓
c)	$P = F \times v = 400\,\mathrm{N} \times 25\,\mathrm{m\,s^{-1}} = 10\,000\,\mathrm{W}$ (10 kW)	✓
d)	Electrical power input: $P_{in} = IV = 80\,\mathrm{A} \times 144\,\mathrm{V} = 11\,520\,\mathrm{W}$ (=11.5 kW) Efficiency $= \dfrac{P_{out}}{P_{in}} \times 100\% = \dfrac{10.0\,\mathrm{kW}}{11.5\,\mathrm{kW}} \times 100\% = 87\%$	✓ ✓
e) i)	1 kW h = 1000 W × 3600 s = 3.6×10^6 J (3.6 MJ)	✓
ii)	Time to travel 100 km $= \dfrac{\text{distance}}{\text{speed}} = \dfrac{100 \times 10^3\,\mathrm{m}}{25\,\mathrm{m\,s^{-1}}} = 4000\,\mathrm{s} = 1.1\,\mathrm{h}$ Energy per 100 km = 11.5 kW × 1.1 h = 13 kW h	✓ ✓
iii)	The car uses more energy when accelerating, climbing hills or travelling along bumpy roads etc. than when moving at a constant speed along a level road.	✓
f) i)	Kinetic energy (of the car) is transformed to electrical energy then to chemical energy in the battery.	✓ ✓ ✓

Tip

In Year 2 Student's book you will study the principles of the electric motor and generator but you should already be familiar with the energy transformations in each device.

Tip

Think about the manufacturing process and where the energy to charge the battery comes from.

Tip

You must include the −sign; it indicates that the velocity is decreasing.

ii) Kinetic energy transferred $= \frac{1}{2}mu^2 - \frac{1}{2}mv^2$

$$= \frac{1}{2}1600 \, \text{kg} \times [(25 \, \text{m s}^{-1})^2 - (15 \, \text{m s}^{-1})^2]$$

$$= 3.2 \times 10^5 \, \text{J} \, (320 \, \text{kJ})$$

g) Energy from a fossil-fuelled or nuclear-powered power station may have been used:

- to provide the electricity to charge the battery
- and in the manufacturing process in mining the lithium for the battery.

h) i) Laminar flow occurs when the paths of adjacent particles do not cross or there is no change in the velocity of the particles.
Turbulence occurs when the paths of the particles overlap and eddy currents are produced.

ii)

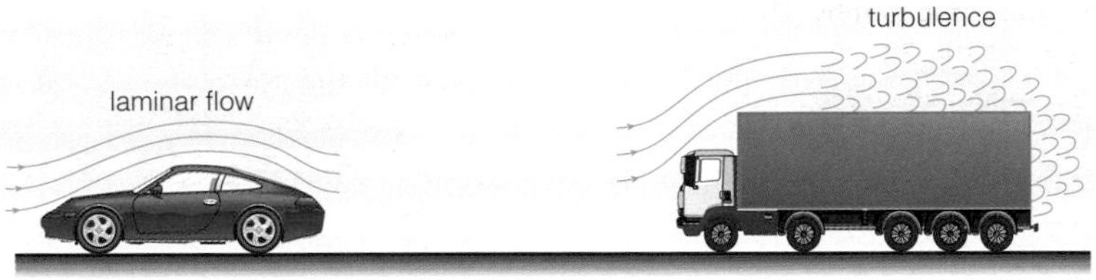

Figure 19.7

Streamlines shown with arrows.
Regions of laminar flow and turbulence correctly labelled.

> **Tip**
>
> The diagrams are not definitive. A clear indication that there will be much more turbulence around the van is needed.

Index

Free online resources

Answers for the following features found in this book are available online:

- Prior knowledge questions
- Test yourself questions
- Activities

You'll also find an Extended glossary to help you learn the key terms and formulae you'll need in your exam.

Scan the QR codes below for each chapter.

Alternatively, you can browse through all chapters at:
www.hoddereducation.co.uk/EdexcelPhysics1

How to use the QR codes

To use the QR codes you will need a QR code reader for your smartphone/tablet. There are many free readers available, depending on the smartphone/tablet you are using. We have supplied some suggestions below, but this is not an exhaustive list and you should only download software compatible with your device and operating system. We do not endorse any of the third-party products listed below and downloading them is at your own risk.

- for iPhone/iPad, search the App store for Qrafter
- for Android, search the Play store for QR Droid
- for Blackberry, search Blackberry World for QR Scanner Pro
- for Windows/Symbian, search the Store for Upcode

Once you have downloaded a QR code reader, simply open the reader app and use it to take a photo of the code. You will then see a menu of the free resources available for that topic.

uantities and units

3 Rectilinear motion

2 Practical skills

4 Momentum

5 Forces

12 Fluids

6 Work, energy and power

13 Solid materials

7 Charge and current

14 Nature of waves

8 Potential difference, electromotive force and power

15 Transmission and reflection of waves

9 Current-potential difference relationships

16 Superposition of waves

10 Resistance and resistivity

17 Particle nature of light

11 Internal resistance, series and parallel circuits, and the potential divider

18 Maths in Physics